HISTORY, PHILOSOPHY AND SOCIOLOGY OF SCIENCE

Classics, Staples and Precursors

HISTORY, PHILOSOPHY AND SOCIOLOGY OF SCIENCE

Classics, Staples and Precursors

Selected By

YEHUDA ELKANA
ROBERT K. MERTON
ARNOLD THACKRAY
HARRIET ZUCKERMAN

A HISTORY OF ELECTRICITY

BY
PARK BENJAMIN

ARNO PRESS
A New York Times Company
New York — 1975

Reprint Edition 1975 by Arno Press Inc.

Reprinted from a copy in
   The University of Illinois Library

HISTORY, PHILOSOPHY AND SOCIOLOGY OF SCIENCE:
Classics, Staples and Precursors
ISBN for complete set: 0-405-06575-2
See last pages of this volume for titles.

Manufactured in the United States of America

Library of Congress Cataloging in Publication Data

Benjamin, Park, 1849-1922.
   A history of electricity.

   (History, philosophy and sociology of science)
   Reprint of the 1898 ed. published by Wiley, New York.
   Includes index.
   1. Electricity--History. 2. Magnetism--History.
I. Title. II. Series.
QC507.B45 1975   537   74-26249
ISBN 0-405-06579-5

A

HISTORY OF ELECTRICITY

Dr. William Gilbert

From the Clamp Engraving made in 1796 from the Original Portrait, then in the Bodleian Library, Oxford.

A

HISTORY OF ELECTRICITY

(THE INTELLECTUAL RISE IN ELECTRICITY)

FROM ANTIQUITY TO THE DAYS OF BENJAMIN FRANKLIN

BY

PARK BENJAMIN, PH.D., L.L.B.

Mem. Am. Inst. Elec. Engrs.; Assoc. Mem. Soc. Naval Architects and Marine Engrs.; Foreign Member Br. Inst. of Patent Agents; Editor-in-Chief Appleton's Cyclopædia of Applied Mechanics and Modern Mechanism; Author of The Age of Electricity The Voltaic Cell, etc., etc.

NEW YORK
JOHN WILEY & SONS
53 East Tenth Street
1898

Copyright, 1895,
By PARK BENJAMIN

All rights reserved.

Braunworth, Munn and Barber,
Printers and Binders,
Brooklyn, N. Y.

PREFACE.

IN this work I attempt to show how there came into the world the knowledge of the natural force, which we call electricity; a force which, within the memory of many now living, has found its most important applications to the needs of mankind, and which exhibits a promise and potency of future benefit, the full extent of which no one can safely venture to predict.

The research has taken many years, has necessitated the gathering of a large collection of ancient, and now exceedingly scarce, writings, not commonly found even in great libraries, and the sifting of an immense mass of recorded facts and theories, often arising in fields far removed from those in which it might naturally be supposed the requisite data would be discovered. The Greek and Roman classics, the results of modern investigation into the old civilizations of Phœnicia, Egypt and even of people of prehistoric epochs, the Norse histories, the ancient writings of the Chinese and Arabs, the treatises of the Fathers of the Church, the works of mediæval monks, magicians, cosmographers and navigators, the early poetry of modern France and Italy; these, mentioned at random, are some of the sources which have been drawn upon, together with the records of the experiments and discoveries of the natural philosophers of all ages. I have made it a rule to note the original founts wherever it seemed to me that such references would be of benefit to others desiring to verify facts or to go over the same ground, and as providing a useful bibliography; while, at the same time, I have endeavored to avoid a multiplicity of annotations relative to immaterial points, which impose only needless labor and uncertainty upon the student.

Above all things, I have sought to write a straight, plain, simple, and, I hope, fairly logical and interesting story. I have rigidly excluded technicalities and scientific demonstrations, which, however interesting to the professional electrician, are as Greek to the general reader; for I address this no more to the wise men of the wires and the dynamos and the batteries, than to the great public whom we all serve, and for whose good we all labor. Popular science, so called, is too often dilute science. Scientific discussions of a didactic or abstract nature, or involving a Babylonish terminology, and requiring minds trained to understand them, cannot be rendered any easier to the mental digestion of intellects engrossed in other departments of the world's work, and, hence, not so educated, by mechanically mixing them with the water of an engaging rhetoric. The facts and the arguments based on them must be digested and brought into true solution, so that the food offered will be easily assimilable; and that is what I have tried here to do.

Perhaps this work may usefully tend to show that electricity, at the present time, is *not* "in its infancy." It has undoubtedly a vast amount of work yet to do, and—I am patriotic enough to believe at the hands of our American inventors, first of all—will yet accomplish things undreamt of in our philosophy; but it will do this not with the feeble uncertainty of the nursling, but with the vigor and might of maturity. Moreover, although in ancient days electricity, in common with all other natural manifestations, was regarded as a mystery, none the less the knowledge of it, as these pages seek to prove, forced its way through the clouds of ignorance and superstition with the unerring directness of a projectile driven through the mist from a modern gun. Electricity is not now occult, it is not mystic, it is not magic, its workings are no more wonderful than are the rise and fall of the tides; in fact, it may be safely said, that we know more about its laws and their consequences than we do about those which determine the fall of a stone to the ground.

I end this essay—which has been more of pleasure than of toil—fully conscious of the errors and inconsistencies which must be in it. At every turn there have been tangled skeins to unravel, whereof the true clews have, no doubt, often been missed; diverging roads, where one selects a path never without misgivings. But with all due submission, I venture to believe that a faithful effort, even if misdirected, is better than none at all, although in that consciousness may well lie the only justification for this book.

PARK BENJAMIN.

CONTENTS.

	PAGE
Introduction.	11

CHAPTER I.

Ancient sources of amber.	15
Amber legends	16
The Syrian women and their amber spindles	17
The lodestone	19
Lodestone legends.	22
Greek knowledge of the lodestone and the Samothracian rings.	23
The Magnetes.	26
Egyptian knowledge of the lodestone.	28
Magnetic knowledge of the Hebrews	29

CHAPTER II.

The opening of the Egyptian ports	30
Greek Nature-worship	31
Thales of Miletus and the beginning of Greek philosophy	32
The Magnet-soul	33
Diogenes Laertius on Thales	34
Aristotle and the foreshadowing of the inductive method	38
Theophrastus, and the first physical description of the amber effect.	39
The mythical Lyncurium	41
The University of Alexandria.	44
Legends of magnetic suspension—Mahomet's coffin	45
Lucretius' De Natura Rerum and its description of magnetic effects.	47
Ancient medical uses of amber	52

CHAPTER III.

The polarity of the magnet.	53
Unknown to ancient Greeks	54
Or to ancient Phœnician navigators.	55
The Betulæ.	56
No knowledge of polarity among ancient Egyptians.	57
Or among the Etruscans.	59
Polarity possibly known to the prehistoric Nomad races	61
Relations of Akkadians and Chinese	63
Ancient China and Chinese chronology.	64
The Chinese south-pointing carts.	67

(vi)

CONTENTS.

	PAGE
Ancient Chinese knowledge of amber and of the geomancer's Compass.	75
The Chinese not natural navigators	77
Nor reliable astronomers	79
Nor competent inventors.	80
The Mariner's Compass probably not of Chinese origin.	85

CHAPTER IV.

The Dark Ages and the rise of Scholasticism.	86
First distinction between magnetic and electric effects drawn by St. Augustine.	87
Patristic references to the lodestone and amber	90
Old medical uses of the lodestone.	93
Claudian's Idyl	93
The Fables of the Magnetic Rocks	96
Ancient Arab navigation	102
The Compass not used in early voyages on the Indian Ocean	103
Nor by the Spanish Saracens.	108
Nor on Spanish ships until 1403	111

CHAPTER V.

The Northmen and their early voyages	112
Physical science among the Anglo-Saxons	115
The Norman invasion and the poem of William Appulus of Amalfi.	116
Scholastic philosophy	118
Alexander Neckam	120
His treatise de Natura Rerum	122
The doctrine of similitudes.	124
And of virtues.	125
Applied to the lodestone.	127
The first European description of the Mariner's Compass	129
And the remarkable magnetic discoveries preceding	131
The Compass points.	133
Gottland, the great nautical rendezvous.	134
Wisbuy and its laws.	135
The Finns and Lapps	137
Their sorcery and relationship to Chinese	139
And possible ancient knowledge of Compass	141
The garlic myth.	142
The punishment for tampering with the Compass	144
The Compass possibly of Finn origin and emanating from Wisbuy.	145

CHAPTER VI.

Thirteenth century thought	148
William the Clerk on the Compass	149
The Bible of Guyot de Provins.	152

	PAGE
And other mediæval poems referring to magnetic polarity	154
The spurious treatise of Aristotle	157
Mediæval lodestone myths and fables	159
Roger Bacon and his discoveries	160
Ancient conceptions of the universe	163

CHAPTER VII.

Peter Peregrinus	165
His perpetual motion	167
His marvelous magnetic discoveries	169
His development of the Mariner's Compass	184
Flavio Gioja and his Compass card	187
Plagiarists of Peregrinus	191

CHAPTER VIII.

The revival of literature in Europe	193
Henry the Navigator and Portuguese voyages	194
Christopher Columbus and his magnetic discoveries	196
Attempts to account for Compass variation by the Magnetic Rocks	202
The voyages of Vasco da Gama and Magellan	205
Peregrinus' disclosure of the magnetic field of force	207
Hartmann partly recognizes Dip of the Compass needle	209
Norman's discovery and explanation of Dip	211
Magnetic deceptions of the period	219
Paracelsus and his magnetic nostrums	220

CHAPTER IX.

Fra Paolo (Pietro Sarpi)	224
His treatises on the magnet	225
Cesare observes magnetism by earth's induction	227
The Jesuits dispute Sarpi's discoveries	228
John Baptista Porta	230
His Society	231
His relations to Sarpi	232
His treatise on natural magic and the magnetic discoveries therein recorded	234
And especially the magnetic field of force	235
And telegraphic communication by magnets	239
Jerome Fracastorio	241
Jerome Cardan	243
And his differentiation of magnetic and amber effects	249
The physicians as physicists	255

CHAPTER X.

William Gilbert	258
The object of his work	268

CONTENTS.

	PAGE
His errors	274
His mode of thought	275
His Terrella and his magnetic theories	277
His magnetic discoveries	288
The inception of his study of electricity	294
The discovery of the Electrics	299
Gilbert's electrical experiments	303
His electrical theory	307
His electrical discoveries recapitulated	313

CHAPTER XI.

Gilbert's treatises	315
Francis Bacon and his suppression of Gilbert's later work	318
Bacon's criticisms on Gilbert	321
Bacon's studies in magnetism and electricity	324

CHAPTER XII.

Physical science in England in time of James I	332
The great Universities	333
William Barlowe and Mark Ridley, and the controversy between them	336
Physical Science in Italy	341
Galileo and his indebtedness to Gilbert	344
Galileo's magnetic researches	347
The electric discoveries of Nicolaus Cabæus	349
The magnetic and electric theories of René Descartes	356
The amber and the magnet in English literature	367
The Rosicrucians and Van Helmont	372
Sir Kenelm Digby, and the rise of physical science in England	377
Sir Thomas Browne, destroyer of errors	380
Some early notions of telegraphy	382
Otto von Guericke	389
His theory of virtues	392
His extraordinary electrical discoveries made with the sulphur globe	396

CHAPTER XIII.

The founding of the English Royal Society	404
Science at the Court of Charles II	406
Robert Boyle	414
His philosophy	416
His electrical discoveries	420
Physical observations in America, and Madam Sewall's sparkling skirt	425
Robert Hooke	426
Isaac Newton and the reduction of electricity under the reign of law	435
Halley's magnetic theories	447

CHAPTER XIV.

	PAGE
The founding of the French Royal Academy of Sciences.	450
Picard's observation of the electric light.	453
Francis Hauksbee and his experiments on light and electricity.	457
Foreshadowings of the identity of electricity and lightning	469
Stephen Gray, charter-house pensioner	470
His remarkable discoveries in electrical conduction and insulation.	474
Charles Dufay.	478
His recognition of resinous and vitreous electricities.	484
Dr. Joseph Desaguiliers	489

CHAPTER XV.

Electrical Progress in Germany. 490
George Matthias Bose, a "modern wizard" 493
Krüger and his theories 501
The medical uses of electricity 502
Supposed combustion of the human body by electricity. 504
The discoveries of Gordon and Winkler. 506
Dean Von Kleist and the discovery of the Leyden jar 512
Abbé Nollet and his experiments. 516
The first electrical circuit and the experiments of Lemonnier and
    Watson 531

CHAPTER XVI.

Benjamin Franklin 537
Discovers discharging effect of points. 541
His experiments on the Leyden jar 543
William Watson and his electrical circuit across the Thames . . . 549
His electrical theory and that of Franklin. 552
Ancient notions concerning lightning. 561
Freke and Winkler's suggestions of the identity of lightning and
    electricity. 571
Abbé Nollet on same subject 573
Franklin's theories of atmospheric electricity 575
The announcement of the Lightning Rod. 582
Kinnersley's lectures on electricity 585
The French experiments on aerial rods 587
Franklin's kite experiment, and the proof of the identity of electricity and lightning. 589
Conclusion 592

Index. 597

ILLUSTRATIONS IN THE TEXT.

	PAGE
Chinese south-pointing cart	73
Norse punishment for tampering with the Compass	145
Peregrinus' floating Compass	180
Peregrinus' pivoted Compass	181
Peregrinus' Compass, plan view	182
Norman's Dipping needle	217
John Baptista Porta	231
Jerome Cardan	243
Gilbert's Terrella	277
Gilbert's armed lodestones	288
Magnetizing hot iron by hammering	290
Gilbert's Orb of Virtue	291
Gilbert's Electroscope	303
Galileo Galilei	346
Cabaeus' picture of magnetic Spectrum	353
René Descartes	358
Descartes' magnetic field	363
Guericke's electrical machine	398
Dalance's title page	448
Hauksbee's mercurial fountain	459
Hauksbee's electric machine	461
Hauksbee's electric glow	464
Nollet's experiment on electrified boy	518
The Leyden experiment	520
Nollet's experiments on electric light	528
Nollet's experiments on electrifying animals	529
Franklin's illustration of his experiments	561

INTRODUCTION.

The intellectual rise in electricity is worthy of historical investigation, not merely because of the material results, actual and potential, which have come from it, but because it shows clearly anew the marvellous power of the human mind as an instrument of discovery, capable of correcting its own errors. Beginning with a single phenomenon, afterwards including effects all, for long periods, seemingly fortuitous and uncorrelated, this rise has involved questions of an interest second only to that which mankind has yielded to the great issues of life and eternity; questions which challenged the human understanding and compelled it to measure itself against them. From one fact it came to include many facts, from one conception grew many conceptions, coincidently with the increase in human learning, the broadening of human thought, and the development of human intelligence.

The initial idea—the germ—found its lodgment in some brain existing at an epoch far beyond the limits of history. The discovery of amber in the ancient lake dwellings of Europe suggests the possible perception of it by pre-historic man. The accidental rubbing against the skins with which he clothed himself may have caused an attraction by the resin, thus electrified, of the light fur in sufficiently marked degree to arrest his attention. Between such a mere observation of the fact, however, and the making of any deduction from it, vast periods may have elapsed; but there came a time at last, when the amber was looked upon as a strange inanimate substance which could influence or even draw to itself other things; and this by its own apparent capacity, and not through any mechanical bond or connection extending from it to them; when it was recognized, in brief, that nature held a lifeless thing showing an attribute of life.

This was more than a mere impression. It was an enigma demanding resolution, and thus endowed with inherent and eternal vitality.

At some other time, perhaps not until after the advent of an Iron Age, a similar power to that of the amber was seen in the attraction of the lodestone for iron. Because of this similitude the ancients somewhat hazily imagined both effects to be essentially one. Progress in discovery concerning either was therefore progress in knowledge concerning both. This is also true from our modern point of view, for not only are the phenomena of magnetism and of electricity directly correlated and interconvertible, but the concept of magnetism perhaps most widely accepted at the present time, holds it to be merely an electric state; the condition of electricity in whirling or vortex motion.

The attempt to account for magnetic attraction as the working of a soul in the stone led to the first attack of human reason upon superstition and the foundation of philosophy.

After the lapse of centuries, a new capacity of the lodestone became revealed in its polarity, or the appearance of opposite effects at opposite ends; then came the first utilization of the knowledge thus far gained, in the mariner's compass, leading to the discovery of the New World, and the throwing wide of all the portals of the Old to trade and civilization.

The predominance of the magnet in human thought was yielded to the amber, when the strange power of the latter was found to exist also in other things. The keen-eyed discoverers saw this new force annihilate time and space, and flash into light; pursued it even to its hiding-place in the clouds; beheld it grow from the feeble amber-soul into the mighty thunderbolt; watched it until the whole universe showed itself pervaded with it.

This was a true intellectual rise. It was the Intellect at work building the universe of which it is the key; finding anew that Nature also is working in every detail after the laws of the human mind.

"It is not, then, cities, or mountains, or animals, or globes that any longer command us, but only man; not the fact, but so much of man as is in the fact."[1]

So in this research, I have felt that it is not so much the trials and the discoveries made in this great and new field of Nature which attract us, instructive and useful, even momentous as they are—for after all to many they are but abstractions—not these, so much as the breathing human beings, who in the far past saw them and deciphered them in the light of those other days, and of whose life they formed a part; who thought of them, and whose thoughts lived on, and became immortal, and moved downward through generation after generation to us; even as our thoughts, joining theirs, will pass through the ages to the generations yet to come.

[1] Emerson : Natural History of Intellect.

CHAPTER I.

THE use of amber begins with the dawn of civilization. The discovery of beads in the royal tombs at Mycenae and at various places throughout Sardinia and the territory of ancient Etruria, proves that trade in it existed in prehistoric times; while the identity in chemical constitution of the amber ornaments of Mycenae and the Baltic amber from the Tertiary formation of the Prussian Samland, the coasts of southern Sweden and the northern Russian provinces, indicates the far distant source from which the resin was anciently derived.[1] Who first brought the resin from the Baltic Sea to the Levant is an undetermined question, since it is known to have come southward across Europe by land as well as around the continent by water.

The Phœnicians—those far-sighted and consummately keen traders, whose commercial and maritime supremacy is still unrivaled by that of any modern nation—extended their voyages past the gates of the world into the unknown ocean in search of both the amber of the Northern Sea and the tin of Cornwall; for to obtain the latter the makers of bronze from all quarters flocked to the great metal market of Sidon. Both commodities also came by way of the Rhine and the Rhone to Marseilles and across the Alps to Etruria and chiefly to the valley of the Po, besides elsewhere by other land routes, along all of which stores of tin and amber have been found as they were ages ago hidden when the caravans were attacked or fell victims to the natural perils of the road. While these ways are known to have existed, and the amber trade over them to have been maintained before Rome or Carthage were

[1] Schliemann : Mycenae and Tiryns, 1876, 203, 245 ; Tiryns, 1886, 369.
Simcox : Prehistoric Civilizations, 1894.

founded, it may be that the Phœnician voyages to the Baltic were of still greater antiquity, for the beads of Mycenae date from at least two thousand years before our era.

The amber was used by the ancient world as a jewel and for decoration. Its color and lustre reminded the fanciful Greeks of the virgin gold which glistened in the sands of Pactolus, even as the brilliant metal had itself recalled to them the yellow sunshine. Afterwards they applied the same name to the compounds of metals which, when burnished, gave a golden glow. They were all children of the sun "Elector"—reflecting in miniature his radiance. Thus, in common with native gold and the silver-gold alloys, the amber, in Hellenic speech, came to be called "electron."[1]

Throughout Greek literature, even from the time of Homer and Hesiod, the mention of it is frequent. It is inlaid in the royal roof of Menelaus, it bejewels the bracelets of Penelope, the necklaces of Eumoeus,[2] and the shield of Hercules.[3] Legends cluster thick about it. Through the lost tragedy of Æschyius, the Hippolyta of Euripides and the Metamorphoses of Ovid comes the myth of Phaeton, recounting his death by the thunderbolt and fall into the river Eridanus, and the transformation of the weeping Heliades into poplars ever sighing and shedding their amber tears beside the stream. The Greek traders coming to the mouth of the Po for their cargoes, easily believed the story—perhaps told to conceal the true source—that the resin had been gathered under the poplar trees along the banks, or on the Electrides—the islands at the outlet of the river. Long afterwards, so firmly did the

---

[1] The ancient Greek poets called the sun ἠλέχτορ and Homer repeatedly so terms it (Iliad. Z.′ 513: T.′ 398). "Electron" is used very indefinitely by the Greek classic writers—and in fact has no permanent gender, though commonly neuter. See Rossignol: Les Métaux Dans l'Antiquité, 345. Paris, 1863.

[2] Odyssey.   [3] Hesiod : Scutum Herculis.

legend persist, men came to search the shores of Eridanus for amber, as the Spanish adventurers sought the Eldorado in the new world.

"Dost thou think that we would tug against this torrent for two oboli a day?" laughed the boatmen of the Po to the discomfited Lucian, "could we find riches under the poplar trees for the picking up?"

To the mythical tales set afloat by the traders, became added the fancies of the poets. Amber is gathered, so ran one fable, by the maiden guardians of the golden Hesperides as it falls from the poplars into Lake Electrum; it is the slime of drear Lake Cephisis, the sweat of the laboring soil under the fierce rays of the sun, the tears of the Indian birds for the death of Meleager, said others. And the sailors told of other Electrides islands in the German ocean and off the Calabrian coast where grew the tree "Electrida," and of stones in far-off Britain "purging thick amber."

It often happens that historical facts become embedded, as it were, in the names of things, and thus preserved, and the knowledge of them so passed down through centuries. Just as we find now locked in the yellow depths of the amber, bodies of insects which lived ages ago, so in one of the designations which the people of ancient times gave to it is embalmed, perhaps, the story of how electricity first became known to the civilized world.

The Syrian women, Pliny says,[1] called the amber "harpaga" or "the clutcher;" which is obviously based on a peculiarity of it altogether different from that which caused it to be likened to an embodied sunbeam. This name, in turn, came from its use in spinning, the oldest handiwork known to the race, and in the mode of spinning which has been employed since the very beginning of civilization. So that we may conjecture that the name came down from the old Phœnicians, and that the amber which they

[1] Pliny: lib. xxxvii. c. 1; Aldrovandus: Musaeum Metallicum. Milan, 1648, 404.

worked into beads and ornaments found its place in the hands of every woman who spun with the distaff, and who could afford the luxury of a spindle made of the much-prized substance. The way in which the spinning was done by distaff and spindle, Catullus tells:

> "The loaded distaff in the left hand placed,
> With spongy coils of snow-white wool was graced,
> From these the right hand lengthening fibres drew,
> Which into thread 'neath nimble fingers grew.
> At intervals a gentle touch was given,
> By which the twirling whorl was onward driven.
> Then, when the sinking spindle reached the ground,
> The recent thread around its spire was wound,
> Until the clasp within its nipping cleft
> Held fast the newly finished length of weft."

As the spindle descended, and at the same time whirled around, it rubbed against the loose feminine garments; thus it became electrified, as amber always does when rubbed, so that on nearing the ground, it drew to itself the dust or bits of leaves or chaff lying there, or sometimes attracted the light fringe of clothing. The spinner easily saw this, because the chaff would leap up to the excited resin, or the fringe filaments extend themselves toward it, and moreover, unless she were careful, the dust and other substances so attracted would become entangled in her thread. Therefore, she called her amber spindle, the "clutcher;" for it seemed to seize these light bodies as if it had invisible talons which not only grasped, but held. This was probably the first intelligent observation of an electrical effect. It is singular that it should have become apparent through the earliest practical, in contradistinction to merely ornamental, use of the amber, though perhaps nothing strange that it is due to the keener perception of woman.

The lodestone or magnetite is an ore of iron[1] which sometimes crops out as a rock above the surface of the ground. The accidental bringing of an iron object into the neighborhood of the outcropping stone probably caused the first observation of the attractive power of the rock for the metal, and thus furnished the basis for the legend which Pliny copies from the poet Nicander (who wrote it two centuries before his time), concerning the Shepherd Magnes, who, while guarding his flock on the slopes of Mount Ida, suddenly found the iron ferrule of his staff and the nails of his shoes adhering to a stone; which subsequently became called after him, the "Magnes Stone," or "Magnet." This legend, in various forms, retained its vitality up to comparatively recent times. As masses of magnetite were discovered in various parts of the world, the stories of its attractive power became greatly exaggerated, especially, as I shall hereafter show, during the Middle Ages. In fact, magnetic mountains which would pull the iron nails out of ships, or, later, move the compass needle far astray, did not lose their place among the terrors of the sea until after the seventeenth century had become well advanced.

The phenomena of the lodestone are, however, two-fold. It not only attracts iron objects, but it has polarity, or, in other words, exhibits opposite effects at opposite ends; by reason of which, when in elongated form and supported so as freely to turn, it will place itself nearly in the line of a meridian of the earth—that is, nearly in a north and south direction. This is its directive tendency, or, as William Gilbert called it in 1600, its "verticity," and upon this quality, as is well known, depends the use of the magnetized needle in the mariner's compass.

We may conclude that whoever gained the first knowledge of the attractive power of the lodestone, was also acquainted with iron, if he had an iron object to present

[1] $FeO.F_2O_3$, sp. gr. 5.2, contain2 72.41 per cent. of iron. Osborn: Metallurgy of Iron and Steel.

to the stone and in this way perceived its attraction. Iron, however, is never found in a metallic state in nature, except in meteorites. Excluding this infinitesimal supply, the metal is obtained from its ores, by means usually involving the development of intense heat, so that to devise modes of attaining the necessary temperatures, let alone the even more complex mental work of contriving apparatus and processes for separating the metal, requires advanced powers of observation and invention. Hence modern ethnological and geological authorities unite with Lucretius[1] and other ancient writers in affirming that the Age of Iron has always followed that of brass or bronze. So far, therefore, as establishing the probable time of the discovery of the attractive force of the lodestone is concerned, it is immaterial whether we consider that the phenomenon was first remarked as an effect of outcropping magnetite upon iron brought near to it; or as one exerted by fragments of magnetite in an iron mine upon other fragments of the same substance, or upon extracted iron. In any case, the observation of the fact seems necessarily to have followed the advent of an Iron Age, and therefore may not extend indefinitely back into prehistoric times.

On the other hand, with regard to the directive tendency of the lodestone a different conclusion is reached. To suspend an elongated piece of the stone and see it turn itself in a definite direction; or to do this repeatedly and with different pieces and thus learn that the phenomenon is true of this particular stone and not of other stones, obviously involves no necessary knowledge of its attractive effect on iron. Therefore, if we admit the possibility of sufficient intelligence in the race then living, we may conjecture that an acquaintance with magnetic polarity may have existed among the earliest peoples of which we have any tradition. I shall show hereafter that reason for such conjecture is by no means absent, which if ac-

[1] De Natura Rerum, v.

cepted, places human knowledge of the directive tendency of the lodestone not only far beyond the limits of history, but even suggests the utilization of that knowledge by wandering hordes for their actual guidance over the wildernesses of the earth, at the same extremely remote epoch.

For the present, however, it is necessary to deal with modern civilization and periods within historical times, and therefore, to begin with an inquiry into the familiarity of the western world with magnetic attraction; for whatever the Asiatic people may have known concerning magnetic polarity, there is no trustworthy evidence that the nations of Europe had the slightest acquaintance with it before the twelfth century of our era.

It is especially difficult to determine the positive date when any nation made the transition from the bronze to the iron age, and practically impossible to do so in the cases of people who either inhabited countries where iron does not abound, or who never acquired the art of obtaining it. In such event, the substitution of implements of iron necessarily imported from other countries for the native ones of bronze, to which the population had become accustomed by ages of use, was an exceedingly slow process, retarded by the mental inertia of the times, and often by national pride in home customs and handiwork. Hence arises the seeming anomaly that among people far advanced in civilization, the general use of iron can be recognized only at a comparatively late period in their history; while among barbarians, incomparably below them in intellectual attainments, we find evidence of its employment at immensely earlier periods. In Denmark, for example, the age of iron corresponds to that of the beech tree. Hesiod, writing in 850 B. C., speaks of the time when "men wrought in brass, when iron did not exist;" and Homer, although frequently referring to weapons and implements of bronze, mentions iron but rarely. The Aztecs, at the time of the Conquest, knew nothing of the metal, although their soil was impregnated with it. The Peruvians, under the same natural conditions, were equally ignorant.[1]

The traditions of magnetic attraction, however, date from periods far earlier than the days of Nicander. The iron of antiquity was mined chiefly on the islands and coasts of the Ægean Sea, and on Elba and Crete, although some came even from distant Ethiopia. That found on the slopes of Mount Ida or on the Mediterranean islands was famous. Its strange hunger for other iron, which it seized and drew unto itself, was to the superstitious Greek a mystery, concerning which the uninitiated might not even think for fear of the anger of the gods : the anger of Celmis, and Damnamenus and Acmon the irresistible, and later of Azieros, Aziokersa and Aziokersos, whose very names were mystic and dangerous to speak.

In far-off ages, so said the legend, Rhea, the earth goddess of Phrygia, sent to Ida, and thence to Samothrace, in the Ægean, those of her children who were skillful underground, and wise in their knowledge of the ores, and where they lay hidden in the cracks and crevices of the rock. And, because of their skill, these emissaries received the name of "Dactyls"—fingers; for they were "the fingers of Rhea." Some of them went to Crete ; but wherever they journeyed (and Samothrace became their main abode), they dug into the earth and brought out the iron ore ; and when the people saw them heat this, and melt it and produce the black, hard ringing metal, they believed them to be gods, and their art a mystery.

As a matter of fact the Idean Dactyls seem to have been merely a roving band of Phrygian miners,[2] who carried

[1] Prescott : History of the Conquest of Mexico. 1865, i., 139, and works there cited.
Lyell, Sir C.: The Geological Evidences of the Antiquity of Man. London, 1873, 8.

[2] Rossignol, cit. sup., refers to the Scholiast of Apollonius of Rhodes on the Phoronid, an ancient and fragmentary poem which he considers as old as the works of Hesiod and Homer. This, concerning the Idean Dactyls, says, "they first found in the mountain forests the art of the

their metallurgical knowledge to places where the ore existed, but like knowledge did not; and who taught mining and iron-working to the Hellenes, or to those who occupied the land before them.

Following the Dactyls came the Cabiri, a second and more skillful band of iron-workers, who were indeed more handicraftsmen than miners. Concerning these, all records are most obscure and conflicting, and they are, besides, inextricably entangled with the myths of several nations. Like the Dactyls, the Cabiri came from Phrygia to Samothrace, Lemnos and Imbros. Their cult seems to have attained its greatest vigor, however, at Samothrace, and ultimately to have spread to Macedonia and Phœnicia. It possessed great vitality, since as late as the fourth century of our era it was in a flourishing existence.

The Samothracian Cabiri became combined with the Dioscuri, Castor and Pollux, the twin sons of Heaven, who presided over the mariners; and with the Egyptian Phtha-Sokari and the Greek Haephaestos; and later with the Corybantes and Curetes, which appear to have been other bands belonging to the same family. Their worship frequently changed form, so that even the mystic recitals of the Orphic hymns relating to it, now ascribed to the false Orpheus or Onomacritus, who lived as late as 514 B. C., are a confused jumble of forgeries, to which even the Christian philosophers are said to have added their quota.

From the various legends and traditions, however, the probable fact appears that the first iron miners of Greece came from Phrygia, which abounded in the metal, and settled in Samothrace. Here they instituted the mysteries which so long afterwards prevailed, and in the beginning, as a proof of their supernatural skill, they exhibited the attractive phenomena of the lodestone through the mystic working of the so-called Samothracian rings.

The first mention of the magnet in the Greek classics is

cunning Vulcan, the black iron, carried it to the fire and produced wonderful work."

apparently that made in the fragmentary Oeneus of Euripides, which Suidas[1] quotes, and which distinctly refers to the attraction of the lodestone for the iron. The subject takes definite form, however, in the Ion of Plato; and there, in the following words, Socrates describes the famous rings :

"The gift which you have of speaking excellently about Homer, is not an art," says the sage, "but, as I was just saying, an inspiration: there is a divinity moving you, like that in the stone which Euripides calls a magnet, but which is commonly known as the stone of Heraclea. For that stone not only attracts iron rings, but also imparts to them similar power of attracting other rings : and sometimes you may see a number of pieces of iron and rings suspended from one another, so as to form quite a long chain ; and all of these derive their powers of suspension from the original stone. Now, this is like the Muse who first gives to men inspiration herself, and from those inspired, her sons, a chain of other persons is suspended, who will take the inspiration from them."[2]

Plato lived between the years 429 and 348 B. C., and from his time forward the rings of Samothrace are described again and again. Lucretius, writing three centuries later, refers to them as still potent. Their well-established existence shows that the Samothracian wonder-workers not only were familiar with the attractive power of the lodestone, but with its capability of inducing a similar power in iron. The popular belief that everything which produces wonderful effects must have wonderful properties, and the converse popular tendency which seeks a cause for any effect not understood, in things concerning which prevailing ignorance is still deeper, were fully as strong in those ancient days as they are now. For precisely the same reason that the modern "magneto-therapist" plays upon the imagination of the patient, or

[1] Suidas : Lex. Graec. et. Lat. post T. Gaisford, Halle, 1853, 658.
[2] Jowett, B. : The Dialogues of Plato.

the charlatan sells to the credulous so-called "magnetic" panaceas for every ailment, so the priests of Samothrace drove a thriving trade in their magnetized iron rings as amulets and cure-alls. They were worn by the worshippers of the Cabiri, later by the Roman priests of Jupiter, and in Pliny's time they became the usual pledge of betrothal.

The Cabiri were remembered long after their individual cult had disappeared. They became converted into the gnomes and the elves of the legends and folk-tales of the Middle Ages, and in the first modern treatises on mining we find them still depicted as dwarfs with their picks and shovels and attended by their dogs, searching for the metals in the depths of the earth. Even so skillful a miner as George Agricola,[1] whose great work begins the present science of metallurgy, cannot divest himself of a half-belief in them; for in his quaint pictures he always shows them at work in the mines, although often amid machinery which the old Greeks who worshiped at Samothrace might well have regarded as the handiwork of higher gods than those which they there adored.

There were many near-by sources for the lodestone which supported and magnetized the Samothracian rings; for iron mines existed not only on the slopes of Mount Ida, and on Elba and Crete, but on the island of Samothrace itself. It was because the magnetic ore was found in the same deposits as the ordinary ores of iron, that the Greeks at first called it "Siderites" or ironstone. Later because of its power of overcoming iron, and of forcing that hard and intractable metal to come to it, they termed it the "Hercules stone," and later still they gave it the name which it still most commonly bears, the magnet, which as Lucretius says comes "from its country, for it had its origin in the native hills of the Magnesians." This, of course, is widely at variance with Pliny's fanciful derivation of the same name.

[1] Agricola: De Re Metallica, 1556.

Lucretius, however, who wrote many centuries after the event, is probably in error, for there is little, if any, magnetic iron ore in the hills of ancient Magnesia—the narrow and mountainous strip of land on which rise Mounts Ossa and Pelion, and which formed the most easterly province of Thessaly. The Magnetes—as the inhabitants called themselves—were, in fact, hemmed in between sea and mountains. The last formed a serviceable barrier against the Thesprotians when this tribe made its irruption into Thessaly; but when, through natural increase of population, the territory of the Magnetes became too restricted for their needs, there was no alternative but to cross the Ægean and seek new footholds on the Asiatic continent, where, Pliny says, they founded the city of Magnesia in Ionia. But a later arrival of Æolians drove them northward, and they established a second city, also named Magnesia, beside Mount Sipylus in Lydia. It is conjectured that their national pride caused them to retain the name of their old home for both settlements: a theory which gains support from the fact that the Æolians and Ionians, in founding new towns, were accustomed to adopt for them local designations.[1] It is this second Magnesia which is most reasonably supposed to have given its name to the magnet, because of the large deposits of magnetic ore, similar to that found at Elba, which still exist in its vicinity and which were probably the ancient source of supply. The town itself was destroyed by an earthquake in the time of Tiberius.

If this emigration of the Magnetes ever occurred, it happened before 700 B. C., and possibly before 1000 B. C., the latter being generally regarded as the period when the colonizing movement of the ancient tribes ended; but, like all such traditions, it is unsafe to accept it as a historical fact. Another version of the same story is that the Magnesians settled in both Lydia and Ionia on their return from Troy; still another makes them out, not willing

[1] Abbott, E. A.: History of Greece, New York, 1888.

THE ORIGIN OF THE MAGNET. 27

emigrants, but fugitives flying from Greece, and a third brings them, not from Thessaly at all, but from Delphoi.[1]

Divested of speculation, there remains simply the fact that there was a town of Magnesia close to a large bed of magnetite.[2] Klaproth[3] notes that this same settlement was called "Heraclea," whence the Greek term "stone of Heraclea" for the magnet; but there was also a town of Heraclea near the first Magnesia, and several other settlements, similarly named, in widely separated parts of Greece and Asia Minor, so that this derivation is also in doubt. Indeed, Pliny[4] regards the name "stone of Heraclea" or "Heraclea-lithos," not as based on locality, but as meaning "Herculean stone," for the reason already given, namely, the conquering power of the magnet over iron; and Professor Schweigger,[5] with labored ingenuity, goes even further, and asserts that "Herculean" and "magnetic" mean the same thing, and that the entire ancient myth of Hercules merely symbolized the natural strength of the magnet.

To these early traditions of the Greeks and Syrians, research into the dim historical annals of other peoples, existing at that far distant time, adds nothing of importance. A familiarity with electrical (or magnetic) effects is often attributed to the Egyptians of the Pharaonic periods; but this seems to be without trustworthy foundation. No legends of magnetic rocks or mountains on Egyptian territory have been encountered. But one Egyptian iron mine shows any signs of having been anciently worked, and, there the ore is of the specular or red, and not of the magnetic variety.[6] Lepsius considers that iron or steel do

[1] Cox, G. W.: A History of Greece, London, 1874.

[2] Trans. Phil. Soc., Cambridge; and Athenaeum, Jan. 4, 1834. See, also, Wilkin's Ed. of Works of Sir T. Browne. London, 1883.

[3] L'Invention de la Boussole, Paris, 1834.   [4] Lib. xxxvi.

[5] Ennemoser: History of Magic, London, 1854.

[6] Wilkinson: Anc. Egyptians, Boston, 1883, ii., 250. Rawlinson: Hist. of Anc. Egypt, London, 1881, 93.

not occur at all in the old empire, but only in the new.¹ Rawlinson,² on the other hand, while conceding the strength of the theory that iron was first introduced into Egypt by the Ptolemies, notes that some implements of the metal have been found in the tombs, with nothing about them indicative of their belonging to a late period; and that a scrap of iron plate was discovered by Vyse in the masonry of the Great Pyramid. He also points out that the paucity of such instances may be partially, if not wholly, accounted for by the rapid decay of iron in the nitrous Egyptian earth, or when oxidized by exposure to the air; so that, as he says, the most judicious of modern Egyptologists seem to hold that, while the use of iron in Pharaonic times was at best rare and occasional, nevertheless the metal was not wholly unknown, and may have been brought into the country from Phœnicia, in a manufactured state.

In such circumstances it is hardly possible to assume any Egyptian knowledge of the lodestone, due to direct discovery of it. The only apparently explicit evidence which has been encountered is the statement of Plutarch, that the Egyptian priest Manetho, who lived about three centuries before our era, and who wrote a history of his country for the Greeks who had recently settled there, reported that the Egyptians of a far distant period called the magnet the "bone of Horus," and the iron the "bone of Typhon." But Manetho's work, when Plutarch wrote about it, was six centuries old and existed only in the form of epitomes which were mutually conflicting, while his chronology is now known to be unreliable.³

It has been suggested that such iron as has been found in Egypt, and referred to Pharaonic times, may have been

---

[1] Lepsius: Die Metalle in den Aegyptischen Inschriften, 1872, 105, 114. Peschel: The Races of Man, New York, 1876, 488.

[2] Hist. of Anc. Egypt, i., 505.

[3] Rawlinson, cit. sup., ii., 6, 8. Cox: History of Greece, i., 614, Appendix D, wherein Manetho's chronology is fully discussed.

made and used by the Hebrews during their servitude, and that when they left the country they carried their knowledge with them. That they were familiar with the metal, at the period of Moses, and hence at about 1,500 years B. C., and possibly had known of it then for a long time, is shown by the mention of Tubal Cain,[1] "an instructor of every artificer in brass and iron," as a personage of great antiquity, at the very beginning of the Pentateuch. Their continuing knowledge of it, over many centuries, is further shown by the biblical references to the bed of iron of Og, the iron chariots of Javin, the miraculous floating ax-head of Elisha, the question "shall iron break the northern iron and the steel" in the Jeremiad, and many other instances, easily found. There are Jewish writers, moreover, who assert that not only were the Hebrews thus fully acquainted with iron, but that they were equally well aware of the magnet and its attractive force. The famous Rabbi Mosheh ben Maimon (Maimonides),[2] who wrote at the end of the twelfth century, mentions not only an image of the sun, in the Babylonian Temple of Belus, as maintained in suspension in the air by means of magnets, but avers that Jeroboam suspended the golden calves, which he commanded Israel to worship, in the same way.[3] No proof, however, seems to support this tradition, which, if true, would show the Hebrew acquaintance with the magnet to have existed at about 950 B. C. Kircher[4] quotes Rabbi Isaac Abarbanel, who wrote late in the 15th century, as authority for the statement that the Israelites knew of the magnet while wandering in the wilderness, and even used it in the construction of the tabernacle; but this again is yet more vague and doubtful than the ascription to Jeroboam.

[1] Genesis, iv. 32.
[2] Moreh Nebukhim (Guide to the Perplexed). Talmud, Tract, Senedrin, c. 3: Gemarah, c. Aegel.
[3] 1 Kings, xii. 28.
[4] Kircher: De Arte Magnetica. Rome, 1654.

CHAPTER II.

The Egyptian ports were, for the first time, opened to general foreign commerce by Psammetichus I., in 640 B. C. Thereupon a stream of immigrants from all parts of Hellas came pouring into the Nile land. Up to this time, Egypt had been a hermit nation, discouraging intercourse, restricting trade and prohibiting the circulation within her territory of foreigners, whom she regarded as cannibals and pirates. Nevertheless there had come to the outer world, reports of her magnificent cities, her great temples, and of a people so ancient and so learned, that, to the barbarians of the North, these stories seemed like legends of the gods. The curiosity of all men concerning her was keen and whetted with the expectation of centuries.

The Egyptian king had triumphed in the civil war against his colleagues by the aid of Greek mercenaries. The unbarring of the country to the men to whom he owed his throne was a political necessity, regardless of the involved violation of customs and traditions hoary with age. The change in national policy was radical, and, once made, the logical consequences followed. Not merely the Ionians, but the people of all Greece, and, in fact, of all states, flocked to the Delta of the Nile, and the swarthy and black-haired builders of the obelisks saw, for the first time, the red-haired and blue-eyed barbarians from the huts of the far north.

The Greek who came then to Egypt lived in a world greater than that which was included within the shadowy boundaries of Hellas, conterminous only with Greek speech and Greek customs. For he abided in one of his own creation, and it abided with him: a world peopled by his own fancy with deities, whose imaginary doings were part and parcel of his life, and which controlled his every action. Every phenomenon of nature to him was the work, voluntary or involuntary, of a personal agent. If the earth quaked, imprisoned giants were struggling against the bonds of the higher gods; Zeus wept in the rain-drops, and the tears of Niobe fell in the snowflakes. Every wood and every stream had for him its divinities. They ushered in the dawn and at night he saw them wandering through the sky. All nature was alive—all things were conscious things. There was no distinction between his mythology and theology, none between the latter and his system of religion, no question which the fictions of his brain could not answer, and no doubt which his imagination could not solve. If limits to his speculative faculty existed, they were to be reached only when it wearied of its own exuberance—a logical impossibility, perhaps, when the creator was the worshiper of his own creations. Equally were there no bounds to the theories which might be evolved to account for natural facts, provided each fact were fitted with its own theory, and the supernatural were open to constant invocation; but when it came to traveling outside of the ratiocinative circle, and to knowing things in themselves and formulating theories which would stand the test of explaining exactly ascertained facts, such conceptions in the mind of the Greek who lived six centuries before our era, had no more place than they have in that of the child who dwells in the happy world of the fairy books.[1]

The Egyptian of the same period claimed a national existence extending back for millenniums. His religion was of double aspect: a strict monotheism combined with a speculative philosophy on the two great subjects of the nature of God and the destiny of man, and a gross and multitudinous polytheism.[2] The intelligent, the learned

[1] Cox: History of Greece, cit. sup., 127.
[2] Rawlinson: History of Egypt, i., 505.

and the initiated were invited to contemplate in the first, a divine nature essentially unitary, pure spirit, perfect, allwise, almighty, supremely good; the ignorant masses in the second, a variety of gods ranging from heroes to bulls, cats, and apes, and a worship teeming with rites unspeakable.

Of science properly so called the Egyptian had none.[1] He claimed to have made records of natural facts for ages, such, for example, as astronomical observations, which, as he boasted, had been kept up for six thousand centuries. But out of this vast storehouse of accumulated data not a single theory explanatory of the motions of the heavenly bodies ever emerged. He heaped up facts as he did the stones of the great pyramid, with infinite labor, and over a great interval of time, but the mountain of facts was as lifeless as the mountain of stone. It was dead, it held the dead, and there was no health in it.

There lived at this time, a keen young Milesian,[2] of an intelligence far above the ordinary mental level of his countrymen; in character uniting the astuteness of the Phœnician, whence he sprang, with the impressionable temperament of the Hellene; one of those "souls born out of time extraordinary prophetic, who are rather related to the system of the world than to their particular age and locality."[3] Upon this phenomenal mind reacted an intellectual environment wherein the most diverse elements were commingled; conceptions of the spirit gods of Egypt, jarring with those of the anthropomorphic deities of Greece; dawning notions of physical astronomy jumbled together with the sports of the shining gods and goddesses in the blue vault, and no straight thought anywhere. The result was the beginning of philosophy; for when Thales of Miletus saw how the machinery given to man to understand facts could neither make the facts nor control them,

---

[1] Buckle : History of Civilization, i., 36.
[2] Plutarch : De Placet. Phil. 1, 3.   Clem. Alex. : Strom 1, 15, § 66.
[3] Emerson : Wealth.

how the most it could do was to react upon itself endlessly in endless circles of myths and shadows, he, for the first time in the history of the human mind, insisted upon finding, not in figments of the imagination, but in the things themselves, a theory intended to account for the phenomena observed. There was a great difference between doing this, however imperfectly or illogically, and referring the same happenings to the interference of the immortal gods. Thus, speculation disengaged itself from theological guidance, the effects of nature became no longer the sport of unseen beings, and the causes of all change were sought in the conditions of things themselves.[1]

Now the particular natural effect upon which Thales pondered, and for which he endeavored to account by a theory, physical through its connection with the thing itself—and not based upon supernatural influences—was the attractive power of the lodestone. And thus it came about that the mystery of the magnet gave the first impetus to philosophic thought.

Aristotle reports the sayings of Thales only by hearsay, and then with extreme caution: the first being that everything is full of gods, and the second[2] (and it is this which is of especial importance in our present research) that "Thales too, as is related, seems to regard the soul as somehow producing motion, for he said that the stone has a soul since it moves iron."

Thus we find the magnet at the very foundation of the world's philosophy. Refusing to account for the attraction of the lodestone by supernatural interposition, as the priests and worshipers at Samothrace had undoubtedly done centuries before, Thales assumed a soul or a virtue i..erent and existing *in the magnet itself*, whereby it was enabled to move the iron. Herein he perceived the manifestation of a first principle, common to all nature, which

---

[1] Lewes: Histy. of Phily., London, 1871, vol. I, 5.
[2] De Anima, I. 2; I. 5.

he conceived to be water—probably, says Aristotle, deriving his opinion from observing that the nutriment of all things is moist, and that even actual heat is therefrom generated and animal life sustained.

Writers of every age, from Aristotle onward, have agreed in regarding Thales as the father of philosophy, and yet very little is known of his life: Herodotus and Aristotle are nearest to him in point of time, and they furnish all that is even measurably trustworthy concerning him. Herodotus[1] describes, first, his prediction of an eclipse of the sun which brought to a sudden end one of the interminable series of battles which the Lydians and Medes were waging, and also that when the advance of the army of Croesus was impeded by a river, he caused a new channel to be made for the stream in rear of the camp, so that the water becoming divided into two branches became sufficiently shallow to be fordable. Modern re-calculation of the eclipse fixes its probable date, and hence the period when Thales lived, at 585 B. C.[2]

If so minute and cautious an investigator as Aristotle could obtain nothing more definite concerning Thales than such as is contained in the meagre statements which he gives, it is hardly to be expected that the commentators who came afterwards could have had any better sources for trustworthy information, especially as time has not brought to light a single writing which can be shown to be the Milesian's production. Nevertheless modern reviews of electrical progress seldom fail to ascribe to Thales the conception of a soul in the amber as well as in the lodestone. The doubtful foundation of this resides in a single sentence in the so-called life of Thales with which begins the "Lives and Opinions of Eminent Philosophers," written by Diogenes Laertius. Laertius is supposed to have been a native of Laerte in Cilicia, and the time when he lived, judging from the periods of the

---

[1] Herod. : I. 74, 75.
[2] Todd : Total Eclipses of the Sun, Boston, 1894.

writers whom he quotes, appears to have been during the last part of the second century of our era—or in other words, about as far distant from the age of Thales as we are from that of William the Conqueror. If, according to other opinions, he did not live until the time of Alexander Severus, and wrote the book for Julia, the consort of that emperor, who was of a philosophical and platonic turn of mind, there is still a wider gap between him and the ancient Greek.

The sentence which he gives is:

"But Aristotle and Hippias say that he attributed souls also to lifeless things, forming his conjecture from the nature of the magnet and the amber."

As a matter of fact, Aristotle says nothing about the amber, and that he should have knowingly omitted mention of it in the passage above quoted is difficult to believe. On the other hand, while Plato, in the Timæus at a later period, speaks of the "marvels that are observed about the attraction of amber and the Heraclean stone," he does not connect Thales with them. Hippias was a traveling Sophist, and a contemporary of Protagoras and Socrates, but none of his writings are extant.

It is necessary merely to glance at the remarkable collection of stories which Laertius has gathered about Thales to see that he has simply brought together items of gossip and tradition which had been accumulating for centuries.

Apuleius,[1] who lived either contemporaneously with Laertius or nearly a century earlier, gives another and different category, in which the amber-soul theory is ignored. Add to this that Laertius refers to no less than five "other men of the name of Thales," including at least one "painter of Sicyon, a great man," and none unknown to fame, a not unnatural suspicion arises that the biographies of all these may have been laid under contribution for the delectation of the fair Julia. "All those letters which are attributed by Laertius to the Philosophers," remarks Julius

[1] Apuleius : Floridor, 361.

Scaliger,[1] as usual, savagely, "I am able to prove, by many arguments, were concocted by the Greeks, in whom the will or faculty for lying never failed."

Let me now recapitulate. We have found lack of evidence to prove that the Egyptians, at the time of Thales, were cognizant of the magnet. Therefore it may be assumed that Thales did not acquire whatever knowledge he may have had concerning this substance from Egyptian sources. We have also found that the working of iron mines in Phrygia was of great antiquity, that magnetite ore existed there and in Lydia, and probably was abundantly disseminated through Asia Minor. So also it appears that the magnet was exhibited as a part of the Samothracian mysteries, which were also of extremely ancient origin. It is not unreasonable, therefore, to conclude that Thales' knowledge of the magnet was home knowledge, and that his doctrine of the soul inherent therein, was intended to be in direct contrast with the prevailing theories fostered by the priests of the Cabiric mysteries, namely, that the stone was supernaturally influenced.

If the tradition of the Syrian women is older than the time of Thales, it may be presumed that the amber attraction was not unfamiliar to him; otherwise I have encountered no direct evidence of earlier knowledge of it than exists in the Timæus of Plato, and Plato lived nearly two centuries after Thales.

The explanation given by Plato excludes all idea of attraction. "Moreover," says the philosopher, "as to the flowing water, the fall of the thunderbolt, and the marvels that are observed about the attraction of amber and the Heraclean stone ; in none of these cases is there any attraction, but he who investigates truly, will find that such wonderful phenomena are attributable to the non-existence of a vacuum, taken in combination with the fact, that these

---

[1] Ep., 306. See also Blount : Censura Celebriorum Authorum. Geneva, 1710, 158.

substances are forced round and round and are changed and pass severally into their own place by composition and divination."[1]

It must be admitted, however, that even if Thales had been cognizant of the amber phenomenon, it was not logically necessary, from his point of view, to include it specifically under his theory based upon the attraction of the lodestone : and hence lack of mention does not, on his part, imply lack of knowledge. All physical philosophy as it stood before the age of Socrates was an obscure, semi-poetical speculation as to first principles. It neither sought to explain nor to clear up phenomenal experiences, but often added new difficulties of its own, frequently contradicting or discrediting experience. In the words of Grote, "Thales and his immediate successors (like their predecessors, the poets), accommodated their hypotheses to intellectual impulses and aspirations of their own, with little anxiety about giving satisfaction to others, still less about avoiding inconsistencies or meeting objections. Each of them fastened upon some one grand or imposing generalization (set forth often in verse), which he stretched as far as it would go by various comparisons and illustrations, but without any attention or deference to adverse facts or reasonings. Provided that his general point of view was impressive to the imagination, as the old religious scheme of personal agencies was to the vulgar, he did not concern himself about the condition of proof or disproof."[2]

Plato while denying the attraction of the amber nevertheless links its effect with that of the magnet; but as to what it acts upon or wherein its action differs, if at all, from that of the Heraclean stone, he is silent.

---

[1] Plato: Timæus, 80. Cicero refers to this in the De Natura Deorum, and so does Timæus of Locri, reputed to have been Plato's teacher, but whose sole extant work is probably an abridgment of the Platonic Dialogues. (Timæus Locrensis, ed. Serrani, p. 102. See, also, Smith: Dict'y of Greek and Roman Antiquities, art. Timæus.)

[2] Grote: Aristotle. London, 1872; Vol. II., chap. XI., p. 154.

Where then does first explicit proof of the amber phenomenon, in fact, exist?

In the spring of B. C., 334, Alexander of Macedon crossed the Hellespont and began the famous campaign which left him master of all the countries between the Danube and the Ganges. At about the same time, Aristotle, who had been his preceptor, established a school at the Lykeum at Athens, and began to gather collections of plants, animals and minerals, wherewith he illustrated his lectures, delivered while walking up and down the leafy paths which wound through the adjacent gardens. In this undertaking he found in his powerful disciple a most willing ally—for Alexander not only contributed a vast sum of money for the purchase of rare objects, but employed thousands of men to collect and transport to Athens all that was strange to the Greeks in the distant countries which had yielded to his arms.[1]

To the gathering of this stupendous mass of material may be traced three results of the highest import; first the acquisition of the multitudinous physical facts which fill the Aristotelian treatises on natural sciences. Second, the foreshadowing of the inductive method of reasoning. Third, the production by Theophrastus, the Lesbian, of a history of stones, probably based directly upon the study of Aristotle's collections.

I have said that Aristotle foreshadowed the inductive theory. As any intellectual rise, coincident in time with that of this great principle, must have been more or less controlled by the mightier mental advancement, some explanation of this statement is perhaps here necessary. Because Aristotle gathered as has been stated a vast mass of facts, it has been frequently maintained that the process which Bacon calls that "double scale or ladder, ascendent and descendent, ascending from experiments to

---

[1] Grote : Aristotle, I. I. 12.

the invention of causes and descending from causes to the invention of new experiments,"[1] was not only foreshadowed but conceived by the Stagirite; even more than this, elaborated into a logical tool ready for the world's use. This view I have not taken. Although the duality of the complex operation, whereof induction is the first and deduction the second half, as well as the especial necessity for the inductive part, was recognized by Aristotle both in actual declarations and by his unwearied industry in collecting facts; although, moreover, he perceived that all science or theory must rest upon this foundation as a whole, nevertheless he devotes himself only to the analysis and to the formulating of the rules of the deductive part. Thus it was, as Grote[2] points out, that science afterwards became disjoined from experience and was presented as consisting in deduction alone, while everything not deduction became degraded into un-scientific experience. Of this last, abundant examples in the field under study will hereafter be encountered, while on the other hand, we shall find the true inductive method practically applied in the same field long before Francis Bacon trumpeted its importance to the world.

Theophrastus was born B. C. 372, and died B. C. 287, surviving Aristotle by thirty-five years, and succeeding him as teacher at the Lykeum. His history describes what he calls the stones and the earths, in contradistinction to the metals; the first, as he supposed, being derived from the earth itself, and the last from water. He refers not merely to stones indigenous to Greece, but to others, of foreign origin, such as the alabaster of Egypt, the pumice of Sicily, the carbuncle of Carthage, Massilla, and of the Nile cataracts and Syene, the emeralds of Tyre, Cyprus, and Bactria, the pearls from the Indies and the shores of the Red Sea, the gypsum of Syria, the cinnabar of Spain, and so on, through a category so extensive, and represent-

[1] De Augmentis, vii. 1.
[2] Grote: Aristotle 1. c. 289; c. 160.

ing the minerals of so many different and distant countries, that little doubt can remain that he wrote the book with the collections of Aristotle directly before him. Here, for the first time, is given definite information concerning the amber attraction. "Amber," he says, "is a stone. It is dug out of the earth in Liguria, and has a power of attraction. It is said to attract not only straws and small pieces of sticks, but even copper and iron, if they are beaten into thin pieces."

Then, bringing the amber and the lodestone into the same attracting class, he adds:

"But the greatest and most evident attractive quality is in that stone which attracts iron. But that is a scarce stone and found in but few places. It ought, however, to be ranked with these stones, as it possesses a like quality."

It is a significant circumstance that there is no suggestion of the soul animating the stones contained in Theophrastus' terse and practical account of their qualities. Their concretion, he says, is due to heat or cold, some kinds of stones being occasioned by the one cause, others by the other; they differ likewise in the matter and manner of the affluxes of the terrestrial particles from which they are formed, and likewise they have "powers" of their concreted masses, which are different from their qualities of hardness, color, density, etc., and which include their capacity for acting upon other bodies or being subject or not subject to be acted upon by them. Thus, he points out, some are fusible, others not so, and others can color water or cause petrifaction, and among these powers is included the attractive quality.

There is no regarding this as anything but a strictly scientific and material view of the subject, which if taken in Aristotle's time, may perhaps account for that philosopher's doubtful and cautious dealing with Thales' theory of the prevailing soul. The calm and terse enumeration of physical characteristics, and the theories and classifications based thereon, are as far distant from the crude spiritual conception of Thales as the last is removed from the older belief in the direct interposition of the gods. It is not difficult even to imagine that Theophrastus looked upon the Milesian doctrine with something of the disdain with which the modern astronomer regards the planetary speculations of the astrologers, or the modern chemist the theories which once gave rise to the hope of achieving the transmutation of metals.

Besides referring to the attractive qualities of the lodestone and the amber, Theophrastus, for the first time, announces the existence of a third substance having identically the same properties as the amber, which he calls Lapis lyncurius or lynx stone. He describes this as used by engravers as the emerald is used, and that it has a very solid texture, in confirmation of which, and also of the statement of the identity of its attractive quality with that of amber, he appeals to Diocles, an eminent physician of Charysta, who is said to have ranked second only to Hippocrates, but of whose works only a few fragments are known.

It is, he says, pellucid, of a fire color, and is found by digging; and then, with some detail, he declares it to be derived from the secretions of the lynx—whence its name.

The precise nature of the lyncurium has long been a bone of contention, and speculations concerning it have been voluminous. The wrangle, occurring as it did in the Middle Ages, is representative of the intellectual condition of the times. From discussions as to what Theophrastus meant, the commentators fell to arguing about what they themselves meant, and the gloss writers of one century expatiated upon the signification of the language of gloss writers of the preceding century, and words were heaped on words, until all sight of the original subject-matter seemed to be lost. This continued until the end of the seventeenth century, when the tourmaline and its attractive effect, when heated, became known, and thereupon the contest ended as illogically as it had continued, in the generally accepted notion that it was the tourmaline to which Theophrastus referred.

Nevertheless there is nothing in the statement of Theophrastus to warrant any such inference. He says that the stone has the same attractive properties as the amber, but not that these are excited by heating instead of by attrition. The amber, he states, comes from Liguria, one boundary of which was the Eridanus or Po river, on the banks of which, as we have seen, the Greeks, from the time of Herodotus, erroneously supposed the resin to be found. Long before the time of Theophrastus, the Ligure or Ligurian stone was well known. In both the original Mosaic version of the Scriptures and in the Septuagint, the "ligure" is the seventh stone in the breastplate of the high priest,[1] and it is likewise the seventh stone in the covering of the King of Tyre[2] in the Septuagint, though not in the original. It may be, therefore, that confusion was caused by the similarly sounding names of the Ligure or Ligurian stone, which was the amber, with the Lyngurian stone derived from the lynx—a substance which Pliny denounces as wholly mythical and non-existing.[3]

---

[1] Exodus xxxiii. 17–20.   [2] Ezek. xxviii. 13.

[3] Pliny: lib. xxxvii. c. 13. Marbodeus, Archbishop of Rennes, has on the title page of his poem on Gems, attributed to the ancient Arabian author Evax, a picture of the Jewish high priest wearing the breastplate, one stone of which is marked "lincurius," and in his commentary he gives the word as "lyngurius" (Marbodeus Gallus, Cologne, 1539, p. 39). Erasmus in his commentary on St. Jerome, says that "lyngurius" and "ligurius" are the same thing, and so does Dioscorides (Lib. 37. 3). Camillus Leonardus (The Mirror of Stones, Venice, 1502, Eng. Trans., London, 1750) notes the "lychinus" or "lychnites" as an "Indian gem red in color," and mentions two species, one of which, purple in color, being heated by the sun or by friction, attracts straws. This suggests of course the tourmaline. But to the "lyncurius" or "lyncis" he attributes no attractive quality, and he further notes the "ligurius," which he says is "like the electorius and draws straws." Iolinus (lib. iii., Utrecht, 1689, p. 59) agrees with Leonardus in defining the "lychnites,"

THE LYNCURIUM. 43

The weight of opinion of the old writers is to the effect that the lyncurium and the amber were the same thing. And so the lynx stone may be relegated to a place in that cloud of delusions which always has darkened and probably always will obscure the path of science. For the long dispute concerning it, the antiquarian may find some pleasure in substituting the question whether Theophrastus erred or whether the stone had its true origin in the ignorance of that ancient bibliophile, Apellikon of Teos, who found the original manuscripts of the philosopher nearly destroyed after some two centuries' exposure to the damp and worms of the cellar of the heirs of Neleus, and proceeded to fill up the gaps after his own fashion.[1]

but says nothing about its attraction when heated. De Boot (Gem. et Lap. Hist., Leyden, 1636) declares that "lychnites" is a kind of marble, and ascribes no attractive power to it, and gives the "lyncurius" as clear like amber, drawing straws and light bodies in the same way. See Aldrovandus, Musæum Metallicum, Bologna, 1636, p. 405; also Agricola, Della Natura de le Cose Fossili. Lib. IV., Venice, 1549, p. 236.

[1] Strabo, xiii., 609.

NOTE.—If a third substance, having the same attractive quality as the amber, was known to the ancients, it was probably jet—a species of lignite resembling cannel coal, but harder and susceptible of a high polish. It does not seem possible, however, to resolve that doubt, owing to the many kinds of coal and other fossil deposits which not only old writers but even modern commentators constantly confuse. Theophrastus speaks of a material which is plainly anthracite coal, and Pliny (xxxvi. 18), of the Gagates, his description of which answers generally to that of jet; but neither author mentions any phenomenon similar to that of the amber as pertaining to it. Later writers apply the word "gagates" to almost any black bituminous material, though they commonly mean "jet" by the term. Leonardus regards the gagate as another species of amber—"black amber"—in contradistinction to yellow, and he describes it as "black, light, dry and lucid, not transparent, and if put into fire has, as it were, the smell of pitch. Being heated with rubbing it attracts straws and chaff." Marbodeus gives almost the same account and states that it is found in Britain, where it is still obtained in the tertiary clays along the Yorkshire coast. This unfortunate confusion of yellow amber and jet, probably first due to Leonardus, has rendered it impossible to tell, from the references to amber attraction by the writers of the sixteenth and even of the seventeenth century, which substance is meant.

Mythology, the controlling factor in the world's intellectual progress, had given way to philosophy, and now philosophy in its turn was beginning to yield its power into the hands of science.

The first great university of Alexandria, begun under Alexander the Great, flourished under the patronage of the Ptolemies for nearly four centuries. It was the gathering place for philosophers from every part of the world. Its students at one time numbered fourteen thousand souls and its libraries contained seven hundred thousand volumes.

Here were made the discoveries of Archimedes in mechanics, of Euclid and Apollonius Pergæus in mathematics, of Hipparchus in astronomy and with the ælopile of Hero, here began the steam engine. All of this great work was done before the year 150 B. C. We need only compare the category of Hero's inventions with the single material notion of Thales, to perceive the radical change in thought which had occurred. It is the contrast of the force-pump and the water-soul. It was not the crude and imperfect classifications of Aristotle which accomplished this. The inductive theory in that stage of the world's history could not have established itself, not merely for want of knowledge of a sufficiency of facts which would demonstrate its truth in any particular instance, but also because there was no group of natural facts which could be clearly seen, unobscured by mists of attending speculation and superstition.

Amid all this activity the progress which was made in knowledge of the amber and of the lodestone was very small. Pliny[1] has the dubious assertion that the architect Timochares began to erect a vaulted roof of lodestone in

---

Singularly enough, as we shall see in dealing with the first-named period, it appears not at all unlikely that the English were then much more familiar with the attraction of jet than they were with that of amber.

[1] Pliny: lib. xxxiv. 42. Vitruvius: De Archit., lib. iv.; time, circa 31 B. C.

MAGNETIC SUSPENSION. 45

the Temple of Arsinoe (wife and sister of Ptolemy Philadelphus) at Alexandria, in order that the iron statue of the queen might have the appearance of hanging suspended in the air. But this work was never accomplished, says the historian, because both the king and the architect died.

This is the same story which, as we have seen in the preceding chapter, the Jewish writers tell of the suspended golden calves of Jeroboam, and the world has never been able to get rid of it. Again and again has it been pointed out, for a thousand years and more, that no piece of iron can be balanced in the air by magnetic attractions oppositely exerted; but the vitality of the falsehood seems even greater than that of the refutations. At the same time there can be little doubt that in some temple, and probably one in Egypt, and at about the time of the University of Alexandria, there was an object held up apparently by no other support than magnetic attraction; and very probably held down by a wire or cord invisible to the spectators. Ausonius[1] directly disputes the statement of Pliny that the construction of a magnetic vault was abandoned. St. Augustine,[2] St. Isidore,[3] and Cedrinus[4] all affirm the existence of the iron statue suspended between ceiling and pavement. Clement[5] of Alexandria causes the Sibyl to sing of "thou, Serapis lying amid rude stones, thou fallest most miserable in the ruins of Egypt," and his scholiast, Clycas, interprets the "lapides rudos multos" as magnets, of which, he says, "many were used in the temple of Serapis on all sides of an iron sun." So that the statue of Arsinoe, in her own temple never completed, may have become confused with an iron sun which did

---

[1] Eidyllum x, Mosella, vers. 314, 320; time, circa 390 A. D.

[2] De Civ. Dei, lib., 21, 6; time, circa 415 A. D.

[3] Originum, lib. xvi., cap. 4; time, circa 595 A. D.

[4] Geo. Cedrinus : Compend. Hist., c. 267 ; time, circa 1057 A. D. Also Suidas : Lex. cit. sup. Art. Magnet; time, circa 1081 A. D.

[5] In Protreptico, 15; time, circa 192 A. D.

exist in the Serapeum; and that there were, in fact, two such different things Ruffinus[1] and others assert.

But note the expansive character of the tradition, and the variety of its transmutations. The horse of Bellerophon, on the island of Rhodes, says the venerable Bede,[2] weighed 5000 pounds, and was suspended by magnets. Martial[3] says that the effigy of Mausoleus was held over his tomb in like manner. As the story grew older, King Theodoric,[4] in a letter to Boesius, applies it to a statue of Cupid in the temple of Diana of Ephesus. And then last, but not least, it reached its final resting place in the legend of Mahomet's coffin. Since this myth furnishes the substance of one of the most common metaphors in use, the facts on which it rests, or rather does not rest, are worth stating.

After Mahomet's death, the Meccans and Medinans disputed possession of the body. Still another faction wished the sepulchre to be in Jerusalem, as the proper place of burial for all prophets. Finally Abu-Bekr interfered and announced that he had heard Mahomet himself during his life direct the selection of Medina. Thereupon a vault was dug beneath the spot where stood the bed on which the prophet slept, in the house of Ayesha. In order to keep the coffin clear of the floor of the vault, it was supported on nine bricks, the earth being heaped about the sides. That is the entire extent to which the coffin was suspended in the air—namely, by nine bricks put under it.[5]

[1] Ruffinus : Aquil. lib., vi. Histor., c. 22; time, circa. 390 A. D. S. Prosperus : De Praedicatione, 3, c. 38; time, circa 446 A. D.

[2] Beda.: de Sept. Mirac. Mundi ; time, circa 703 A. D.

[3] Lib. De Spectaculis, time, circa 78 A. D.

[4] Cassiodor.: Variat Lib., 1, Ep. 45; time, circa 500 A. D.

[5] Gagnier: Histoire de la Vie de Mahomet.

Gibbon's note (the Decline and Fall of the Roman Empire, chap. 50) as to this is as follows: The Greeks and Latins have invented and propagated the vulgar and ridiculous story that Mahomet's iron tomb is suspended in the air at Mecca ($σῆμα\ μετεωριζόμενον$, Laonicus Chalcondyles: De Rebus Turcicis, 1. iii. 66) by the action of equal and potent

The Mahometans have always ridiculed the tradition, and certainly it is exceedingly difficult, short of assuming it to have been made out of whole cloth, to find any basis for it in the facts above stated. There is, however, another version, credited to one Bremond,[1] an indefinite "traveler of Marseilles," who asserts that he saw "above Mahomet's tomb a magnet, two feet long and three fingers thick, from which is suspended a golden crescent enriched with jewels, by means of a big nail in the middle;" but this obviously lacks the essential feature of the something being held floating in the air by magnetic attraction.

Meanwhile, the knowledge of the magnet had spread beyond the confines of Greece and Asia Minor, in other directions than to the southward. It had moved to the west and to Rome. The Roman, Lucretius,[2] in that greatest of all didactic poems, "On the Nature of Things," tells of the Samothracian rings as still existing (95 to 52 B. C.), and as having been seen by himself.

"You may see, sometimes," he says, "five or more suspended in succession and tossing about in the light airs, one always hanging down from one and attached to its lower side, and each in turn, one from the other, experiencing the binding power of the stone: with such a continued current its force flies through all."

Here is the first suggestion of a moving current traversing a conductor, in contra-distinction to a soul or virtue merely pervading the object. The distinction between the

lodestones (Dict. de Bayle. Mahom. Rem. E E. F F.). Without any philosophical inquiries, it may suffice that, 1. The prophet was not buried at Mecca; and 2. That his tomb at Medina, which has been visited by millions, is placed on the ground. (Reland: de Relig. Moham., 1. ii., c. 19, p. 209-211.)

[1] Azuni: Dissertation sur la Boussole, Paris, 1810, p. 27.

[2] Lucretius: De Natura Rerum, Book 6. Translated by H. A. J. Munro. Cambridge, 1866.

magnetic current flowing through the rings and its effect exerted upon the space around the magnet is also drawn, for in addition to the continuing current, Lucretius says that there streams from the stone "very many seeds, or a current, if you will, which dispels, with blows, all the air which lies between the iron and the stone," thus producing, as he imagines, a vacuum in front of the iron, into which the air pressure "thrusts and pushes it on, as the wind a ship and its sails;" and on this theory he accounts for attraction. Furthermore, as Lucretius describes his "streams" as continuously circulating around the lodestone, the vortex magnetic theory of Descartes is here curiously foreshadowed, if not actually suggested.

Up to this time, as we have seen, there is nothing in the ancient authors indicating any knowledge by them of the repulsive effect of the magnet. It is always spoken of as drawing the iron. When, however, two magnets are brought together, attraction occurs only when their *un*like poles are presented to one another—the north pole attracting the south, and *vice versa*. But if like poles are approximated, just the opposite result happens, and the magnets mutually repel. It is immaterial whether two lodestones, or one lodestone and a magnetized piece of iron, or two magnetized pieces of iron, such, for instance, as two compass needles, be employed; the result is always the same. Hence, as iron that has been brought into contact with the lodestone (as was the case with the Samothracian rings) very readily becomes magnetized by induction from the stone, it is evident that there was a possibility of two rings having become magnetized in this way, being accidentally approximated with their like poles facing one another, and under conditions when one or the other of them might be free to move under the repulsive force. Whatever may have been observed as to this at an earlier time is not known; but an unmistakable and, probably, the first recorded recognition of the phenomenon appears in the poem of Lucretius.

"Sometimes, too," he says, "it happens that the nature of iron is *repelled* from this stone, being in the habit of flying from and following it in turns."

The allusion is now, not to the current which flows through the rings, but to the influence of the stone upon the iron, merely placed in its neighborhood—or, as we now say, in its "field of force" and not in contact with it. He is describing the turning of the ring, so as first to present one pole to the lodestone and then the other, for a ring usually has its poles located diametrically opposite each other. If the ring were supported so that its poles could be thus alternately presented to one and the same pole of the lodestone, then, whenever the ring pole was of the same name as that of the lodestone [as north pole to north pole, or south pole to south pole], the ring would be repelled, and would swing away from the lodestone; but if the ring pole were of different name from that of the lodestone [as north pole to south pole, or south pole to north pole], then the ring would be drawn to the lodestone, and if the latter were moved, the ring would follow it. Hence, by turning the ring to and fro, as on an axis, it could thus be made to swing or vibrate backwards or forwards in front of the lodestone, or, as Lucretius explains, the ring will fly from or follow the stone "in turns." Here is the first foreshadowing of the motion of an armature—for such is the ring—before the pole of a magnet, by change in relative polarity of magnet and armature; in the light of present knowledge we might even regard this as the advent into the world of the conversion of the energy of electricity into mechanical motion, and the germ of the electric motor.

Lucretius says, further, that he has seen the Samothracian rings "jump up" when the magnet stone had been "placed under." It is unquestionably true that in a suspended chain of rings, as he describes, the pole at the bottom of the lowest ring would be of the same name as that of the pole of the supporting lodestone—say, north. If now, the same or north pole of a second lodestone were brought up to the lower part of that last ring, then that ring would be repelled and "jump up"—exactly as Lucretius says.

Even more remarkable than this is his statement that iron filings "will rave within brass basins" when the stone is placed beneath. This was the first perception of the field of force about a magnet by noting not merely the effect of its attraction or repulsion exerted upon the pole of another magnet brought into it, but upon loose iron filings free to dispose themselves therein along the lines of force. Then, under the astonished gaze of the poet, the particles of metal arranged themselves in the curious curves of the magnetic spectrum, and rose like bristles in front of the poles. And as he moved the stone beneath the brass basin which held them, he saw them fly from one side of it to the other, sometimes grouping themselves for an instant in dense bunches, then leaping apart and scattering all so incoherently and so wildly, that it is small wonder that he regarded them as raving in their frantic desire to break away from the mysterious force. We shall find the performances of these raving iron filings astonishing the philosophers of the sixteenth century and remaining always a puzzle until Faraday and Maxwell found the key to it within our own time.

The explanation which Lucretius gives of magnetic attraction is repeated by Plutarch[1] who wrote a hundred and fifty years later and who applies it also to the amber attraction. He says, "that amber attracts none of those things that are brought to it, any more than the lodestone. That stone emits a matter which reflects the circumambient air and thereby forms a void. That expelled air puts in motion the air before it, which making a circle returns to the void space, driving before it towards the lodestone, the iron which it meets in its way." He then proposes a

[1] Plutarch : Platonic Quaest., tom. 2.

difficulty "why the vortex which circulates around the lodestone does not make its way to wood or stone as well as iron," and, again like Descartes, answers, that "the pores of the iron have an analogy to the particles of the vortex circulating about the lodestone which yields them such access as they can find in no other bodies whose pores are differently formed."

Plutarch also refers to magnetic repulsion and says that "like as iron drawn by a stone often follows it, but often also is turned and driven away in the opposite direction, so also is the wholesome good and regular motion of the world."

It must not be assumed, because of the interpretations which it is possible to make at the present time of the magnetic phenomena mentioned by Lucretius, that any actual knowledge of the polarity of the lodestone existed in his day. Not until centuries later did this come to the civilized world.

Even when in course of time the recurrence of the repelling effect of the magnet attracted attention, no conception of polarity resulted. On the contrary, it was for a long time believed that the stone which repelled was a totally different stone from that which attracted iron. This supposed repelling stone is described for the first time by Pliny,[1] who calls it the "theamedes" and says that it comes from "Ethiopia, not far from Zmiris." For the first thirteen centuries of our era, belief in its existence was implicit. It served conveniently to explain magnetic repulsion, and hence, as frequently happens in such circumstances, it prevented investigation of that effect.

For discoveries concerning the amber, search may now be made through many centuries in vain. Plato, as has been stated, had linked together the attraction of the amber and the Heraclean stone, and Epicurus had attributed both to the same cause, namely, atoms and invisible

---

[1] Pliny: lib. xxxvi. 25.

bodies outwardly projected from the attracting body combining with and bringing back the body attracted. That seems to have convinced the Greeks and Romans then, and the rest of the world for the ensuing two thousand years, that the amber and the magnet were interrelated; or, at all events, that they both attracted for exactly the same reason, and therefore nothing was to be gained by looking into the subject further. As for the Egyptians, it is doubtful whether they ever brought amber into extensive use at all, before quite a late period of their history. Only a few amber beads have been found in their tombs, and these last were of the 2d and 3d centuries of our era.[1]

The great Greek physician, Asclepiades,[2] recommends pills of amber as a specific for hemorrhages, and that seems to be the first medical use of the resin. His equally eminent brother of Rome[3] has scant mention of it in his great work on materia medica.

All that the civilized world had learned concerning the lodestone and the amber has now been in substance stated. It is briefly summed up in the knowledge of the attractive capacity in each, of the ability of the magnet apparently to transfer its powers to iron, and of the existence of (supposedly) a kind of lodestone by which iron is repelled.

[1] "An amber necklace, about 22 inches long, was also found in a grave here—one-third of it—the small beads only were kept at Bulak, as amber was almost, or quite, unknown in Egypt before." Tanis. 2d Memoir. Egypt. Explorat. Fund. W. F. Petrie. London, 1889. *Per contra* Clemens (Clem. Alex. Paedagog. iii. c. 2,) speaks of the sanctuary in Egyptian temples as shining "with gold, silver and amber." Possibly the word "amber" here is a mistranslation of the similar term for the electrum alloy. See Wilkinson: Anc. Egypt, i. 246, Boston, 1883.

[2] Lib., vii., de Comp. Med. Time, circa 200 A. D.

[3] Lib. de Simp. Med. See for this and preceding reference, Aldrovandus, Musaeum Metallicum, Bologna, 1648, p. 415.

CHAPTER III.

How or when the tendency of a freely-suspended magnet to set itself in a nearly north and south direction was first discovered is a question, the answer to which is probably forever lost. The civilized world remained in ignorance of the fact for nearly eighteen centuries after the attractive effect of the lodestone had become well known. Although, as I have already stated, it is not impossible to conjecture that the phenomenon was familiar to the ancestors of primitive civilization, who, from the highlands of Central Asia, dispersed in many races over the earth; yet the knowledge came to the people of the Middle Ages anew, through the invention of the first and greatest of electrical instruments—the mariner's compass; first, in its utilization of the mysterious force existing in the magnet; greatest, in that it has contributed more than any other product of human intelligence to the progress and welfare of mankind.

The obscurity which veils the discovery of the underlying principle of the compass in the remote past seems to extend to all the circumstances in which that contrivance originated. It has been ascribed to the Greeks, the Phœnicians, the Etruscans, the Egyptians and the Chinese. It is said to have first appeared on the ships of mediæval Italy, and yet to have been first known in mediæval France. It is also claimed as German, Arabian, English and Norse.

It is necessary to examine briefly the principal arguments advanced in behalf of these several nations. In this way we shall best perceive the conditions which caused progress or checked it, and so trace through its many channels the rise which we are following.

The review of Greek knowledge of the magnet, already made, is, perhaps, in itself sufficient to show how slight must be the basis for any hypothesis that the compass is of Hellenic origin. The commerce of ancient Greece was of limited extent, and did not involve long voyages—her ships, in fact, entering the Turrhene seas in constant fear of the Etruscans. They were held as interlopers on the west coast of Italy even up to 533 B. C.,[1] the carrying trade meanwhile being mainly confined to the Carthaginian and Etruscan fleets. Nevertheless much has been written in support of the theory that Homer was familiar with the compass because, in the Odyssey, he speaks of the Phocian ships which sailed "tho' clouds and darkness veil the encumbered sky"—the argument being that ships could not possibly "fly fearless" through darkness and clouds, unless provided with a binnacle and its appurtenances.[2] Such contentions are hardly worthy of serious consideration. The application of similar reasoning to the passage in the same poem which mentions

> "Wondrous ships, self-moved, instinct with mind,
> No helm secures their course, no pilot guides,
> Like man, intelligent they plough the tides,"[3]

might with equal propriety be taken to show the familiarity of the bard with steam, and possibly electric, propulsion, or even with the still unsolved problem of automatic steering.

The long voyages of the sailors of Sidon and Arvad have led many to regard the compass as of Phœnician origin, under the assumption that such journeys could not have been made without its help. The writers of the seventeenth century are fond of asserting that the Phœnician

[1] Gray: History of Etruria, I, 173.

[2] W. Cook: An Inquiry into the Patriarchal and Druidical Religion. London, 1874. Cook's argument is upheld by Salverte: Philosophy of Magic (trans. by Thomson), N. Y., 1847, vol. II.

[3] Odyssey, viii, 610.

ships sent out by King Solomon must have been equipped with it, because it is no more than reasonable to assume that Solomon's wisdom included such valuable knowledge. On the other hand, remarks the old chronicler,[1] relaxing his gravity for the sake of the pun, "Solomon had all the knowledge necessary to Morall, Politike and saving wisdom, and to the end for which God gave him so large a heart. But the sea hath bounds, and so had Solomon's wisdom. Somewhat was left for John Baptist to be greater than he, or any borne of women. Neither was the knowledge of the compass necessary to Solomon, who, without it, could and did compass the gold of Ophir."

The fact that Phœnician vessels went to this Ophir was also deemed another good reason for believing the needle to have been on them; this, mainly, because no one could say definitely where Ophir was, and hence nothing was easier than to insist that its situation lay at the very ends of the earth, whither ships could not possibly find their way unaided. Thus, some writers place Ophir in Peru, others at the extremities of India, from which last place the traveler Bruce removed it.[2] The geographer D'Anville[3] subsequently found a suitable situation for it in "the Kingdom of Sofaula," in Africa.

Finally, however, the chroniclers concluded it to be safer to rest upon the tradition that it took Solomon's ships three years to go to Ophir (wherever it was) and return; hence, on the chronological argument only, they insisted that the distance must have been vast. But Huet,[4] Bishop of Avranches, disposed of this inference by explaining that the first year was used for the outward voyage and the second for the return, and the third for laying up and repairing the ships; and then he adds with much wisdom,

---

[1] Purchas, his Pilgrims. 1, ¿ 8.
[2] Bruce: Travels in India. Book II., Chap. IV.
[3] Venanson: De l'Invention de la Boussole Nautique, Naples, 1808.
[4] Huet: Des Navigations de Solomon, c. 8, 3.

"It is great error to judge of ancient navigation by present. To-day sailors go on at night and in cloudy weather, while anciently they came to anchor. The ancients followed every angle and sinuosity of the coast. The author of the Periplous of the Red Sea proves that the Egyptians got to India only by following the coast in little ships," and he closes with Pliny's even more sagacious remark—"The desire for gain rendered India less distant than the rest of the world." The appearance of Phœnician ships in the Persian Gulf in 697–695 B. C. gave, however, great impetus to commerce with the far East, for they were much larger, better built and more sea-worthy than the vessels of the Babylonians and Assyrians. Voyages in the Indian Ocean in search of new markets then became longer, and finally the southern shores of Shantung (East China) were reached in about 675 B. C.[1]

There is no trustworthy evidence, however, that the Phœnicians, despite their skill as iron workers, had any knowledge of the directive property of the magnet. Their most ancient book, written by Sanconiathon, "the philosopher of Tyre," deals with the progress of the human mind and the discoveries made by man, and, in accounting for these last, says that "it was the God Ouranos who devised Betulae, contriving stones that moved as having life." On this passage the theory that the betulae must have been the lodestone has frequently been based, and Sir William Betham asserts unequivocally, though none the less inconsequently, that this statement is quite sufficient to prove the acquaintance of the Phœnicians with the compass.[2] On the other hand, it has been elaborately demonstrated by one author that the betulae were not animated stones at all, but merely stones figuratively so considered, or, in other words, idols;[3] while other writers

[1] De Lacouperie: Western Origin of Early Chinese Civilization, London, 1894.
[2] Sir W. Betham: Etruria-Celtica, London, 1842, II., 8, et seq.
[3] Fourmont: Reflexions sur les Anciens Peuples, Paris, 1747.

have argued in support of the conclusion that the stones were probably pieces of magnetic iron from meteorites, worn as divining talismans by the priests of Cybele, who supposed them to contain souls which had fallen from heaven.[1]

I have already alluded to the lack of evidence tending to show that the Egyptians of the Pharaonic period had knowledge of the lodestone, whence it necessarily follows that they could have known nothing of the compass. Nevertheless, upon a contrary assumption, it has been frequently maintained that the orientation of the Great Pyramid is such as to indicate, with reasonable probability, that the compass needle was used in establishing the positions of its faces.[2]

The difficulty with this supposition is that the Pyramid is, in fact, placed with too great accuracy for the work to be done even by the best modern compass. Its sides face astronomically the north, south, east and west; not to the cardinal points of the compass, but to the azimuthal direction of the earth's axis and to a line at right angles thereto. The compass, however, is subject to variations, due to regular daily, monthly, yearly and centennial changes in the earth's magnetic field, which controls it. Hence, the task of figuring backward the probable position of the needle at the time of the building of the Pyramid—a period which is in doubt—might well cause despair in the most skillful investigator of terrestrial magnetism; for, in the least interval which has elapsed, the needle has probably swung over large angles from the true north, back and forth many times. But, granting such a possibility, still it may be safely questioned whether the most accomplished surveyor or topographical engineer of to-day could run the lines of the pyramid faces, by the aid of the best modern compass, with no greater error than 19′ 58″,

[1] Ennemoser: History of Magic, II., 27.
[2] Gliddon: Otia Ægyptiaca, London, 1849.

which the French Academy, in 1799, determined to be the entire amount of variation of these faces from the true astronomical direction.[1] Accidental mechanical imperfections in pivoting the needle, or in the shape of the latter, might easily result in far greater error. The assumption that an instrument free from fault existed in such remote antiquity is, of course, untenable.[2]

The spirit of maritime enterprise which animated the Phœnicians and Carthaginians, and even the Greeks, was never rife among the Egyptians of early eras, and, at later epochs, they were content to await the importation of goods by the foreign merchants, and to do their bartering on their own territory. They had no timber for ship-building, and dreaded the sea. It was only after the ports were opened and commerce was forced upon her that Egpyt became a maritime state, and obtained her timber from Syria, and then Necho (610 B. C.) built his navy, part in the Mediterranean and part in the Red Sea, and expended 120,000 lives in trying to cut a canal which would enable him to unite his fleets. This failing, he sent the Red Sea squadron to discover a route around the African continent, which it did, rounding the Cape of Good Hope and entering the Mediterranean; but, as the ships sailed from point to point along the coast, they expended three years in making the trip, and so the king decided the undertaking of no value.[3] It is hardly necessary to add that a mari-

---

[1] C. Piazzi Smyth: Our Inheritance in the Great Pyramid, 3rd Ed., Lond., 1877, 67.

[2] It has been argued that the Egyptian *bàa-n-pe, celestial iron,* signifies magnetic iron: and that the expression *res-mehit-ba, south-north iron,* in the inscription of the pyramid of Unas (last Pharaoh of the 5th dynasty), *if correctly read,* would indicate an Egyptian knowledge of polarity. This, however, seems to be unsupported conjecture. De Lacouperie: Chinese Civilization, cit. sup. Deveria: Le Fer et l'Aimant dans l'ancienne Egypte, 1870.

[3] Rawlinson: Ancient Monarchies, ii; History of Egypt. Draper: Intell. Dev. of Europe, i., 78 et seq. Kenrick: Anc. Egypt under the Pharaohs, N. Y., 1853, vol. II, 36. Plutarch: Isis and Osiris, 363, c. 32.

time showing such as this affords no help to the inference of a knowledge of the mariner's compass.

Of the ancient Mediterranean nations, there still remains to be considered that strange people which came by thousands and tens of thousands from Lydia, and with their great fleet descended upon the astonished Umbrians, as unexpectedly as if they had fallen from the sky. The Rasenna, as they called themselves, or as we now term them the "Etruscans," "were not like any other nation," says Dionysius, "in either speech or manners," and modern ethnology brings them into the great Finno-Ugric family, and makes them relatives of the Finns, the Tartars and the Mongolians.

Here was a nation which, if it did not undertake the long voyages of the Phœnicians, for which there was no need—since, as we have seen, it got its amber by a much more direct road, and probably acquired its other foreign supplies by the simple and convenient process of piracy—fostered the sailor and all his arts certainly from a period thirteen centuries before our era. The Etruscans invented the anchor and the cutwater or prow, and stamped the latter on their coins. Likewise they placed on the bows of their ships, small idols pointing the way in advance, and we retain them still in the modern figure-head.[1] Their augurs consecrated the spot on which a temple was to be built by marking on the ground and in the air, lines at right angles indicating regions called "cardines," and hence our word "cardinal," and our denomination "cardinal points." These regions were subdivided so that the ground occupied by the building had sixteen points, each giving its peculiar augury.[2] They laid out their roads in straight lines, and built great sewers and tunnels for irrigation, water-supply and drainage throughout their territory; and under such

---

[1] Dempster: De Etruria Reg., Florence, 1723, lib. vii., c. lxxxi. 441; Suidas: Lexicon, verb. Pattaeci. Herod: lib. iii., 37; Gray: History of Etruria, i., 317, 411.

[2] Gray: History of Etruria, cit. sup.

conditions, especially in subterranean works, that it is difficult to perceive how the alignments could have been made without the aid of the magnetic needle.

But there is nothing tangible to suggest Etruscan knowledge of the compass, except a single object found in the tombs, which bears an incoherent inscription concerning "steering on the ocean by night and day," a bas-relief of a man holding a rudder, and an eight-pointed star, very like the similar star which has been on the compass card ever since the latter appeared in Europe, and commonly known as the "rose of the winds." It also exhibits, at the end of the ray corresponding to the north, a figure closely resembling the "fleur de lis" or "Lilly," which also appeared upon the very earliest compasses. The terminals of the rays corresponding to N. E., S. E., N. W., and S. W., are similar and rounded, and thus differ from the sharp apexes corresponding to the cardinal points.

It was originally argued that the object was in fact a compass dial above which the needle was suspended by a fine thread or wire,[1] but with the refutation[2] of this theory by the Italian antiquaries who showed it to be a lamp, archæological interest in it ceased. Nevertheless the conjecture is still possible that the dial which first appeared in Italian compasses may have been copied by the mediæval navigators from some such Etruscan design.

With this brief survey, we may lay aside as unproved by the evidence outlined, the various hypotheses which attribute the invention of the compass to one or the other of the ancient nations bordering upon the Mediterranean. With regard to the Phœnicians and Greeks, there is no apparent ground even for reasonable conjecture that they had any knowledge of the magnet beyond its attractive power; while as to the Egyptians it is extremely doubtful that they knew anything of the lodestone at all.

[1] Sir W. Betham: Etruria-Celtica, cit. sup.
[2] Dennis: The Cities and Cemeteries of Etruria, London, 1878. II., 105.

Whether the Etruscans, however, were completely ignorant of magnetic polarity is open to question—not merely because of the considerations relating to them and already noted, but for another and broader reason; their race connection with the Mongolians. Consideration of this is a natural prelude to the discussion of the alleged Chinese invention of the compass—and hence to that of the part which Asiatics have taken in the intellectual rise under review.

Among the races of mankind which are included in neither the Aryan nor the Semitic nations, there is a group termed the Turanian, which comprises all those which can be philologically proved to have a genetic connection, and which therefore constitute a true linguistic family. The most important branch of the Turanians is made up of original inhabitants of the great Asiatic tableland, and in these are included the Finnic, Samojedic, Turkic or Tartaric, Mongolic and Tungusic tribes, or as they are sometimes collectively termed, the Ugric or Altaic nations.

These people have certain well-marked peculiarities, which distinguish them from all other races. While the Aryan and Semite nations are found inhabiting large areas of continuous territory never separated by any great interval from others of their own race, and moving by land by a system of lateral extension, so that they colonize by individuals and families, rather than by tribes or by the migration of an entire community, the Ugrics, on the other hand, present characteristics of an opposite description. They are found, so to speak, in isolated patches. There are Finns in Sweden, in Hungary, in Russia, in Persia and in Siberia; Mongols on the Don and Mongols two thousand miles distant on the slopes of the Altai, and congeners on the shores of the Arctic Ocean and on the Bosphorus. These people migrated in bodies with their herds and their flocks. They came upon desired territory and took it by conquest; they multiplied rapidly, and when their land became inadequate to the support of its population, the excess again migrated and the process was repeated. In this way the Mongolic hordes originally conquered China and penetrated to Moscow and Poland. In this way, the dynasty of the Great Mogul was founded in India, and that of the Manchoos established itself in modern China, where it still exists, as the long queues of the Celestials bear witness.[1]

The genesis of the Etruscans has always been a disputed point among ethnologists, who have assigned them to the Greeks, to the Egyptians, to the Phœnicians, to the Canaanites, to the Libyans, to the Armenians, to the Cantabrians or Basques, to the Goths, to the Celts, and to the Hyksos. There are persuasive arguments, however, which connect them with the great Ugric family. Their language has been shown to be very similar to that of the Finns and the Tartars, and their pictures exhibit them with high cheek-bones and oblique eyes, such as the Aryans and Semites never have; and again, unlike these last, they were unemotional and stubborn and conservative. They reverenced ancestors, and built tombs and cared for the needs of the dead as if they were living, all of which is foreign to the thoughts and feelings of the Aryan or the Semite, who bade farewell to his dead at the brink of the grave and proclaimed his own vitality in his palaces and temples.

They came either directly or after a sojourn in Egypt from Lydia in Asia Minor, where the magnetite is abundant; still earlier from that cradle of the human race, the Asiatic highlands, whence still earlier again, others of their kin wandered off, even before the old ice was gone, into the caves of Aquitaine and to the Swiss lakes, where their bones are still found mingled with those of the reindeer and the cave-bear, and with their stone axes and bone needles; while their characteristic tombs and mounds extend over Europe and Asia.

[1] Taylor: Etruscan Researches. London, 1874.

Whether this great Ugric family, before its dispersion, became familiar with iron and the lodestone, we can only surmise.

Nor is the hypothesis incredible. The deserts and steppes of western and northern Asia, over which these races wandered, were as trackless as the deep, and perhaps that same necessity which is "the mother of invention" may as well have operated to suggest the lodestone as a means of guidance to the nomad of prehistoric times as to the venturesome sailor of the Middle Ages. We should thus naturally seek traces of such ancient knowledge among the Etruscans, Mongols and Finns, rather than among the people of the Aryan and the Semite families; in fact, among these we have failed to find it. The Etruscan tombs have yielded suggestive but slender evidence. When we turn, however, to the Mongols, the presumptive proofs multiply.

Modern research establishes a connection between the prehistoric Akkadians and the Chinese. The language and the legends, the written character, the astronomy, the arts, agriculture and domestic economy of China, all show traces of a prehistoric community of origin with those of the first inhabitants of Babylonia. M. De Lacouperie, who regards the Bak tribes, which migrated eastward from the last named region during the twenty-third century B. C., as the first civilizers of China, especially suggests that the early Chinese names of the four cardinal points much resemble those given to the same points by the Chaldeans. The same authority collates an extraordinary number of instances in which the results of Chaldean culture are found embodied in earlier Chinese civilization, showing, for example, that from the Chaldeans the Chinese obtained knowledge of the solar year, of their metrical system, of divination, of their musical scales, of the gnomon and the clepsydra, of decimal notation and local value of figures, of the transit instrument, of the fire drill, of brick-making, canal digging, river embankments and irrigation works, of the use of metals and the art of casting them, of skin boats, of war chariots, and of so many other items as to afford ground for his belief that everything in Chinese antiquity and traditions points to a western origin.[1]

I have now to consider the knowledge of the ancient Chinese concerning the magnet and the amber, and their oft-reputed invention of the mariner's compass.

Lying to the south of the steep declivities of Gobi, on the Asiatic continent, there is a fertile lowland where a profuse semi-tropical vegetation exists, in abrupt contrast with the sparse and rugged growth of the desolate northern steppes. Here the warm and dry weather of the spring months, followed by the abundant monsoon rains of early summer, cause the bamboo and the wheat to flourish with equal luxuriance, so that the products of the soil combine the hardy character of those of the temperate zone with the rapid advance to maturity of the tropical yield. This territory was the nucleus of the Chinese Empire. Its situation being entirely inland, its inhabitants, under the favorable conditions of soil and climate, became of necessity, and above all, an agricultural people.

From the adjacent dwellers in Thibet, India and Central Asia, the Chinese were separated by a difference in language, by natural barriers, and, artificially, by the great wall which they built along the edge of the northern cliffs. It was not until a comparatively late period in their history that their boundary advanced, by conquest, to the sea-coast.

Endowed, therefore, originally with a territory situated geographically to advantage, with a soil capable of providing for all their needs, surrounded by neighbors of the same descent as themselves, whom they surpassed in civil-

[1] Simcox: Primitive Civilizations. N. Y., 1894, 16 et seq.

ization for thousands of years, comparatively unmolested by invasion, and, even when overcome by the Tartar hordes, absorbing their conquerors, and thus converting subjugation into a mere change of governing dynasty, there prevailed, among the Chinese, conditions which infallibly tended to the promotion of peaceful self-evolution and also the development of an intellectual and material independence of the rest of the world; an independence which finally hardened into national conservatism of an intolerant type.

In seeking to discover the chronological periods when events even of great national moment occurred in the history of such a people, the difficulties encountered are by no means trifling. When it comes to fixing, with any degree of certainty, the time of happenings of a specific or less important character, they are practically insurmountable. No epoch can be assigned as certainly that of the beginning of Chinese history. The national annals, in one form or other, are claimed to extend back through the Kingin-Chan era to the reign of Yao, 2357 B. C. Tradition still more vague reaches to the ascent of the throne by Hoang-ti in 2704 B. C. But there are Chinese authors who gravely assert periods of national existence as elapsing prior to the death of Confucius (479 B. C.), ranging from 276,000 to 96,961,740 years.[1]

In China there are no great structures, such as the Egyptian pyramids, which can serve as proof of the civilization and attainments which existed at any period prior to that of the building of the great wall. The enlightened ruler of the Tsin dynasty[2] who constructed not only that wonderful work (B. C. 204), but provided the country with those potent civilizing agents, good roads, conceived that the services he had rendered were amply sufficient to

---

[1] Azuni: Dissertation sur la Boussole. Paris, 1809. Quoting De-Guignes: Discours. prelim. au Shoo-king.

[2] Williams: The Middle Kingdom, New York, 1883, ii. 92.

justify his assumption of the title of "Emperor First," and the consignment to oblivion of all annals which could preserve traditions of any earlier reigns. Therefore he constructed, metaphorically speaking, another wall which has been even a more effectual barrier to historical research than was the great pile of masonry to the incursions of the northern barbarians—that is to say, he burned every book he could find excepting those treating on agriculture and medicine; and lest their contents should be remembered or reproachful comment should be made upon his act, he buried alive five hundred of the most learned scholars. The intention was to completely blot out every trace of preceding emperors. Some thirty years later, when Wăn-te, of the Han dynasty (B. C. 178), wished to revive literature, even so venerated a classic as the Shoo-king could not be found; so that it was re-constructed from memory by one Fuh-sang, then ninety years of age, who in the reign of the Emperor First, being one of the principal literati, had put out his own eyes and feigned idiocy in order to escape death. A few years later it was claimed that a number of books had been found in pulling down a former abode of Confucius, and on this alleged discovery some of the existing Chinese classics are based.[1] At the present time, if the latter were destroyed, scores of Chinese scholars could undoubtedly be found capable of reproducing them verbatim from memory; but the fact that the version of the Shoo-king repeated by Fuh-sang was considered far inferior to that of the supposed old book, discovered as before mentioned, seems to indicate that the extraordinary education, which the Middle Kingdom now requires of its people as a condition precedent to social and official honors, did not, in those ancient days, reach its present degree of minute thoroughness.

While the beginning of Chinese history is placed by De Lacouperie at the 23d century B. C., other Chinese annal-

[1] The Shoo-king, or the Historical Classic. Trans. by Medhurst. Shanghae, 1846.

ists regard it as impossible to rely upon any records dating back more than 800 years before our era.[1] Legge[2] fixes the beginning of trustworthy chronology at 826 B. C., and Plath, at 841 B. C. It is apparent, therefore, that in dealing with the legends and traditions which form the basis for the assertion of knowledge of the magnet by the Chinese at very ancient epochs, the doubt whether they properly belong to mythology or to history is unavoidable.

The most ancient of these legends relates to the victory of the Emperor Hiuan yuan, or Hoang-ti, over the rebel Tchi yeou, or Khiang, an event supposed to have taken place in the year 2634 before our era. Khiang, having been defeated, "excited a great fog in order to put, by the obscurity, disorder in the ranks of his adversary. But Hiuan yuan made a chariot *which indicated the south*, in order to recognize the four cardinal points," and by the aid of this he overtook and destroyed Khiang.[3]

This legend is so clearly mythical that it would deserve no attention, were it not constantly quoted by pro-Chinese advocates in support of their favorite claim that the invention of the compass by the Chinese extends back to the remotest antiquity.[4] In the form in which they present the story, it perhaps warrants Klaproth's conclusion that there is nothing so plainly fabulous about it as to render it certain that it has no historic foundation; but the anti-Chinese writers have unearthed various ancient works in which the tradition is very differently stated. In one of these Khiang is destroyed by a monster-winged dragon, sent after him by Hoang-ti, which threw him into a valley

[1] Azuni, cit. sup.

[2] Chinese Classics.

[3] Thoung Kian Kang Mou, imperial edition of 1707, fol. 22. Quoted by Klaproth: l'Invention de la Boussole. Paris, 1834, 72.
Also, Kou tin tchou, quoted by Biot. Comptes Rendus, vol. xix., 823.

[4] Arriot : Abregé Chron de l'Hist. Univ. de l'Empire Chin., vol. 13. Memoirs concerning the Chinese, p. 234, No. 3. Martini : Historia Sinica, 106.

full of devils; while, in another, Hoang-ti gains his victory by the aid of arms obtained from a celestial virgin, and only by that means overthrows Khiang, who "had the wings and body of a beast."[1]

The tribes which began the settlement of China are believed to have maintained a jade traffic with western Asia, the trade route of which was also a channel for the trans-continental flow of intelligence. This commerce, which had gradually decreased, appears to have revived after the conquest of the country in 1100 B. C., and the establishment of a new dynasty therein by the Tchoü, an energetic and powerful race of Kirghiz origin, which had occupied for centuries the territory bounding China on the northwest. Not only did new learning arrive through the increased traffic, but the Tchoü themselves had probably already acquired much astronomical and astrological lore from Khorasmia, where a focus of such knowledge had been established by a branch of the Aryan race in about 1304 B. C.[2]

An interval of fifteen centuries separates the legend of Hoang-ti from the one next in chronological order, wherein a supposed reference to the magnet is contained, and which according to one Chinese authority ascribes knowledge of polarity to Tchoü-Kung, the founder of the Tchoü dynasty, who is supposed to have obtained it from the sources above mentioned.[3] A later and more complete version is found in an historical memoir[4] written in the first half of the second century of our era, a production which is, in fact, an attempt to collect such fragments of ancient annals as were believed to have survived the wholesale burning of a thousand years before. It does not appear that this work

[1] Azuni, cit. sup., 102.   [2] De Lacouperie, cit. sup.

[3] De Lacouperie, cit. sup., noting an amplified version of the lost 56th chapter of Shoo King, written by Kwei Kuh tze in 4th century B. C.

[4] The Szu Ki or Historic Memoirs of Szu ma thsian quoted in Thoung Kian Kang Mou, Ed. of 1701, vol. I, fol. 9. Reproduced by Klaproth, cit. sup., 79.

now exists, except in the form of extracts quoted in a book issued during the last century, so that the story may well be regarded as not only an exceedingly doubtful tradition, but one which has certainly undergone two modern attenuations. Its period is 1110 B. C., when the Cochin-Chinese are alleged to have sent ambassadors to offer white pheasants to the Emperor, and to do him homage, because there had been no particularly annoying convulsions of nature for the preceding three years. Three envoys were dispatched over different routes, because the "road was very long and the mountains high and the rivers deep," and if a single individual should go astray, the others might succeed in reaching their destination. As it happened, all arrived safely and made their fferings, but when the time came to return they conclu ted that they had forgotten the way back. The Emperor then presented them with five carts, or chariots, which always indicated the south, whereupon they set forth, but instead of steering a straight course back to Cochin-China they seem, somewhat inconsequently, to have made their way to the seashore, and to have followed the coast to their native land; and what reflects still more upon the efficacy of the carts is that it took them a whole year to make the journey.

"The Mirror of Chinese History," a native commentary illustrative of the facts related in the Shoo-king, tells the story with some variations, the final statement being that "the duke gave them five close carriages, each of which was so constructed as to point to the south; the ambassadors mounted these, and, passing through Foo-nan and Lin-yih to the seashore in about a year, they arrived at their country. Hence the south-pointing carriages have always been used to direct the way and to show the submission of distant strangers, in order to regulate the four quarters of the world."

Another work[1] gives a sequel to this story to the effect

[1] Ki kin chu, written by Tsui-p'au during the Tsin dynasty. Jour. N. C. Branch, Roy. As. Soc., n. s., xi., 123.

that "the officers who accompanied the ambassadors to their country then returned. They came back in the same carriages in a direction opposite to that which they pointed, and occupied a year as the journey out had done. The axles and protruding axle-ends were originally of iron, which was completely rusted away when they returned. The chariots were entrusted to officers to be kept for use of the envoys of subject states located at a distance." This was written centuries after the events described, and is probably wholly imaginary.

But in the Shoo-king itself, in the account given of the funeral of the King of Chow, which occurred at about the same time (1102 B. C.), there is described the placing of the royal vehicles about the palace—and "the great or pearly carriage is to be on the visitors' or western stairs facing the south: the succeeding or golden carriage on the eastern stairs facing the south"—and so on for the category of chariots, each successive one being made of less valuable material, and the last being of wood. It will be noted here, that the chariots were merely placed or installed so as to face the south, and the south in China has always been regarded as the honorable quarter. The emperor takes his position facing that point, and all important buildings are similarly placed. Whether the south-pointing chariots of the legend (as the commentaries and alleged translations, made many centuries later, assert) actually indicated the south by some contrivance contained in them, though not described; or whether they were merely chariots of honor, which, like those of the King of Chow, were placed ceremonially facing the south, is thus a debateable question. It is a noteworthy fact that the commentary on the Shoo-king, written in 1200 A. D., is elaborate on astronomical, musical and geographical topics, even to the details of the armillary sphere and the minute proportioning of cords for producing musical tones. It is, therefore, exceedingly significant that both text and commentary—the latter written long before the

A LOST ART. 71

invention of the compass became a matter of international dispute—should be completely silent on the subject of the magnet, if it were in common use.

The fact that the tradition of the ambassadors persisted in itself, does not render it any the less mythical. Besides, like the older legend, it is encountered in bad company. Azuni[1] quotes from the Chinese work, in which he finds it, an equally grave narration concerning men "with bodies of beasts and heads of bronze, who ate sand and invented arrows and frightened the world." And the "Mirror of Chinese History," whence I have transcribed the verbatim recital here given, likewise solemnly records the appearance of a yellow dragon and of a flame which presently "changed into a red bird having a soothing voice."

The most ancient historical record of chariots indicating the south is that found in the work of Han-fei-tsu, a Tao philosopher who lived in the fourth century B. C. His work is non-existent, but, as usual, is quoted in a comparatively modern Cyclopædia, Iu-hai, as follows:

"The ancient sovereigns established indicators of the south (See-nan) to distinguish the morning side from the evening side."[2]

A later writer Liu-hiang (80–89 B. C.) ascribes the chariots to an earlier date, asserting that the Duke Hien of Tsin, who lived between 822 and 811 B. C., attempted to construct them and failed, and that the Duke Huan of Tsi, a century and a half later, succeeded.[3] If the art was lost and recovered at this early epoch, it is a curious fact that

---

[1] Dissertation sur la Boussole, cit. sup. Legge (Chinese Classics, Shoo-king, Vol. III., 535-7) rejects both the Hoang-ti and the ambassadors' legends.

[2] Biot: Comptes Rendus, cit. sup. Klaproth, *contra*, says that the earliest work containing a like reference dates only from the fourth century A. D., and that merely fragments of it have come down.

[3] De Lacouperie, cit. sup.

history should have repeated itself in the same particular, thirteen hundred years later;[1] for during the fifth century, and, although some chariots still existed, a skilful workman, after a year's study, was unable to reproduce one, and thereupon poisoned himself "with the feathers of the bird *ming*, macerated in wine." The task was finally accomplished by one Ma-yo, whose method "was found perfect."

The commentary on the Hoang-ti tradition says that nothing was known as to the ancient form of these chariots, but that they were devised by the Emperor Hian-tsoung, who reigned from 806 to 820 A. D. We are told that they had four gilded dragons on the corners which held up a feather canopy, and that a wooden figure on the top pointed southwards; but nothing is vouchsafed about the magnet. And that is the case[2] with every one of the Chinese descriptions of these south-pointing chariots antedating the introduction of the compass in Europe. It is true that Klaproth, Duhalde, Biot and other sinologists conceive that the *a posteriori* inference that a south-pointing chariot is one containing a magnet needle may fairly be made; but this cannot overcome the force of the omission above noted, especially in view of the further fact that no direct statement of Chinese knowledge of the magnet exists of a date earlier than 121 A. D.,[3] a period when the Europeans had been conversant with the lodestone and its attractive properties for six hundred years, and probably longer. And this statement consists of but six Chinese characters in the dictionary Choue-Wen, where the character "Tseu" is defined as "the name of a stone with which the needle is directed." Even this is known only by citations in later works.

The mediæval and modern Chinese encyclopœdists de-

[1] Klaproth, cit. sup., 89. Biot notes the annals of Wei (235 A. D.); the official history of the Tsin dynasty (265 to 419 A. D.); of Chi hou (335 to 349 A. D.), and of the Soung dynasty (420 to 477 A. D).

[2] China Review: 1891, Vol. XIX, 52.

[3] Biot: cit. sup., p. 824; Klaproth: cit. sup., p. 66.

pict the south-pointing cart or chariot as represented in the accompanying illustration, which appears in the so-called great Japanese encyclopædia of 1712, and originally in a Chinese work of similar character of 1341. The figure, some sixteen inches in height, was made of jade. Within the right arm, extended in front, was concealed a magnet, the directive force of which is supposed to have turned the manikin on its pivot, and thus to have caused it always to point to the south. This arrangement, however, the Chinese concede to have been unknown before the 5th century A. D., when they assert that it replaced a magnet hanging within the chariot.[1]

CHINESE SOUTH-POINTING CART.

Iron was extensively worked in Shensi in B. C. 220, for at that time there was a heavy excise duty on it, and there is a tradition that such imposts were laid as far back as 685 B. C. Hence, as magnetite is known to exist in the iron deposits of the above locality, it has been argued that sufficient evidence is thereby afforded of Chinese knowledge of the properties of the lodestone at the earliest named date. But the same argument would bring home a like acquaintance to the Syrians, for example, and therefore it is of no value in a determination of priority in invention between the different iron-working nations.

So far, nothing has been adduced showing any cognizance by the ancient Chinese, of the attractive quality of the lodestone, nor any knowledge at all of the amber.

---

[1] The *Ku yu tu* (Illustrations of Ancient Jades), first published in 1341, copied into a Chinese encyclopædia of 1609, and then into the Japanese encyclopædia. Klaproth: cit. sup. De Lacouperie: cit. sup. Obviously the dimensions of cart and figure, in the picture, are out of proportion.

Klaproth states that abundant deposits of the resin exist in the empire, but also records that it was imported, in various manufactured forms, as presents to the emperor from Rome and western countries, during the first and second centuries B. C. De Lacouperie says that the knowledge of amber came to the Chinese from the west—Kabulistan—and points out the similarity between the Chinese and Persian names for it. The earliest reference to its attractive property is also apparently the first mention of the like property of the magnet, and appears in a "Eulogy of the magnet," written by Kouo pho in 324 A. D., in the following words:

"The magnet draws the iron, and the amber attracts mustard seeds. There is a breath which penetrates secretly and with velocity, and which communicates itself imperceptibly to that which corresponds to it in the other object. It is an inexplicable thing."[1]

But this is nothing more than a restatement of the European notion of the flow, or virtue, or current, or soul, emanating from the stone or the amber, with which theory the western civilized world was then familiar, and which, it is safe to say, involves a power of abstract conception which the Chinese mind has never possessed. In fact, the originator of such an interpretation of a physical happening, of necessity finds in it an explanation satisfactory at least to his own mind; and it does not seem logically possible, as a part of one and the same mental process, that he could regard the effect as "inexplicable."

The attraction of the lodestone is referred to in a later Chinese work on natural history, in which the magnet is said to draw iron "like a tender mother who causes her children to come to her, and it is for this reason that it has received its name."[2]

It is necessary to distinguish clearly between the land

---

[1] Klaproth, cit. sup., p. 125.

[2] Pen-thsao-chy-i of Tchin thsang khi, published 727 A. D., noted by Klaproth, cit. sup.

use of the compass—as for directing carriages, locating buildings, etc.—and its employment for finding the way at sea, the latter being by far the more important.

So far, it will be noted, no marine use of the compass by the Chinese has been suggested. The first passage, remotely capable of such interpretation, appears in the official history of the Soung dynasty, which, after mentioning the carts, says that "under the Tsin dynasty (265 to 419 A. D.) there were also ships indicating the south." During the same period Shih-hu is said to have built a boat provided with a south-pointing magnet, and to have used it on the "Pond of the Cackling Crane," but this seems at most to have been but a toy.[1] No definite statement, however, is found until the end of the 11th century is reached, and then, in a work entitled Mung-Khi-pi-than,[2] we meet the following extraordinary passage:

"The soothsayers rub a needle with the magnet stone, so that it may mark the south; however, it declines constantly a little to the east. It does not indicate the south exactly. When this needle floats on the water it is much agitated. If the finger-nails touch the upper edge of the basin in which it floats they agitate it strongly; only it continues to slide, and falls easily. It is better, in order to show its virtues in the best way, to suspend it as follows: Take a single filament from a piece of new cotton and attach it exactly to the middle of the needle by a bit of wax as large as a mustard seed. Hang it up in a place where there is no wind. Then the needle constantly shows the south; but among such needles there are some which, being rubbed, indicate the north. Our soothsayers have some which show south and some which show north. Of this property of the magnet to indicate the south, like that of the cypress to show the west, no one can tell the origin."

[1] De Lacouperie: cit. sup., noting the Tsin Kung Koh Ki of the 4th century. See also note 2, page 76.

[2] Thsa-chi, book 24, cit. by Biot.

Almost exactly the same recital appears in a medical history composed in the years 1111 to 1117.[1] And it has been claimed that a record exists of the use of the compass on board the ship which carried a Chinese ambassador from Ning-po to Corea during the year 1122.[2]

Two very significant facts may here be noted, namely, that exactly the same knowledge (the variation of the needle excepted) existed in undoubted connection with the nautical compass in Europe at a closely approximate period; and second, that in the before quoted description of two instruments no nautical employment of either of them is suggested. The latter fact is fully recognized by Klaproth, who admits that he can find "no indubitable use" of the compass in the Chinese marine until toward the end of the 13th century, at which time, as will hereafter be abundantly proved, it had been on European ships for a hundred years.

The tendency of the magnetic needle to depart from true north (commonly termed its variation), appears to have been observed by the Chinese geomancers in the compasses used by them, long before any marine use of the instrument was made. A so-called life of Yi-hing, a Buddhist priest and imperial astronomer, undertakes to show that the variation in the 8th century was nearly three degrees to the right, or west of south. Later, we find the geomancers adding special circles of symbols to the compass card; such as a circle of nine fictitious stars, a circle of sixty dragons, and so on; and, among these, circles of points especially constructed to allow for variation. This was done in the year 900 by Yang Yi when the variation was 5° 15′ east of south, and again three centuries later when it had increased to 7° 30′, in the same direction.

Such, in brief, is the evidence which the Chinese re-

[1] Pen thsao yan i, quoted by Klaproth, 68.

[2] Trans. Asiat. Soc. of Japan, 1880, viii. 475. Jour. North China Branch Roy. As. Soc., New Ser., xi., 123. Shanghae, 1877.

cords have yielded. Let us now turn to the characteristics and achievements of the people themselves, and endeavor to ascertain therefrom the probabilities of the existence of their claimed early knowledge of the magnet, and whether circumstances favored their invention of the compass or discovery of electrical effects.

The Phœnician traders and other navigators of the Indian Ocean reached the Shantung peninsula in the 7th century B. C. and monopolized the sea traffic of the coast. This maritime intercourse appears to have terminated before the end of the 4th century, the more convenient route through Indo-China having diverted the trade. From these hardy seafarers the Chinese seem to have learned little or nothing.[1] Agriculture, as I have already noted, was the chief pursuit of the Chinese in the beginning of their history, and has so remained. In nautical belief, the farmer is always the opposite of the sailor; or, in other words, his is the calling which the seaman regards as furthest removed from his own. The maritime powers of a nation are always the last in reaching maturity; and those of one which is pre-eminently agricultural in its pursuits either never attain that point, or else, if the Chinese be taken as typical, require a greater time for development than is included at present within historical limits. The Chinese, moreover, have been united for ages in one inflexible system of manners, letters and polity, and have dwelt upon land capable of supporting them; so that there has been little natural inducement to them to enter into communication with the rest of the world. The bordering nations were, for centuries, far lower in the scale of civilization, and could offer nothing to barter but raw materials, of which China had either an abundant natural supply, or for which she had no use. True, navigation of the great rivers which irrigated the country be-

---

[1] The eyes on the bows of Chinese junks (also present on modern Dutch boats) are said to have been copied from ancient Phœnician vessels. Perrot-Chipiez, Hist. de l'Art., iii., 517.

gan at an early date; but river navigation does not make and never has made deep-sea sailors. The Chinese streams, like the Egyptian Nile, were merely highways (and, in some cases, practically streets, whereon the dwellings of the inhabitants floated about, as they still do), and, so far from the aquatic life to which they give rise, evolving seamen, the greater facilities for interior communication afforded by the rivers and canals rendered it easier for the dwellers on the seaboard to draw upon inland sources of supply, than to seek foreign ones across the unknown waters.

Nor were their coasts or the adjacent seas favorable to navigation; while the Chinese ships, to the sailors of the western world, have always seemed the very opposite of what sea-going vessels should be. The huge junks, with bulging hulls and high sterns, were modeled after popular notions of sea monsters, the teeth and eyes of which were depicted on the bows, and the fins imitated in the shapes of the sails. The typhoons upset them or drove them upon the reefs, or blew them helplessly far out to sea. Yet, with singular ingenuity, their builders constructed them with double skins and with water-tight compartments, long before the sea-kings of the west dreamed of such safeguards.

The early voyages of the Chinese were merely coasting trips made by the river boats, which crawled timorously along the shore. No sea-going ships were built until 139 B. C. At the time of the Christian era, the Chinese knew scarcely anything of the nearest islands to the eastward, and in the 2d century it is doubtful whether they ever sailed beyond the extreme point of the Shantung peninsula. At this time a fifteen ton boat was considered enormous. In the 3d century some desultory traffic was carried on with Japan, but after that period the extension of sea commerce was slow. At the beginning of the 5th century Java had not been reached, and not until fifty years later did Chinese junks venture as far as Ceylon and the Persian Gulf.[1] All this is doubly significant as show-

[1] De Lacouperie, cit. sup.

ing first, that at the commencement of this great period of six hundred years there was no deep-sea sailing which called for the use of the compass; and second, that toward the end of it, although voyages were made wherein the guidance of the magnetic needle would have been of great utility, and although the traditions of the south-pointing carts then became more numerous, still no similar records have been encountered showing that ships were steered by the lodestone's aid.

Arguments in support of the presumed knowledge of the Chinese regarding navigation are often based on their alleged attainments in astronomy; for they have undoubtedly studied the phenomena dealt with by that science, since time immemorial. But their calculations of eclipses have been found erroneous; and the astronomer Cassini, in examining an observation of one winter solstice very celebrated in their annals, discovered therein an error of no less than 487 years. They are rather astrologers than astronomers, and their tribunal of mathematics, existing, as it has, for centuries, has found its chief occupation in indicating to the Government fortunate days for national enterprises or ceremonials rather than in gathering the results of observation. In brief, their system of astronomy is rigidity itself, and if its predictions fail they argue that the fault is not in themselves, but in their stars, and settle the matter by deferring further prophecy until after the event.

The student who attempts to glean from the early missionary writers on China any definite information as to the real status of her people in fields of invention or discovery, will find himself confronted by an abundance of exaggerated statements and contradictions innumerable. The later Italian and French authors, who have endeavored to reconcile these, fail to do so, and unite in regarding the missionary reports as generally unreliable. Nor can favorable inferences be drawn from other achievements ascribed to the Chinese. They invented a written character of their own, but only with syllabic, and not phonetic, symbols. They are credited with the invention of gunpowder, but it is an open question whether they did not get it originally from India. They knew of it as early as 250 A. D., but then only used it in fire-crackers. No evidence exists of its use as an agent of warfare earlier than the middle of the 12th century, nor did the Chinese know anything of its propulsive effects until the reign of Yung loh in the 15th century, after it was first employed for festival and ceremonial purposes.[1] They invented the abacus, but not the positional value of figures.

On the other hand, the credit of first printing from carved wooden tablets, or from movable porcelain type, inventing India ink, chop-sticks, silk manufacture and the macadamization of streets is seldom denied to them. Their persistent conservatism, to some, is a potent argument in support of the proposition that whatever they have adopted must be sanctioned by immemorial usage. On this ground many of the pro-Chinese writers take a firm stand. Barrow, for example, considers that the astrological inscriptions on the card of the modern Chinese compass is quite sufficient evidence of an extreme antiquity. They have engrafted upon it, he says, "their most ancient and favorite system of mythology, their constellations and cycles, and, in short, the abstract of the elements. That a people so remarkably tenacious of ancient custom, and thinking so very meanly of other nations, would ever have submitted to incorporate their rooted superstitions by engraving on the margin the sacred and mystical characters of Fo Shu with an instrument of recent introduction and barbarian invention" he regards as incredible. To this may be added the fact that to the magnet the Chinese have always paid divine honors. "An astonishing number of offerings," says the missionary Gutzlaff, "are brought to the magnet; a piece of red cloth is thrown over it, incense is kindled before it, and gold paper, in the form of a

[1] Barrow: A Voyage to Cochin China in the years 1792-3. Lond., 1806.

Chinese ship, is burnt." Barrow also notes that a Chinese navigator not only considers the magnet needle as a guide to direct his track through the ocean, but is persuaded that the spirit by which its motions are influenced is the guardian deity of his vessel.

From the actual Chinese records we have now found that the legends of south-pointing chariots antedating the Christian Era are probably mythical. No reference to the lodestone appears in Chinese literature until 121 A. D. If Chinese knowledge of the magnet dates from about this time, then, certainly, so-called south-pointing chariots existing at an earlier period could not have been magnetic, and the omission of any mention of the lodestone in the descriptions of them follows of necessity. If, after 121 A. D., the magnet was used in them, then it is difficult to reconcile this with the fact that the later writings continued to describe the chariots in the same terms for centuries and until long after the compass had come into general use in Europe, and never contained a word concerning the agency upon which their south-pointing virtue depended. It is, moreover, a curious circumstance that while the first south-pointing chariot known in Japan was constructed by a Buddhist priest in 658 A. D., the lodestone itself was not found in that country until nearly half a century later.[1]

No recorded evidence of the attraction of the magnet or amber appears in the Chinese books of earlier date than the fourth century of our era, and then we find it explained by a physical theory totally out of harmony with Chinese modes of thought and the same as that which had been advanced by the Greeks, eight hundred years before.

Turning to the characteristics of the people themselves, it is undeniable that among them have originated many inventions of great importance. But each achievement is isolated. It cannot be traced in correlation with anything else, nor as the result of any evolutionary process or gradual development. Nothing is more clear than the ab-

[1] Klaproth, pp. 93–94.

sence, in Chinese thought, of the processes incident to inductive reasoning. They possess a sort of inventive automatism; and in the results they have achieved, the environment appears to have been by far the more potent factor than the brain. This faculty they probably have, and always have had, in higher degree than any other people. But they have chiefly expended their brain energy, so to speak, upon a multitude of rites, ceremonies and inflexible customs, governing and restricting every phase of their existence; and upon the acquisition by rote, of the contents of volumes of precepts and historical traditions, which find no practical applications. The consequence is minds of stunted or abnormal growth, capable of great subjective action and the grinding out thereby of many words; but even under the influence of the needs of three hundred and fifty million people, and aided by favorable temperature and abundant physical resources, incapable of taking more than the first inventive step.

Their love for the marvelous and supernatural is fostered by their national customs. Their unwillingness to learn from the outer barbarian, is exhibited in the disastrous consequences of their war with the Japanese. They degraded the science of astronomy into mere astrology. They have produced no great picture, no famous statue faithfully representing nature, although they have handled the brush and chisel with consummate skill for ages. But they are the most wonderfully cunning of imitators in the world.

The data which has now been presented concerning the Chinese, lead to the following conclusions:

If the south-pointing chariots which existed prior to the 12th century A. D., be regarded (despite the doubts suggested) as governed by a south-pointing magnetic needle, or if the traditions of such a needle in the hands of the geomancers be accepted as true, then Chinese annals furnish the earliest recorded proof of the turning of magnetic polarity to useful account. But the same records give no information as to how the discovery was made, or when it was made. As to the first, the Chinese legends are grotesque and incredible; as to the second, the traditions are hopelessly conflicting, save in that all refer to periods in remote antiquity.

The prehistoric people from the western Asia migrated, as I have said, in all directions; the Finns, for example, going northward, and the Mongols eastward, and Etruscans, perhaps, westward or southward. If the hypothesis be accepted provisionally, that the parent race knew of the directive tendency of the lodestone, and that all of its offshoots could thus have used it during their migrations as a means of guidance over the deserts and wildernesses, it follows, of course, that the discovery was not originally made on territory which has ever been recognized as Chinese, or by the Mongols exclusively; but, on the contrary, was a part of the stock of knowledge which the different tribes once possessed in common. Now, bearing in mind the conservative, inelastic, non-progressive character of the Chinese, and their seeming inability to advance beyond the first act of discovery or invention, it apparently follows that the directing needle might well continue among them in its original state, and thus remain applied for ages only to its original uses. Therefore, we should naturally expect to find familiarity with the needle only as a means of land guidance, and used either for indicating a quarter of the horizon, or for establishing lines in definite direction, as in placing buildings, laying out tunnels, etc. This comports with the facts. The Chinese, having a great expanse of territory, would have use for the land compass in traveling over long distances; equally their religious system, as well as their engineering knowledge, called for its employment in the establishing of sites for their edifices. The Etruscans, belonging to the same Altaic group, had but small territory, and, therefore, no need for the guidance of the stone in traversing it; but, as I have already pointed out, their augurs, corresponding to the Chinese geomancers, knew the cardinal points, and were able to fix them, and the alignments of the Etruscan walls, and especially of the tunnels and sewers, are made with an accuracy which it is difficult to explain, unless the use of the needle be supposed. As for the existence of Chinese record evidence, and the absence of like proof concerning a similar employment of the magnet by other nations, this finds a ready explanation in the unchanging permanence of the Chinese community and its customs, over a vast period of time, during which state after state of the Western world has risen and fallen. The noteworthy fact is not that such early records exist, but that their character is so doubtful, and their number so few.

It is true that the actual Chinese record, indicating familiarity with the compass afloat, is somewhat earlier in date than the first European record of similar purport, but on the other hand the latter, as will hereafter be abundantly shown, describes the instrument in use as one which European navigators had long known. Moreover, the construction of the first European compass clearly demonstrates it to be the product of an evolution which, in view of the slow intellectual and inventive progress of the time, must have extended over a period far greater than the interval which separates the two epochs.

Further than this I am impressed by the fact that I have been unable to find any series of connecting links by which Chinese knowledge of the instrument could have been brought to Europe, during the twelfth century or earlier, through channels of navigation and trade, and that in reviewing such possibilities of communication as have been suggested serious doubts have always appeared. On the other hand, it may be said that it does not follow that the intelligence spread through any of the regular channels of international intercourse, but may have come through chance travel between Europe and the far East. Against this hypothesis stands the fact that the presence of the compass in the early European fleets, manned by natural and instinctive seafarers, can be reasonably accounted for, (and this I have yet to show) while the presence of the compass on the contemporary Chinese junks, manned by people having no inborn inclination for the sea, is a circumstance seemingly destitute of ancestry.

The identity of construction of the two instruments, European and Chinese, renders inevitable the presumption that one is an imitation of the other. As between people whose skill lies in originating and people whose skill lies in the wonderful minuteness and accuracy of their copies, few, I imagine, will hesitate in deciding which was probably the re-producer; or fail to reach a reasonable conviction that the mariner's compass of the East is literally a "Chinese copy" of the instrument which led, not the indolent Asiatic, but the daring mariners of England and Spain and Portugal and Italy to the most magnificent achievements of the human race.

CHAPTER IV.

I HAVE now to resume the tracing of progress in the Western world. With the decline of the divine school of Alexandria, which followed the period of the Ptolemies, the inventive thought of civilization became almost stationary for nearly a thousand years. Mankind devoted itself to thinking in circles, and believing before it understood. Gradually the doctrine of faith in things spiritual extended itself to things physical, and the latter, being exalted above reason, became removed from the field of human inquiry. From the acceptance of the theory that the Scriptures contain all the knowledge vouchsafed to man, to the interpretation of phenomena by texts, and the gauging of physical laws by the rules of orthodoxy, was but a natural descent. The downward path from the splendid achievements of Archimedes and Hero and Euclid was broad and easy, and it ended in the slough of the schoolmen and the mystics, wherein the world wandered for centuries, mistaking the fitful corpse lights of dead falsehoods for the clear daybreak of coming truth.

Fortunately for future progress, the mystery, which was regarded as inseparable from the effects of the magnet and the amber, proved the salvation of continuing knowledge concerning them. There can be little doubt but that many of the inventions made by the acute student minds which congregated in Egypt were totally forgotten and lost during the dark ages. It is only recently that the art of portraying the human countenance in colors and with a skill in handling and modeling hitherto supposed to have had its origin with the painters of the Renaissance, has been proved to have been known and practised in the Greek-Egyptian settlements dating from the early centuries of our era. The machine which releases its contents or gives some information on the insertion of a coin, and which only in recent years has invaded our public places, stood at the doors of the Egyptian temples, and automatically doled out its little measure of consecrated water in return for five drachmas dropped into the slot in its receptacle.[1] But there was nothing surprising or mysterious about either mechanisms or portrait painting. On the other hand, the magnet and the amber, both seemingly lifeless, yet animated, formed, as it were, the connecting link between the dead earth and living objects. Short of things divine no greater mystery than this could be conceived. It became an ever-present and always-questioning Sphinx rearing itself above the desert of ignorance and superstition, in which, for generation after generation, men were doomed to strive and struggle.

Through the early centuries of the Christian Era, we shall find this problem dealt with again and again—sometimes purely physically, more often metaphorically; sometimes by the poets, and with greater frequency by the historians and fathers of the church.

"When I first saw it," says St. Augustine,[2] speaking of the attraction of the magnet, "I was thunderstruck (*"vehementer inhorrui"*), for I saw an iron ring attracted and suspended by the stone; and then, as if it had communicated its own property to the iron it attracted, and had made it a substance like itself, this ring was put near another and lifted it up, and as the first ring clung to the magnet, so did the second ring to the first. A third and fourth were similarly added, so that there hung from the stone a kind of chain of rings with their hoops connected, not interlinking, but attached together by their outer surface. Who would not be amazed at this virtue of the stone, subsisting, as it does, not only in itself, but trans-

---

[1] Heronis Alexandrium: Spiritalium Liber, Urbini, 1575, s. 29, xxi.
[2] De Civitate Dei, lib. 21, c. 4. (Dod's Translation.) Edinburgh, 1871.

mitted through so many suspended rings and binding them together by invisible links?"

"Yet far more astonishing is what I heard about the stone from my brother in the episcopate, Severus, bishop of Milevis. He told me that Bathanarius, once Count of Africa, when the bishop was dining with him, produced a magnet, and held it under a silver plate on which he placed a bit of iron; then as he moved his hand, with the magnet underneath the plate, the iron upon the plate moved about accordingly. The intervening silver was not affected at all, but precisely as the magnet was moved backward and forward below it, no matter how quickly, so was the iron attracted above. I have related what I myself have witnessed. I have related what I was told by one whom I trust as I trust my own eyes."

Here is the first statement of the movement of a magnetic body under the control of a moving magnet. Lucretius, as we have seen, had told of the to-and-fro vibration of a magnetized ring as the magnetic poles presented to it were reversed, of the fact that brass intervening would not cut off the magnetic virtue, and of the "raving" of the iron filings, but not of the pieces of iron actually following the lodestone, when the latter was moved from place to place. St. Augustine tells all that was known about the magnet at his time. He could have said no more, and the veriest stickler for didactic scientific accuracy, fresh from his Aristotle and his Euclid, could have said it no better.

"Let me further say what *I have read* about this magnet," he continues. "When a diamond is laid near it, it does not lift iron; or, if it has already lifted it, as soon as the diamond approaches it drops it." That error,[1] for

---

[1] For other Patristic writers referring to this same delusion see Eugenius: Opusculorum, P. ii., xxviii.

    Magnes ferri color ferrum suspendere novit
    Sit praesens adamans, quod tenet ille cadit.

Also Aldhelm: Aenigmata. Lib. de Septenario et de metris 8. De magnete ferrifero.

which he will not vouch, lasted for fifteen centuries. The caution is characteristic of the author, who, at seventy years of age, reviewed all his writings and retracted that which appeared doubtful or extravagant, and sought to harmonize his opinions where they seemed in conflict. Elsewhere he is careful to distinguish between matters of hearsay and things which he knows or which can readily be tested, and among these last he includes quicklime, which burns in water and remains cold in oil, and the magnet; and then he says that he does not know by what imperceptible potion the lodestone *refuses to move straws and yet snatches the iron*.[1]

That was a significant question. It marks the first dawning notion of some possible difference between amber attraction and magnet attraction. Why should the lodestone move iron, and yet be powerless to stir the light chaff? Why should the amber draw the chaff, and yet be unable to attract iron? The querist believed the resin and the stone to be generically the same. Hence, the anomaly which surprises him. The Chinese Kouopho who said, a century earlier, that the amber and magnet effects were inexplicable, had not perceived that the mustard seeds which flew to the amber refused to obey the call of the stone.

When this difference was suggested, then the rise of electrical knowledge, in human thought, began to move in parallel channels. The world waited for a dozen centuries before finally recognizing the distinction and separating the phenomena into those which were electric or amber-like and those which were lodestone-like or magnetic; but the first suggestion of it came, none the less, from the great philosopher and saint of the early church.

It may appear singular that St. Augustine should have referred to the mystery of the magnet and amber, not in any metaphorical way, but in the form of statement of actually-observed physical fact. Yet, on the other hand,

[1] De Civ. Dei, lib. 21, c. vi.

it is evident that in no stronger manner could he have used his knowledge of magnetic attraction in order to accomplish his object: a defense of miracles and to deprecate the demand that the latter should be explained by human reason. If, he argues in substance, man cannot explain these plain natural occurrences, how can man be expected to explain events that are supernatural? "When we declare the miracles which God has wrought or will yet work, and which we cannot bring under the very eyes of men, skeptics keep demanding that we shall explain these by reason, and because we cannot do so, inasmuch as they are above human comprehension, they say that we are speaking falsely."

St. Gregorius Nyssenus [1] also describes the communication of the magnetic virtue from one piece of iron to another with simple accuracy, but in most of the Patristic writings which refer to it the phenomenon is dealt with as illustrative of the attraction of the soul to Deity, of Divine control, or of the permeation of the Holy Spirit.

"For although God appeared to material things," says Tertullian,[2] "yet He did not injure them because of grace, and approached, but did not become of them, like the magnet to iron." "If the magnet and amber have the strength to draw rings and reeds and chaff," says St. Jerome,[3] "how much the more irresistibly can the Lord of all created things draw unto Himself that which He desires." St. Ambrose,[4] after describing how the magnetic virtue most strongly affects the ring next adherent to the magnet, and is weakest in the last ring of the chain, finds in this an illustration of the gradual lapse of the soul from the pure state to sin. St. Gregory Nazianzenus[5] sees in the united rings an illustration of the binding power of the Spirit. Theodoritus, Bishop of Cyrrhus,[6] referring to the capability

---

[1] De Homine, cap. i. [2] Lib. adv. Hermogenem, cxliv.
[3] Comment. in Ev. Matthaei, Lib. 1, cxix., 50.
[4] Epist., xlv., 983. [5] Oratio, de se ipso.
[6] De Curatione Infidelium Græcorum, ser. 5.

of the magnet of selecting iron only and lifting up that metal, holding it without visible prop from beneath or apparent means of suspension from above, and all by some hidden or occult cause, declares that this does no more than does the word of God for all men, if they would only give ear to it. Yet, though many listen, only the faithful are garnered; while not even to these is held out the consolation of earthly happiness below, nor is the bond which unites them to heaven above manifest. Hence, it is something unknown, or rather the hope of it, which supports them, even as the unknown virtue of the magnet raises and supports the iron.

Nevertheless, the tendency of the early teachers of Christendom was to discourage the study of natural philosophy. The momentous questions involved in the new faith, in their estimation, so completely dwarfed all mundane issues, that the search for physical truth seemed but a misapplication of the mental powers, which should be devoted solely to the consideration of moral duties and the future world. "It is not through ignorance of the things admired by them," says Eusebius,[1] "but through contempt of their useless labor that we think little of these matters, turning our souls to the exercise of better things." All physical reasoning was denounced as "empty and false;" and to dispute concerning such matters as the dimensions of the sun, the nature of the heavenly bodies and the magnitude of the earth, "is just as if we chose to discuss what we think of a city in a remote country of which we never heard but the name."[2]

Such being the attitude of the most cultivated minds toward scientific research, it necessarily ceased. St. Isidore, in his encyclopedic "Etymologies,"[3] at the end of the 6th century, adds nothing to the facts reported by St. Augustine, and repeats the same account of the Samothracian rings and the silver plate, with a few additions, mainly derived from Pliny.

[1] Praep. Ev., xv, 61.   [2] Lactantius, liii., init.   [3] Originum, lib. xvi., iv.

The last description of the ancient magnetic contrivances is given by Ruffinus,[1] writing in 390, and he mentions those in the Serapeum in Alexandria, as merely intended to deceive the people. A little window was arranged near the statute of Serapis, so that, at sunrise, a beam would fall upon the lips; and this the priests explained as the sun's morning salutation; while, at sunset, an iron figure of the sun, delicately counterbalanced, was made to rise by the attraction of a magnet concealed in the roof, and that was the sun's good-night. "But there were so many other means of deception," adds the chronicler hopelessly, "that it is impossible to tell them all." It is said that while the image of Serapis was falling under the blows of a battle-axe in the hands of one of the destroying party which Archbishop Theophilus, in his furious zeal, led to the Serapeum, a stray invader wandering through the recesses of the temple, found the hidden magnet in the roof, and removed it; and, thereupon, a four-horse chariot, which had been suspended in the air, came crashing to the pavement.[2]

The last part of the narration is often criticised as fabulous, under the assumption that the four-horse chariot, which probably was the iron image of the sun described by Ruffinus, was caused to float in the air with no support save magnetic attraction; but if, as Ruffinus states, it was carried on the arm of a balanced lever, the improbability is not so manifest.

During the four centuries of undivided Roman empire, beginning with the reign of Augustus Caesar and ending with that of Constantine (306 A. D.), the names of Dioscorides the Cilician,[3] and Galen of Pergamus,[4] stand out most prominently as observers of nature. But Galen tells us merely that the magnet and the Heraclean stone are the same thing, and resemble haematite or bloodstone; and

[1] Hist. Eccles., lib. ii., 294.
[2] Draper: Int. Dev. of Europe, i., 320.
[3] Lib. 5, c., 100.   [4] Lib. Sim. Med., ix.

Dioscorides, after announcing that the best magnet attracts iron most readily, is blue, dense and not too heavy, suddenly exhausts his knowledge with the rather inconsequent remark that three drachms of pulverized magnet, taken in sweetened water, will prevent fat. There was also Alexander of Aphrodiseus, who lived in Caracalla's time, and who, though far less known than the preceding philosopher, nevertheless invented the distillation of seawater, and suggested that the same process might be applied to wine.[1] He said that the attraction of magnet and of amber is inexplicable, in which he agreed with the Chinese philosopher Kouopho; but to him is apparently due the credit of having evolved the theory that the magnet actually eats and feeds on iron, a notion which lasted some twelve hundred years, and was very prevalent in the 16th century. Marcellus Empiricus,[2] physician to Theodosius the Great, wrote that the magnet both attracted and repelled iron, the last property being termed by him *antiphyson*. This, however, was in the 4th century A. D., and hence long after Lucretius and Plutarch had referred to the same phenomenon.

From the grave homilies of Jerome or the sombre lines of Lucretius to the gay and voluptuous idyl of Claudian[3] is a far cry; but the subject which could inspire the saint and the philosopher was equally potent to influence the volatile brain of the singer. He speaks of the magnet as a stone, discolored, dull and vile, unfit to adorn beauty, to shine amid the purple of Cæsar or to deck the bridle of the fiery steed. And yet no gem of Orient is so prized by those who know its power. It lives by iron; without iron it thirsts and dies. Yet in the statue of the god of war is iron, and in that of the goddess of love, the magnet. Wherefore the priest consecrates the union of these two divinities, the sacred torch guides the chorus, the doors

---

[1] Meteorol. Comment. Venet, 1527. Humboldt: Cosmos, 562, 589.
[2] Klaproth: Inv. de la Boussole, 12.
[3] Claudian: Idyl V. Magnes.

of the temple are hung with myrtle, and, amid festoons of purple and garlands of roses, there are celebrated the nuptials of Mars and Venus, while in the love of the magnet and the iron a new metaphor is given to the world which even the greatest of its poets has not disdained to use.

With the decadence of the Roman Empire the commentator waxed more and more in strength, and original thought became correspondingly enfeebled. Men forgot or feared to consult nature, to seek for new truths, to do what the great discoverers of other times had done; they were content to consult libraries, to study and defend old opinions, to talk of what great geniuses had said.[1] Thus no new gold was mined, but the supply on hand was beaten to the last degree of tenuity or twisted into a myriad of forms. The three things which blocked progress were the overshadowing claims of religion to the sole domination of the reflecting mind, the prevalence of the Platonic doctrine that all science may be evolved by the use of the reason, and the disposition to dispute about terms, or to seek new facts by new and subtle collocations of words in the endeavor to read nature through books.

From these there became evolved first, a blind faith in the supernatural, and in its constant intervention in the physical world; second, an imagination capable of conceiving such interference as occurring under any and all circumstances, and as being the one and the sole explanation for everything that was in the least respect phenomenal; and, third, a habit of dealing with all learning at second hand, which quickly obliterated the distinction in value between evidence of the senses and mere hearsay reports of speculations, especially after the latter had permeated down through two or three generations of commentators. The result was mysticism, injected into the Greek philosophy by the Alexandrian school, and then spreading through the whole body of human thought and poisoning it to its very centres. If the ancient Greek was so familiar with

[1] Whewell: Hist. of the Inductive Sciences, i., 312.

his gods that they invaded every motive and act of his life, at least there was a freshness, a fragrance, a childlike quality in his philosophy and in the legends which his own exuberant fancy had interwoven with it, and which turned it all into poetry. To him the universe, though diverse, was yet harmonious and so unitary. He had no system of revealed truth, no need of choosing between the acceptance thereof and perdition, no conception of a fall of man, and hence no doubt of the ability of reason to penetrate to the science of things. His science—his logic, and his geometry—was rational.

Replacing all this, now came the belief in a constant struggle between dual powers, existing not only in the external world, but in the human mind; the belief that every thought, every intelligent effort was the plaything of divine caprice on the one hand, or infernal machination on the other. Philosophy became an alleged imposture of the devil; reason, vitiated and untrustworthy. There was no causality to be sought for, no field for scientific investigation; for what could be fairly determined by the instrumentality of the senses, when they, as well as the reasoning faculties, were liable to deception or distortion to suit the occult purposes of the warring powers of good and evil? Faith became so far independent of thought that it was better to say concerning any myth, mystery or marvel, in the words of Tertullian, "I believe *because* it is improbable, absurd, impossible."

The last of the fathers educated in philosophy died with St. Augustine. In the Dark Ages which followed, science disappeared, and magic took its place. The most intelligent minds became entangled in the subtleties of spiritual relations pervading even numbers and figures, sought occult meanings in every work of nature, and made a degraded superstition the controlling factor in life.

Yet here again the inherent mystery of the magnet became a potent agency in the preservation of the knowledge concerning it. Many centuries went by before the commentators supplemented their transcriptions from Pliny with the absurd notions which finally clustered thick around the lodestone and the amber. In fact, it was only after some intellectual force had been gathered in the general awakening, that men acquired sufficient ingenuity to propose myths which, although absolutely untrammeled by the least regard for the truth, seemed at all capable of deepening a mystery which, by the universal consent of past generations, had already reached the limit of profundity. To these delusions I shall refer hereafter. I have now to note the rise and spread of one which, in the end, advanced science instead of impeding it; for it immensely magnified the powers of the magnet without attempting to ascribe to them any new or different quality. It rested on a natural, if not a reasonable inference, and this perhaps accounts both for its early conception and long persistence; in fact there are few fables which have had so great vitality or have been so widely believed as that of the Magnetic Mountains.

In that marvelous collection of romances, the Arabian Nights Entertainments, is the tale of the third royal mendicant,[1] who ventured to sea. After a long period of calm, the captain of the ship tells him, in great perturbation, that "to-morrow we shall arrive at a mountain of black stone called lodestone ; the current is now bearing us violently towards it, and the ships will fall in pieces and every nail in them will fly to the mountain and adhere to it ; for God hath given to the lodestone a secret property, by virtue of which everything of iron is attracted towards it. On that mountain is such a quantity of iron as no one knoweth but God, whose name be exalted ; for, from times of old, great numbers of ships have been destroyed by the influence of that mountain." On the following morning, as the ship approached the fatal stone, "the current car-

[1] In some editions called Agib, the third calendar.

ried us toward it with violence, and when the ships were almost close to it they fell asunder, and all the nails and everything that was of iron flew from them towards the lodestone."[1]

The germ of that story lies in the legend of the shepherd Magnes, the iron nails in whose shoes held him fast to the magnet rock on Mount Ida, which, as I have said in a former chapter, Pliny copied from Nicander's now lost poem. When Ptolemy wrote his geography[2] in the 2d century of our era, he conceived the notion of enlarging the rock and substituting a ship's fastenings for the shepherd's shoe-pegs ; and, in order to give to it verisimilitude, he proceeded to locate the magnetic mountains in the sea between Southern China and the coasts of Tonquin and Cochin China, on certain islands which he calls Manioles. But, as Nicander's shepherd did not have the nails of his shoes pulled out by the magnetic attraction, Ptolemy, evidently from scruples against venturing outside of the four corners of the tradition, is careful not to say that the iron is torn from the vessels, but only that "ships which have iron nails are stopped, and that is why they are put together with wooden nails, in order that the Heraclean stone which grows there may not attract them." The story-teller of the Arabian Nights is equally wise as to the materials which the lodestone will not attract; for the dome on top of the lodestone mountain is of brass, and so is the horseman thereon which the adventurous calendar brings down with a leaden arrow from a brazen bow, and after the sea magically submerges the mountain it is a man of brass who appears in a boat to row the hero away from the dangerous spot.

The wanderings of the magnetic rocks over the surface

[1] Lane : The Thousand and One Nights. Lond., 1859, 161. This collection was first made known in Europe about the end of the 17th century by Galland, from a manuscript brought from Syria dated 1584. The stories probably date from about the middle of the 15th century.

[2] Geography, lib. vii., c. 2.

of the earth now begin. In a treatise attributed to St. Ambrose,[1] a Theban story-teller recounts exactly the same facts, but says the mountains are on the thousand islands of the Arabian and Persian sea, called "Mammoles." Yet, singularly enough, the Chinese author So Soung,[2] writing between 1023 and 1063, completely corroborates Ptolemy, and says that "at the capes and at the points of Khanghai (the southern sea on the coasts of Tonquin and Cochin China), the waters are low and there are many magnet stones, so that if the great foreign ships which are covered with iron plates approach them, they are arrested, and none of them can pass by these places." So Soung quotes from a still earlier work, appropriately entitled "memoirs on the extraordinary things seen in the southern countries."

Even more curious than the support afforded to Ptolemy is the reference to great foreign ships covered with iron plates: which raises the question of whether the Norse iron-clad vessels made so long a voyage at so far distant a period. The Anglo-Saxons, ordinarily, before going into battle hung their shields along the gunwales of their ships,[3] and the Icelanders did so in stormy weather or in time of peril.[4] Whether, therefore, any Icelandic vessel managed to get as far as China, and being in danger from the reefs mentioned by So Soung, hung out her shields, is a matter of curious conjecture. The Icelanders had explored the American coast many years before; but there is not even a legend of a Viking ship ever sailing to China.

By the 12th century the myth of the magnetic rocks had reached the north of Europe. In one of the earliest of the Dutch poems, which Longfellow characterizes as "the

---

[1] De Moribus Brachmannorum. Ed. Bissaeus. Lond., 1665, p. 59.

[2] Thou King Pen Thsao., cit. by Klaproth (cit. sup.), p. 117.

[3] "Then from the wall, the Scylding warder, who had charge of the cliff, beheld them carrying over the gunwale their bright shields, their material of war ready for use." Beowulf (Anglo-Saxon Epic.) Trans. by T. Arnold. Lond., 1876.

[4] Hakonar Saga. Vigfusson. Lond., 1887, 106, 291.

Divina Commedia of the Flemish school," (The Journey of St. Brandaen),[1] the saint is driven by a storm into the Leverzee (the old German Lebermeer), where he saw a mast rise from the water and heard a mysterious voice bidding him sail to the eastward to avoid the magnetic rocks, which drew to them all that passed too near. Ogier, the Dane, is wrecked upon Avalon Island by the attraction of the lodestone mountain or castle thereon.[2] Although the masts crack and many a sail is stretched by the brisk breeze, the vessels of the Norse-German fleet are held motionless by the magnet rock at Gyfers, says the Lay of Gudrun.[3] Magnus Magnussen, on his voyage to discover Greenland, finds his ship stopped by a lodestone at the bottom of the sea. And there is that most redoubtable of travelers, Sir John Maundevile, Knight,[4] who not only depicts the perils which ships searching for the "Yle of Prestre John" encounter from the "grete Roches of Stones of the Adamant that of his proper nature draweth Iren to him," but gives his flamboyant imagination full play in describing the "Buscaylle and Thornes and Breres and grene Grasse" which spring up from the fragments of the wrecked vessels and clothe the rocks as with a "grete Wode or Grove." "And therefore," he concludes, "dur not the Marchauntes passen there, but zif thei knowen wel the passages or elle that thei han gode Lodesmen."

The magnet rocks are frequently mentioned by the Arab writers of the 12th and 13th centuries. Cherif-Edrisi, the geographer, speaks, with great particularity, of an archipelago in the Red Sea, near the straits of

---

[1] Reis van Sinte Brandaen: Longfellow: Poets and Poetry of Europe, Phila., 1845, 372 : Oudulaemsche Gedichten der xiie xiiie en xive Eeuwen, nitgegeven door Jonkhr, Ph. Blommaert. Ghent, 1838-41.

[2] Keary: Outlines of Primitive Belief. New York, 1882.

[3] Ludlow: Epics of the Middle Ages. London, 1865.

[4] Halliwell: The Voiage and Travaile. London, 1866. See also Olaus Magnus: Hist. de Gent. Sept., l. ii., cxxvi. H. Von Valdeck: Herzog Ernest von Bayern's Erhötung, 1858. Spenser: Faerie Queene, ii., canto 12. Also Encyclo. Metropol., XXI. article, Magnet.

Bab-el-Mandeb, wherein rises a mountain called "Mouru-kein," which is, however, partly submerged by the tides: and this, he says, (as usual on the authority of some one else, one Hhasan ben al-Mondar,) seizes and holds ships built with iron nails. He adds, however, on his own authority, that there is still another magnetic mountain, which he is careful to locate as far off as possible, to-wit: in a gulf near Cape Zanguebar, which is very high, and over the sides the waters fall with a frightful noise, and it is named Adjoud.[1]

The finishing touch to the romance which set the nails flying from the ships, was given by one Bailak, a native of Kibdjak,[2] who wrote, in 1242, a treatise on stones, in which he quotes from a pretended work[3] on the same subject by Aristotle, which, in fact, was an Arab concoction. His story is worth quoting in full, because it not only rounds out the fable to completion, but also, as is very likely to happen in such cases, contains the explanation which leads to it own destruction. He says:

"According to Aristotle, there is a mountain of this stone in the sea. If the ships approach it, they lose their nails and their iron, which detach themselves from the ships, and fly, like birds, toward the mountain, without being retained by the cohesive force of the wood. Hence the vessels which navigate in this sea are not fastened with iron nails, but cords made of palm fibres are used to unite them, being secured by soft wooden nails which swell in the water. The Yemen people also fasten their ships with strips detached from the branches of the palm tree. It is said that there is a similar mountain in the Indian Sea."

It may be remarked, in passing, that when the Domin-

---

[1] Klaproth, cit. sup., 119.   [2] Ibid, 57.

[3] As is well known, there is much doubt concerning the actual works of Aristotle. Most of those now accepted are not included in the full Aristotelian catalogue given by Diogenes Laertius, nor were they known to Cicero. Grote: Aristotle, v. i., c. ii.

ican monk Vincent de Beauvais[1] repeated the story in 1250, he put the mountain squarely on the shores of the Indian Sea, and gave as his authority a Book of Stones written by Galen, who was just as innocent of any such production as Aristotle was. And when John Taisnier,[2] arch-plagiarist, in turn told it in 1562, he divided the mountain into several pieces, and put it in the "Aethiopian Sea;" while he changed the Yemen people into Cantabrians, and, regardless of the baldness of the fiction, made them construct their ships of wooden blocks fastened together by glue.

The evolution of the legend is characteristic of the times. The outcropping magnetite of Phrygia probably attracted the iron tools of the ancient miners, and Nicander transferred its drawing power to the shoes of Magnes. Ptolemy made the rock a mountain, set it on the seashore and caused it to arrest ships; St. Ambrose and Soung So multiply the mountains, the Norsemen and the Arabs distribute them widely over the earth and cause them not merely to hold the ships but to pull out the iron nails, and thus we reach the story of the Arabian Nights.[3]

But, to return to Bailak's recital, at the end of which is revealed the probable key to the myth. The Arabs had no iron, or so little of it that vessel fastenings of that material could not be obtained.[4] The great majority of their ships were mere fishing crafts, intended to keep in

[1] Constantinus, in Libro Graduum.

[2] Taisnier: De Nat. Magnet, 1562.

[3] Lane (cit. sup.), says that the Arab author El-Kazweenee (in Ajàib el-Makhlookàt) in his account of minerals, places the mine of lodestone on the shore of the Indian Sea, and reports that if the ships which navigate this sea approach the mine or contain anything of iron it flies from them like a bird, and adheres to the mountain; for which reason it is the general custom to make use of no iron in the construction of vessels employed in this navigation. Note 72.

[4] Agatharcides affirms that iron in ancient Arabia was twice the value of gold. (De Mari Rubro, 60.)

sight of shore. They were built[1] of wood of so hard a quality that it was liable to split or crack like earthenware, so that nails, even if they had been available, could not be driven into it. The planks, after being bored, were fastened to the stem and stern posts by wooden pins, and were then bound together with ropes made from the fibrous husk of the cocoanut, the cocoa fiber or coir of the present time. Marco Polo,[2] after describing these boats, says that they "are of the worst kind and dangerous for navigation, exposing the merchants and others who make use of them to great hazards." Being unfit to venture upon the open sea, these vessels were of necessity kept near land. Hence they were constantly exposed to danger from reefs and shoals, and especially from such currents as the Arabian Nights story-teller mentions, which swept them irresistibly upon the rocks, so that it might easily seem that the ships were dragged to the latter by some mysterious attractive force. A few tales of shipwreck of that sort were easily elucidated by simply picking out from Pliny's Natural History (which the world implicitly relied upon for centuries) a convenient explanation. The story of Magnes furnished one which fitted neatly to the facts, and the tropical imagination of the Orient needed no access of fervor to add the flying forth of the nails like birds, and the breaking of the ill-fated ship into a thousand pieces.[3]

We have now to return to the study of the world's progress in knowledge of magnetic polarity. We have examined, briefly, the reasons which render the claim of origination of the mariner's compass by the Chinese as one to

---

[1] The Three Voyages of Vasco da Gama. Hakluyt. Soc. London, 1869, 240.

[2] Travels of Marco Polo. London, 1854, 20-21.

[3] See Thevet: Cosmog. Univ., p. 445. Azuni: Dissert. sur la Boussole (cit. sup.)

EARLY ARAB NAVIGATION.

be regarded with much doubt, if not to be wholly denied. Even, however, if we concede the credit of this great achievement to the Celestials, there still remains the problem of how to account for the (necessarily assumed) transmission of intelligence concerning the compass from them to the western nations. The limited extent of their voyages, due to their ignorance of geography, navigation and seamanship, militates in advance against any hypothesis of direct communication by the arrival of a Chinese junk in a western port; and, in fact, that supposition is seldom ventured. Perhaps the most favored theory is that the Arabs, during the 12th century, brought the instrument from China to the Mediterranean.[1] It is probably true that Arab travelers found their way into China long before any Europeans did so; and it is said that the knowledge of silk was by this means brought to the western world during the latter period of the Abbasides, and fully five hundred years before Marco Polo's famous voyage.[2] The discovery of ancient Chinese oil bottles, bearing on them quotations from the Chinese poets, in Egypt and Asia Minor, is considered proof of commercial connection between the Arabs and Chinese, prior to the middle of the 13th century.[3] And the known fact that Arabian vessels did constantly sail from the Persian Gulf to the Chinese coast, has been deemed in itself sufficiently indicative of the presence of the compass, without which so long a voyage, it is argued, could not be made.

But the greater strength appears to lie in the considerations which support the opposite conclusion. I have already pointed out the structural weakness of the Arabian ships and their unsuitable construction for ocean navigation.[4] Consequently, their long voyages to China were always along

[1] Klaproth, (L'Invention de la Boussole) and Humboldt, (Cosmos) both so argue, and most cyclopædias follow them.
[2] Peschel: Races of Man, 363.
[3] Williams: The Middle Kingdom, ii., 27.
[4] Azuni: Dissert. sur la Boussole (cit. sup)

the coast by way of Cape Malabar, the shore never being lost sight of at any time.[1] Trips to India, however, being shorter, were frequent, and these were made even by the Egyptians long before the Christian era. After studying the periodic direction of the winds and the monsoons, one Egyptian navigator, Hippalus, was bold enough to venture into the open sea and to trust to the steady-blowing west monsoon to waft him to the port of Musiris on the Malabar coast.[2] His success was regarded as so extraordinary that the wind was named after him, and thenceforward, on Indian voyages, both the Egyptians and the Arabs, when they did leave the shore, risked the venture only when they could rely on the persistence of the wind, and when it was fair for the course they desired to sail.[3]

It is true that the Arabs had astrolabes and other astronomical instruments, which their pilots used for finding their positions at sea; and obscure descriptions of these have frequently been taken as referring to the compass. The Arabs, however, never were inventors, and their early knowledge was mainly derived from Greek books, from which they could have learned nothing of navigation, much less concerning the compass. Azuni[4] refers, moreover, to a planisphere stated to be in the Treasury of St. Mark in Venice, copied from one brought back by Polo, on which is the explicit inscription that the ships traversing the Indian Ocean "navigate without the compass, for an astrologer is stationed aloft and apart from every one else, with the astrolabe in his hand," and thus vessels are piloted. Nicholas Visconti,[5] who made the Indian tour in the middle of the 15th century, asserts positively that the Indian navigators "never navigate with the compass, but

[1] Renaudot: Anc. Rel. des Indes et de la Chine. Paris, 1718.
[2] Robertson: Hist. Dis. of Anc. Ind., § 2.
[3] Arrian: In Periplo Maris Erythrœi. Vellanson: De l'Inv. de la Boussole. Naples, 1808.
[4] Azuni: Dissert. sur la Boussole, cit. sup.
[5] Ramusio: Coll. Voyages. Venice, 1554, vol. i., 379.

guide themselves according as they find the stars high or low, and this they execute with certain measures." Whether the Arab pilots who were met by Vasco da Gama, in 1497, after his famous voyage around the Cape of Good Hope, were provided with compasses is a disputed question. One of da Gama's companions,[1] after stating that the largest ships, encountered did not exceed 200 tons burden and were of very weak construction, adds that "no one ever navigates these seas with the compass, but with certain quadrants of wood, which appears to be very difficult, principally when the weather is foggy and the stars cannot be seen." Contrariwise it is asserted that the Arabs, at the time of da Gama, were instructed in so many of the arts of navigation that they did not yield much to the Portugese mariners in the science and practice of maritime matters.[2]

It is obvious that even up to the end of the 15th century a decided doubt exists as to the use of the compass by any Arab or Indian navigators. Nor can anything be inferred in their favor, even if it be conceded that the Chinese vessels were employing it on the Indian Ocean. The Chinese are not a communicative people, and, whether as a marine or a land device, the magnetic needle has always been regarded by them as animated by a spirit. This is the guardian deity of the ship, and hence, from the beginning, the compass has been shut up in a little cabinet in the stern of the vessel, with other sanctified utensils, and jealously guarded from strangers. The instrument, moreover, is adjusted for the course before the ship leaves port by the ship's owner, and the navigator is therefore especially solicitous that it should not be disturbed *en voyage*.[3] Add to all this the fact that a magnetic needle cannot be recog-

---

[1] Ibid, vol. i., c. 3. Barrow: A Voyage to Cochin China. Lond., 1806, 355. Renaudot: Dissert. sur les Sciences des Chinois, 288–289.

[2] Hakluyt Soc. Three Voyages of Vasco da Gama, 1869. Varthema: Travels, 31.

[3] Barrow, cit. sup.

nized as magnetic by looking at it, and, unless the copying Arab possessed, not merely the knowledge of magnetic attraction, which he might have had, but also that of magnetic polarity, which he certainly did not have, it would be impossible for him to reproduce the apparatus.

Besides, the Chinese mariner was grossly ignorant, and even if he could explain the mysterious little needle, it is unlikely that the haughty Arab, of a totally different race and religious belief, would view other than with contempt, the signs and astrological hieroglyphics, which were part of the Chinaman's religion, and to which he would be sure to attribute much of the marvelous powers of the compass. The Arabs had possessed charts and astrolabes for a long time, and had proved them to be efficient guides at sea, so that it is improbable that they would readily supplant them by any such incomprehensible Chinese contrivance. Nor indeed is it necessary, because the Arabs made long voyages, or because early mariners of the Indian Ocean undertook journeys on which the modern navigator would never venture without the aid of a compass, to assume that such an instrument existed among them. A fairly good means of guidance at sea, known since the days of the Phœnicians, was the flight of birds. Those birds which accomplish the longest flights, and cross the widest oceans, always select, by some marvelous instinct, the shortest ocean routes; and birds which know their way are an invaluable guide to the sailor who has lost his. There is, for example, a kind of falcon which breeds in Southern Siberia, Mongolia, and Northern China, and winters in India and Eastern Africa.[1] It is able to make this long migration by moving from station to station in the Indian Ocean, so that it is plausibly supposed that guided by these birds, the ancient ships might have made voyages from the coast of Malabar even to the far distant Archipelago of Madagascar.

[1] Dixon : Migration of Birds. London, 1892. Simcox : Prim. Civilization, cit. sup.

If, therefore, the eastern Arabs neither invented the mariner's compass themselves nor derived it from the Chinese, the many claims based upon its supposed origin among them must be laid aside. Indeed, it is seldom that recent writers attempt to establish directly the first appearance of the instrument among these people, the usual contention being that the knowledge of it came with the Saracen armies which conquered Egypt, the north African coast, and finally Spain; and that it was during the period of advanced civilization which that country reached during the Arab-Moor domination that the compass found its way into general use on the Arab fleets in the Mediterranean.

It is certainly not an unreasonable supposition that the people whose attainments and culture shine out with the highest lustre against the black background of the dense and all-pervading ignorance prevailing throughout Europe during the dark ages, and to whom the rejuvenation of physical science was chiefly due, should have been, of all others, the one to bring forth an invention of such transcendent importance; but even the strongest advocates of the Spanish Saracens do not pretend that the compass was discovered on Spanish soil, but allege only in a general way that "the Arabians, finding it in their eastern conquests among the treasures of natural magic, brought it into Spain certainly as early as the eleventh century, and used it very generally there in the twelfth."

But if the Arabs of the East, as we have seen, did not have the instrument, it is hardly necessary to remark that the Arabs of the West could not have obtained it from them; and, therefore, the problem is not to be solved by such an hypothesis. Certain eminent Italian historians (patriotically unwilling to relinquish the credit accorded to Italy for many years during which the claims of Flavio de Gioja of Amalfi were favorably regarded), while conceding the invention to the Saracens, deny it to the Saracens of Spain and ascribe it to the Saracens of Apulia. That the compass did probably come to Apulia at an early date in its history, and that there is consequently a certain support to this opinion, we shall see further on; but there is a total lack of record evidence that the Saracens, who dwelt chiefly in Lucera or Nocera, a city of refuge in the Italian province, had any part in its introduction.

Our present inquiry, however, is to determine what, if anything, the Spanish Moors contributed to the science, the development of which we are studying; and to this I now address myself.

By the middle of the eighth century, the Arabs throughout their whole empire, from Syria to the Atlantic, had begun to turn from the study of the Koran to that of science and profane literature. They went to the Greeks for their philosophy, and translated into Arabic the works of Aristotle and Plato, Euclid, Apollonius, Ptolemy, Hippocrates and Galen, undefiled by the distortions of the Christian revisers, and untrammeled by the theological dicta of any religious system. The syllogism of the Stagirite commended itself to the subtle Saracen intellect, and the disputations of the shady walks of Athens, long since silenced, became again heard in the schools which flanked the mosques from one end of the Mediterranean to the other.

In Spain the awakening was even more thorough, and the progress more swift; for the men of action outstripped the men of thought. Cordova produced her unrivaled leather, and pointed to a paved street ten miles long and brilliantly lighted at night. Toledo brought forth her sword-blades, which still laugh the modern armorers' art to scorn. The Arabic numeral, arithmetic, algebra and chemistry came into the world. Rice and sugar and cotton and spinach and saffron and nearly every fine garden and orchard fruit followed the conquerors from the east. The vineyards of Xeres and Malaga then first yielded their famous wines. Art took on new and fascinating forms, and such dreams of beauty as had never before been known appeared amid the groves and gardens of Granada.

If the Spanish Moors built any great ships or made any long voyages on blue water, all have escaped record as completely as has the memory of the mariner's compass, with which their advocates say their apocryphal vessels of the tenth and eleventh centuries were provided.

We shall look in vain through the encyclopædic astronomical work of Ibn Younis—the great Hakemite Tables (1007 A. D.)—for any reference to this instrument, inestimable as is its importance in observations of the heavens. Equally in vain will the works of Cherif Edrisi (1153 A. D.)—the most famous of all Arabian geographers—be searched for it; nor has the grammatical and historical lexicon of the Byzantine Greek, Suidas, full as it is in its reference to the magnet, a word which reveals the slightest knowledge of the directing needle.

If the Saracens had constructed large vessels and had made extensive voyages in them upon the open sea, it is reasonably certain that some clear and indisputable records thereof would long since have come to light. But the only craft of unusual magnitude which they built were flatboats for the transportation of troops or goods over short distances, while their ships were of inconsiderable dimensions. Out of eighteen hundred, which they sent against Constantinople, only twenty were large enough to carry one hundred men each, and all of them were destroyed by the Greek fire showered upon them by the besieged, in a single night. On a succeeding venture most of the seven hundred and sixty ships composing the attacking fleet met a like fate.

Thus it appears that the compass had no early existence among the Arabs of the Persian Gulf and Indian Ocean, nor among the highly civilized Saracens of Spain. There still remain the Arabs who traded in the Eastern part of the Mediterranean; but here our quest is short, for in the work of Bailak of Kibdjak already noted, and written in 1242, the first of all Arabian descriptions of the compass is found.

"The captains who navigate the Syrian Sea," he says, "when the night is so obscure that they cannot perceive any star to direct them according to the determination of the four cardinal points, take a vessel full of water which they place sheltered from the wind and within the ship. Then they take a needle, which they enclose in a piece of wood or reed formed in the shape of a cross. They throw it in the water contained in the vase, so that it floats. Then they take a magnet stone large enough to fill the palm of the hand, or smaller. They bring it to the surface of the water, and give to the hand a movement of rotation toward the right, so that the needle turns on the surface of the water. Then they withdraw the hand suddenly, and at once the needle, by its two points, faces to the south and to the north. I have seen them, with my own eyes, do that, during my voyage at sea from Tripolis to Alexandria in the year 640 (or 1240 A. D.).[1]

Bailak's assertion that the compass was in use at this date is of course, in itself, enough to dispose of the oft repeated statement that the first tidings of it were brought to Europe by Marco Polo, for that traveler did not return from Cathay until 1295. But the theory of its invention by the Eastern Arabs must also fall; for, at the time that Bailak wrote, the Northmen had been steering their ships by the magnet needle for more than half a century, and the compass was well known to the sailors of England, France and Italy.

That the mariners of Christian Spain had like knowledge may perhaps be inferred; but the earliest Spanish record of the compass which has been found is in the famous compilation of laws known as Las Siete Partidas,[2]

[1] Klaproth: 57. From Arab MSS., No. 970, Bib. Nat. Paris.
[2] Las Siete Partidas del Rey Don Alfonso El Sabio. Madrid, 1807. Part II., Title IX., Law 28.

made by order of Alfonso X., King of Castile, in 1263; the 28th statute of which is the following:

"And as the sailors are guided in an obscure night by means of the magnet needle, which is their mediator between the star and the lodestone, and shows them where to go as well in good weather as in bad; so those who have to aid and to counsel the king should always be guided by justice, which is the mediator between God and the world, always giving safety to the good and punishment to the wicked, each according to his deserts."

It appears, however, that the instrument was not then in use in Spanish ships. In the chronicle of Don Pedro Niño, Conde de Buelna, a famous Castilian knight, appears the decisive statement under date of 1403: "The galleys of Conde, it is said, left the island of la Alharina in Barbary... The pilots compared their needles rubbed with the magnet stone and opened their charts." This the distinguished Spanish historian Capmany[1] says is not only the first mention of the use of the compass in a Spanish vessel, but he finds that in the inventories of a three-decked ship fitted out in Barcelona in 1331 against the Genoese, there is no reference to such an instrument; nor yet in the similar schedules of 1364 of the galleys of Don Pedro IV., of Aragon, although all articles, even those of very small account, are noted. On the other hand, he points out that in the galley inventories of Alfonso V., of Aragon, dated 1409, the compass is fully set forth.

[1] Capmany: Memorias Historicas Sobre la Marina, Commercio, etc. Madrid, 1792.

CHAPTER V.

BEING a sea-wolf, and living among sea-wolves, the mediæval Northman was controlled by wolf law, which compelled him to keep his powers, offensive and defensive, in the best possible order, lest he should be eaten. For, when there was no quarry at hand for the common pack, its members fell one upon another and the Danes harried the Saxons, and the Swedes worried the Finns, and the Norwegians came upon any and all of them; and consequently, as this fighting was done chiefly at sea, that nation which had the strongest navy for the time being, at least, was paramount. Therefore, to the building and maintenance of ships everything else was subordinate, and even the lands were divided so as to secure the largest possible contribution of vessels, or the greatest tax levy for their support. In the beginning, the galleys were small, and the seven hundred of them which made up the fleet of Hakon and Harald Bluetooth were little more than canoes; but they grew apace in size until in the 11th century, the Long Serpent of King Olaf Tryggvason went into action against the ships of Norway, Denmark and Sweden, with thirty-four banks of rowers beating the water into foam at her sides. And, in that same battle of the Svold, Eirik, Jarl of Norway had a vessel with beaks on both stem and stern, and covered at her bow with great iron plates which reached to the water.

As larger ships were built, the wonderful energy of the Northmen found a new outlet in overcoming the dangers and hardships of long voyages, even unto regions wherein seamen had never before penetrated; and their trading craft went not merely into the Mediterranean, but to Greenland and along the American coast. The northern Sagas may, however, be searched in vain for any allusion which can be interpreted as referring to the compass. In the middle of the ninth century, Harald Fairhair, of Norway, drove many of the chieftains from the country, and the record of their voyages in search of new lands then begins. Naddod Viking discovered Iceland in 861, and was followed by Floki, the son of Vilgerd, a noted pirate, in 865. Floki sailed from Roga land in Norway, and on reaching Smörsund offered a great sacrifice and consecrated the three crows by means of which he meant to find the way—for, says the Saga, "the magnet was not then in use for the northern sailors." When he thought he was near to the land, he freed the first crow, which returned to the port from which he had sailed; the second crow flew about aimlessly until tired, and then came back to the vessel; but the third crow went onward, and after the manner of its kind, in a straight line, and so laying his course, Floki found the eastern coast of the island.[1]

Following the discovery of Iceland came that of Greenland, by Thorvald and his son Eirik the Red, who made the first voyage[2] in 985, and followed the coast. Later

[1] Landnámabók: i., c. 2, § 7. Wheaton: History of Northmen, London, 1831. Mallet: Northern Antiquities, 188.

This use of crows or ravens for finding the way at sea is believed to have been common among the Northmen, and there may have been a particular variety of these birds trained for the purpose and consecrated thereto by religious rites which fell into disuse on the introduction of Christianity; a probability strengthened by the fact that the raven was the bird of Odin, the raven god, Hrafnagud, as he is called in the Scald poetry.

The Icelandic saga was written in the 11th century, and hence its direct reference to the non-use of the magnet at an earlier period has been cited to establish undoubted knowledge of the compass at the date of the work. The latter, however, was left uncompleted by its original author, and it was glossed by many writers up to the time of Hauk, the son of Enland, who entirely re-made it in the 14th century—so that the reference belongs to that date and not to a period three centuries earlier. See Klaproth, L'Invention de la Boussole, cit. sup.

[2] Flateyjarbók, i., 429; Du Chaillu: Viking Age, cit. sup., 18.

still was the first journey of Bjarni to America,[1] and as to this the Saga says that "after three days' sailing, land was out of sight and the fair winds ceased and northern winds with fog blew continually, so that, for many days, they did not know in what direction they were sailing "— a statement which completely negatives the presence of the compass, even without the aid of the ensuing description of how the ships afterwards sailed in sight of the shore.

We have now to turn to the Anglo-Saxons. A century and a half after the pirate ships of Hengist had appeared off Thanet, the "strangers from Rome," sent by Gregory the Great, marched into Canterbury with censers burning, the silver cross borne aloft, and chanting the solemn litany of the Church: so returned into England the Latin tongue, and with it Christianity. In the reign of Aelfred came peace, long enough for the establishment of order and the beginning of the teaching of the people. Of all the great things which Aelfred did, the most significant with respect to our present research, are the opening of channels of thought and commerce between England and the people of the north countries, and the great impetus which his larger and better ships must have given to the making of long voyages.[2] Thus a more extended knowledge of the art of navigation and of matters pertaining thereto was gained, better conditions of intercommunication were established, and the spread of intelligence among the seafaring nations greatly quickened. Meanwhile, and long before the reign of Aelfred, the magnet was well known in Britain. The Greek and Roman writings with which the clergy were familiar—and those of Pliny especially—contained, as we have seen, abundant references to it; and, as iron had been freely mined before the Roman occu-

---

[1] Flateyjarbók., i., 430.

[2] The Saxon Chronicle and William of Malmesbury, 248. Wright, T. A.: Essay on the State of Literature, etc., under the Anglo-Saxons. London, 1839, 92.

pation, the lodestone which Harrison,[1] in 1577, speaks of as "oftentimes taken up out of our mines of iron," was present in abundance in the country. The first Anglo-Latin epigrammist, St. Aldhelm,[2] writing in the latter half of the seventh century, devotes a stanza to it, mainly with relation to the supposed power of the diamond to cut off its attraction; and the Venerable Bede indirectly alludes to it in his mention of Bellerophon's horse suspended in the air at Rhodes.

The writings of St. Isidore and Bede were the chief text-books of science of the Anglo-Saxons up to the twelfth century. Their dicta were accepted as articles of faith to be learned, and not questioned. Compilations and re-compilations were made from them, often intermingled with spurious treatises, and the whole buried under great masses of commentaries, so that, to determine therefrom the state of knowledge existing at any particular period is at best a doubtful undertaking. The Anglo-Saxon work in which we might expect to find the compass described, if it were known, is the Manual of Astronomy abridged by Alfric from Bede's De Natura Rerum, in the 10th century; but it contains no reference to the instrument, and, on the contrary, alludes to the northern or "ship star," and its fixedness in the heavens.[3]

With the monastic reforms of Dunstan and Athelwold some slight revival of scientific investigation becomes apparent. But it was of weakling growth, and when it was found linked with like progress in Saracen Spain, the great body of the monks looked upon it with suspicion as savoring of witchcraft and heresy. Despite the fame which Dunstan gained by his supposed victory in a per-

[1] Harrison, W.: A Description of England. Lond., 1577. Book iii., c. 12.

[2] Aldhelm: Lib de Septenario et de Metris. Ep. viii. De Magnete Ferrifero.

[3] Wright, T.: Popular Treatises on Science during the Middle Ages. London, 1841.

sonal contest with the Devil, his neighbors threw him into a pond to determine whether he was, in fact, a wizard or not; and when Ailmer of Malmesbury, having invested himself with a pair of wings, jumped from a steeple and broke his legs, they ascribed his failure to evil influences with which he had paltered, and not, as he insisted, to his having forgotten to put on a tail behind.[1]

In none of the chronicles of Saxon England, nor in the old legendary poems of the north, can any definite sign of acquaintance with magnetic polarity be recognized. While the Normans undertook long excursions, the ordinary voyage made by them was merely between points on the narrow seas where the pilots were seldom out of sight of land, and in waters which had been navigated for centuries and wherein all the peculiarities of coasts and currents were intimately known. During the 10th and 11th centuries, however, the forays of the Normans, originally confined to the lands bordering on the sea, were extended into the heart of Europe. They ruined France, placed her monarchs under tribute, and occupied and named one of the fairest portions of the Frankish territory. There, having embraced Christianity, they began pilgrimages to Italy and the Holy Land, with all the fervor of new-made converts. Their acceptance of the new faith does not seem to have extended as far as the doctrine of loving one's neighbors; and any behavior on the part of the latter toward the pious pilgrim on the basis of such a presumption speedily converted the meek wayfarer into an astonishingly skilful manipulator of axe and sword. Their pilgrimages, therefore, were not easily distinguishable from invasions, and aroused resentment, and finally wars, especially with the Greeks and Saracens of Southern Italy, one result of which was the conquest of Apulia, and the transfer to Norman control of the flourishing and opulent Amalfi, a

---

[1] Wright, T.: An Essay on the State of Literature, etc., under the Anglo-Saxons. London, 1839. 64–69.
William of Malmesbury (Scriptores Post Bedam), 92.

city which, for the preceding three hundred years, had been one of the great maritime trading marts of the world.

It is in an original account of Norman prowess in this part of Europe, written by William Appulus,[1] a native of France, in 1100, that a possible trace of the compass appears; and this only in a single line of a poem which, in describing Amalfi and its glories, mentions the many mariners tarrying in the city as "skilled in opening the ways of the seas and the heavens." Gibbon[2] regards these words as relating to the compass; but, inasmuch as the eminent historian himself dwells upon the extension of the Amalfitan trade to the African, Arabian and Indian coasts, they seem more applicable to the general nautical skill which could conduct ships to such distant places, rather than to any specific aid in so doing which the compass might afford.

At all events, if the silence of all written records from the reign of Aelfred to the beginning of the 12th century is to be regarded as disproving the prior or contemporary use of the instrument, the continuance of that same silence after the time of William Appulus and in the face of the great commerce of Amalfi, is even more significant as showing its absence.

On the other hand, it is not safe to accept these premises as controlling, in view of the existing state of European civilization. Hallam tells us that from the middle of the 6th century a condition of general ignorance lasted for a period of about five centuries; and that not until the close of the 11th century began vigorous attempts to retrieve what had been lost of ancient learning, or to supply its place by the original powers of the mind.[3] Then, unfortunately, the newly-developed energy was turned into a path almost diametrically opposite to that which led to the

---

[1] William Appulus (apud Muratori, v.) lib, iii., 267.

[2] Gibbon : The Decline and Fall of the Roman Empire, c. lvi.

[3] Hallam : Literature of Europe, Part I , c. i., § 10.

cultivation of physical science. If ignorance had left that road choked and impassable with weeds, the scholastic philosophy, of which John Scot Erigena began the assertion in Aelfred's reign, conducted inquiring minds altogether away from it. In the early days of Christendom the heathen philosophy was regarded as different from the philosophy of the new dispensation, and, therefore, it was silenced. Now, it was maintained that the heathen philosophy was identical with that deduced from divine revelation, and consequently that theology was inherently and essentially philosophical truth. Wherefore Abelard insisted that logic includes the whole of science, which is the same thing as saying that the key to all knowledge lies in combining and recombining the notions conveyed by words: or that the manipulation of a mathematical formula can result in the discovery of a new mathematical truth. Thus, a universal science was established with the authority of a religious creed. Error became wicked, dissent became heresy; to reject the received human doctrines was nearly the same as to doubt divine declarations.[1]

In the scholastic philosophy so founded, physics had no proper part, as distinguished from metaphysics. "Quiddities" were spoken of as distinct from qualities and quantities. Peter became an individual because of his humanity combined with "Petreity."[2] The nature of angels, their nine hierarchies, their modes of conversing and the morning and evening state of their understandings;[3] the character of the crystalline waters above the heavens wherein the stars are set;[4] the mystical analogies between man and the universe, such were some of the subjects which were discussed and disputed in endless circles until minds became polarized

---

[1] Whewell: Hist. of the Inductive Sciences, i., 315; ii., 151; Tennemann: Geschichte der Philos., viii., 461; Ranke: Hist. of the Popes, i., 502.

[2] Ibid., 321.

[3] Hallam: Literature of Europe, cit. sup.   [4] Whewell, 318.

and incapable of either receiving or understanding physical truths. And they were always such vast topics, such ponderous metaphysical disquisitions; and so momentous were the consequences supposed to depend on them that the modern student heartily joins with old Burton in wondering how his scholastic predecessors "could sleep quietly and were not terrified in the night, or walk in the dark, they had such monstrous questions and thought such terrible matters all day long." Where was there any place in the literature of the beginning of such a period for exact physical descriptions of the magnet and its phenomena? What was substituted for them appears in the very first writing[1] in the Anglo-Norman language—the lingua Romana. This was not a sermon, although the resemblance is frequently strong, but a poem written in 1121 by Phillippe de Thaun, under the high patronage of the Queen of Henry I., Adelaide of Louraine. The work is a Bestiary, founded partly on Pliny's Natural History and partly on the zoölogical classifications of St. Isidore, interspersed with fables—some evidently borrowed from the Orientals. It deals with the subject of magnetism in the following not altogether lucid manner:

"And this know freely, that they break in pieces the lodestone with goats' blood and lead: it signifies a great matter. By the blood of the goat, we understand corruption in our law. By the lead we understand sin by which men are ensnared. But the lead weighs the iron, which draws sinners to hell. And this virtue it has in it, that it draws iron with it: it signifies that Christians draw Pagans to their law when they leave their heresy and believe."

Although these were days, as I have said, when any eccentricity in thought or deed might give rise to suspicions of paltering with the powers of darkness, no charge of sorcery or the compassing thereof could lie against the inspired author of this sort of poesy; but when, in the following paragraph, he proceeds to describe

[1] Wright, T.: Popular Treatises on Science during the Middle Ages, 125.

a mountain in the east where the adamant is found, which mountain (whether because of the lodestone in it or not, he carefully neglects to say) "by night emits a great light and it does not appear in the face of day," he becomes apprehensive. So with entire prudence, he makes it clear that he, the poet, does not aver this, but that it is the dictum of Physiologus—an expedient which many writers of later times found convenient to imitate when discussing prohibited subjects in a way likely to arouse in their behalf the solicitude of the Holy Inquisition.

The intellectual movement in both literature and science gained force rapidly as the 12th century advanced. Schools sprang up over the continent, and letters were cultivated nowhere better than in Normandy. Then the Norman French—the *Langue d'oui*—gradually became the vehicle of literary expression, and, with the reign of Richard—he of the lion heart and poet soul—a new era of literature intervenes, when the trouvères and jongleurs come upon the scene and the indomitable Norman spirit bursts forth in romances of chivalry and honor and love: and when the beautiful legends of Arthur of the round table and the "San Graal" are told by the descendants of the fierce Berserkers, whose delight lay in hearing over and again the bloody sagas of rapine and massacre stridently shouted by the Skalds.

In the month of September, 1157, two infants, born on the same day, the one at Windsor, the other at St. Albans, were confided to the care of good dame Neckam. The first was Richard of England, son of King Henry; the second, she herself, by the best of all rights, named Alexander,[1] and afterwards he became commonly known as Alexander of St. Albans.

[1] Wright, T. Alexander Neckam, lib. ii. London, 1863, quoting MSS. James Coll., vii. 34. Wright's Latin text of Neckam's treatise and his biographical introduction thereto, have been followed in the present chapter.

The boys, after the manner of foster brothers in those days, grew up together until the difference in their stations moved them asunder. The Prince went to the wars—the subject to the schools. Perhaps the royal favor followed the young student, for we find him a distinguished professor at the University of Paris when but twenty-three years of age: and a member of the school which had been established by his countryman, Adam du Petit Pont, which was celebrated for the subtleties of its disputations. Here, he tells us, he both studied and taught the arts—rhetoric, poetry, civil and canon law, Biblical criticism and medicine: an odd combination from a modern point of view. Then he returned to England and became master of the Dunstable school; but he evidently had less taste for teaching than for learning, and the books, the congenial companionship, the literary atmosphere to be found only in the monasteries, became an overpowering attraction to the scholar, who felt, as many another of like kidney has felt since, that he was made for better work than hammering a parrot-like knowledge of the Trivium into boys whom the Assize of Arms was enrolling in the new militia as quickly as they were able to wield a lance.

So he wrote to the Abbot of St. Benedict, seeking admission into that order; for of all monks, the Benedictines were, in pursuance of the injunctions of their founder, most devoted to letters.

"Si vis veniam: sin autem, etc." (If you wish I will come: if not, etc.) ran his missive, with a curtness and a shade of hauteur worthy of his royal nursemate. But the witty Abbot had a pat answer, and a pun besides, ready at hand:

"Si bonus es, venias: si nequam, nequaquam." (If you are good for something, you may come: if not, don't.) The "nequam" and "Neckam" were perilously alike: too nearly so for the sensitive spirit of the would-be monk; so he turned his back on Benedict's house and made favor with that of Augustine at Cirencester. There he seems to have lived out his days uneventfully and to have risen to be Abbot. He died at Kempsey, near Worcester, in 1217, and that is all that is known about him personally.

Neckam was a typical product of the prevailing philosophy of his time. His principal treatise bears the title De Natura Rerum, which was a stereotyped one among the mediaeval encyclopædists, and in it he epitomizes all the scientific knowledge of his day which he has gathered by observation, and proceeds to explain it by the aid of a tropical imagination tempered by theology. He delights in intellectual labor, and detests scholastic methods, yet sees no other mode than these last by which its productions may be utilized. He collects his facts with patient care, all the time thinking that the study of the liberal arts, while useful, leads people into the "vanity of overcurious researches." And then he seeks to reconcile the inconsistency by averring that the arts are commendable in themselves, but those who abuse them are worthy of reprehension : regardless as to whether or not he himself may be found in the latter category. But for scholastic reasoning as such, it is "a thing full of vacuities."

There is no direct statement in Neckam's writings which fixes exactly the time when they were produced ; but John of Brompton, whose chronicle ends with the accession of King John (which occurred during his lifetime), quotes from the De Natura Rerum in a way that shows it to have been well known at the end of the 12th century. It is a treatise constructed very much on the lines of St. Isidore's Etymologies, but is not so categorical as the older work. It is a treasure house of curious folk-lore and legends. In it appears for the first time the fancy of the Man in the Moon; the traditions of the development of the goose from the barnacle;[1] the swan that sings ere it dies ; the unnatural ostrich which starves its young, and

---

[1] This persisted even in the Royal Society at the end of the 17th century. Phil. Trans., No. 137, p. 927, vol. xii., 1677-8.

the pelican which dies to feed them; the nightingale which sings on one bank of a stream and never on the other; the grasshoppers generated by the cuckoo; the parrot which drops dead on hearing the language of its native land; the dog which manages the sails of a boat which its master steers; the wren which hides under the eagle's wing and when the eagle rises in the air above all other birds, slips out and flutters over him and so wins the contest; the squirrel which crosses rivers on a chip with his tail for a sail; the lynx with eyes so sharp that it can see through nine walls: all discussed as Neckam promises in the beginning—"morally."

The portions of Neckam's writings which are of especial interest and importance in our present research are the chapter[1] on attractive strength in his De Natura Rerum, and a paragraph in another treatise De Utensilibus, the last being a sort of vocabulary or series of lists of articles in ordinary use. These show clearly the point to which knowledge of the magnet and amber had progressed, and the curious conceptions and fancies which had become intermingled with it.

In attacking the subject of attraction, Neckam defines the existing doctrine of similitudes, which was very closely like the ancient theory of sympathies and antipathies, by means of which it was sought to explain every phenomenon of nature by a mutual affinity or reciprocal dependence of bodies, whether celestial or terrestrial, organic or inorganic; such as gravity, cohesion, the force we call chemical "affinity," (and for which we still retain the old name though with a different understanding of it) and all movements, natural and instinctive, of living things.

The theory came originally from the Greek, and especially from Galen, who maintained that there was a vital, intelligent and divine power in nature, by virtue of which every substance appropriates that which suits its constitu-

[1] Cap. xcviii.

tion and its needs.¹ This was practically giving to the nature-soul of the ancient Greeks a selective capacity. In the 9th and 10th centuries the Arabs applied the same doctrine to the magnet. Serapion says that a solvent medicine, when it reaches the stomach, then draws with an attractive virtue the humor suitable to itself, but it is not drawn to the humor; just as the magnet moves the iron to itself, but is not moved to the iron.² Ali ben Abbas likewise makes a similar comparison,³ which, in later writers, is repeated over and over again, although it is essentially false, and simply due to the iron being more weakly magnetized than the attracting lodestone.

The doctrine of similitudes is thus a mediæval form of the old canon *similia similibus*, and rests on the same concepts. All compounds, for example, were supposed to derive their qualities from their elements by resemblance, being hot by reason of a hot element, heavy in virtue of a heavy element, and so on. For a long period, medical science rested on these distinctions, disorders being hot and cold, and remedies being similarly classified. One Eastern story teller relates⁴ that the Persian physicians were scandalized by the prescription of mercury by a European brother, for the cure of ill-effects following over-indulgence in cucumbers; for, they maintained, cucumbers are cold, and hence their ill-effects can not be overcome by mercury, which is cold also. "He makes no distinction," complain the oriental practitioners, "between hot and cold diseases and hot and cold remedies, as Galenus and Avicenna have ordered, but gives mercury as a cooling medicine."

---

¹ Martin: Obs'ns and Theories of the Ancients on Magnetic Attractions and Repulsions. See also Atti dell' Accademia Pont. de Nuovi Lincei, T. xviii., 1864–5.

² Steinschneider: Intorno ad alcuni passi di Opere del Medio Evo relativi alla calamita. Rome, 1868.

³ Lib. Practicae, lib. ii. c., 53.

⁴ The Adventures of Hadji Baba. Ed. by J. Morier, N. Y., 1855, p. 98.

Similar notions persisted among the metallurgists until the beginning of the 18th century. Thus the ready combination of metals with mercury to form amalgams was regarded as proof of mutual benignant regard, and the combination of metals in their alloys was similarly explained. Lead is loved by gold and silver, but brass abhors lead.[1] The astrologers claimed that metals exercised a selection in benevolently mixing with various parts of the human body, the gold seeking the heart; silver, the brain; lead, the spleen; mercury, the lungs; tin, the liver, and so on. But to living beings as units, they thought that metals manifested great contrariety, because, as it was gravely pointed out, no animal could subsist on metals, plants do not flourish where metallic veins abound, and in mines the vapors are deadly. Even in preparing pearls as medicine, they must be brayed in marble mortars, because otherwise iron might thus be imported into the body and act malevolently.

Neckam follows these ideas closely. Some things, he says, are drawn naturally, others by accident, and when by accident, either from necessity or chance similitude; from necessity, as when the body, through hunger, attracts so that its famishing members will thrive on insufficient food, such as bran (there were evidently dyspeptics in those days), or even on noxious herbs. Accidental similitude occurs when non-nourishing things are combined with nutriment. Natural attraction takes place, we are told, in many ways, "as by the power of heat, or by a virtue, or by the natural quality of similitude, or by the law of vacuity." Fire, for example, by the strength of heat draws oil for its nutriment.

The concept of an "attractive" virtue is the mediæval modification of Galen's selective vital force. This attraction by virtue, says Neckam, is caused in two ways, either occultly or manifestly. Occult virtue is closely allied to similitude in its effects, and acts as scammony draws bile

[1] Aldrovandus: Musaeum Metallicum, ii.

and hellebore the vapors. But manifest virtue is virtue that is perceptible and—here we suddenly find ourselves within the borders of the particular field which we are exploring—it is seen when the lodestone draws iron and the jet chaff.

This reference to jet is noteworthy. The ancient writers spoke of "gagates," which acted like amber, and left it in doubt what gagates might be. After jet had been certainly determined to possess the amber quality, the word was so interpreted. Neckam, however, is not quoting from any ancient author, but stating his own facts and beliefs; and the frequent later use of the English word "jet" by English writers instead of "amber," in referring to the phenomenon figuratively, renders it altogether probable that the learned Abbot was speaking not of the doubtful substance from Lydia, which he had never seen, but the lustrous black stone which had been mined in his own country ever since the Roman invasion.[1]

"If you ask its value as an ornament," he says, "jet is black and brilliant: if its nature, water burns it and it is extinguished by oil: if its power, being heated by rubbing, it holds things applied to it, like amber: if its use, it is an excellent remedy for dropsy." It was commonly found in Derbyshire and Berwick, and the Romans preferred it to that which was found in Germany. "The old writers," says Harrison,[2] "remember few other stones of estimation to be found in this Island, than that which we call 'geat,' and they, in Latin, 'gegates.'"

The explanation of "the quality of natural similitude not without attractive virtue" is ushered in by an illustration borrowed evidently from the Arabs. A warm stomach draws warm nourishment, and a cold stomach, cold nourishment: and we are to note that, according as by friendly

---

[1] The value of jet and of Kimmeridge coal for ornamental purposes was then well understood, and jet ornaments have been found in graves of the period. Traill: Social England, i. 92.

[2] Harrison: A Description of England. London, 1577.

similitude, attraction occurs, so, by hostile dissimilitude expulsion takes place. So that, for example, if vinegar and water be poured around a tree, the water will be absorbed and the vinegar rejected.

Now comes the first faint suggestion of the polarity of the lodestone. "So," he says, continuing his illustration, "the lodestone attracts by one part by similitude and from another part expels by dissimilitude." This is not the mere statement that a lodestone will repel as well as attract: nor is it, on the other hand, quite the affirmance of "opposite effects at opposite ends," but it is a clear recognition that one and the same stone will repel at one part and attract at another part. Where these parts were situated with reference to the figure of the magnet—whether at its ends or otherwise—Neckam did not know; but that this dual property exists in it, he makes plain. Compare Neckam's statement with that of Aldrovandus written four centuries later; "the lodestone attracts iron by natural sympathy *at one end* and repels it by antipathy *at the other.*"

Continuing, he explains that the appetite virtue draws by friendly similitude, and the expulsive virtue rejects by hostile dissimilitude; but the attracting thing—again he goes back to the Arabs—must act more violently than the attracted thing, for if equal they would counterbalance. Whence it is that the lodestone draws iron and not another lodestone, although it may have thereto greater similitude, because the lodestone opposes to the lodestone an equal and mutual contradiction. The iron yields itself because of weaker virtue.

The entanglement of his mind in the snares of sympathies and similitudes is obvious. On the theory of similitude, a lodestone should attract another lodestone; but that, he holds, is not the fact. *Similia similibus* cannot be at fault; that would be to dispute the hypothesis, which is indisputable. Wherefore, *query*, how can an incontrovertible fact be reconciled with an indisputable theory when they diametrically disagree?

The dialectical subtlety of the problem cannot have been otherwise than fascinating to the intellect skilled in the casuistries of the Petit Pont, and it grappled with the difficulties just as it had perhaps many a time done with

> "Whether angels in moving from place to place
> Pass through the intermediate space "

—and emerged triumphantly.

The similitude is undeniable; so, likewise, the sympathy. One lodestone resembles and sympathizes with the other, even as the other does with it. Therefore, why should *attrahens* act upon *attractum* any more than *attractum* upon *attrahens?* If Sortes and Scholasticus at opposite ends of a rope pull against one another with equal strength, is Sortes drawn to Scholasticus or Scholasticus to Sortes? Certainly not; they remain quiescent *in statu quo;* so do the lodestones. Q. E. D.

Now, this is not setting up Sortes—that favorite straw man of the schools—to be proved a stone, or a rose, or a lily, or what not; nor does it demonstrate that any remark of Sortes is both true and false at one and the same time, nor that he knows something, yet nothing—all favorite quibbles of the mediæval disputants—and, therefore, what Neckam calls "vacuities": this is what a 12th century mind, trying to break away from that sort of reasoning, manages to accomplish in the effort. The reasoning is wrong, of course; but it is physical reasoning, and that, even if wrong, is something better than "vacuities."

Now follows in this old treatise of an English monk probably the first of all known descriptions of the *mariner's* compass. Here it is:

"The sailors, moreover, as they sail over the sea, when in cloudy weather they can no longer profit by the light of the sun, or when the world is wrapped in the darkness of the shades of night, and they are ignorant to what part of the horizon the prow is directed, place the needle over the magnet, which is whirled round in a circle, until, when the motion ceases; the point of it (the needle) looks to North."

The paragraph from the De Utensilibus may be best considered simultaneously with the foregoing. The Latin words[1] present many obscurities, to which it is needless to refer in detail here, since they are considered in the following translation:

"If then one wishes a ship well provided with all things, one must have also a needle mounted on a dart. The needle will be oscillated and turn until the point of the needle directs itself to the East (North), thus making known to the sailors the route which they should hold while the Little Bear is concealed from them by the vicissitudes of the atmosphere; for it never disappears under the horizon because of the smallness of the circle which it describes."[2]

The manner of using the compass described in these remarkable passages is altogether different from that now followed; but is easily interpreted in the light of the in-

[1] "Qui ergo munitam vult habere navem habet etiam acum jaculo suppositam. Rotabitur enim et circumvolvetur acus, donec cuspis acus respiciat orientem; sicque comprehendunt quo tendere debeant nautae cum Cynosura latet in aeris turbatione; quamvis ad occasum numquam tendat, propter circuli brevitatem." Wright, T.: A Volume of Vocabularies, London, 1857.

[2] D'Avezac: Anciens Témoignages Historiques Relatifs à la Boussole. Bull. de la Soc. Geog., 19 Feb., 1858.

See, also, Bertelli ; sulla Epistola de P. Peregrino, Rome, 1868, Mem. ii. p. 41. D'Avezac points out that the statement in the original that the needle directs itself to the East is evidently an error, and translates the somewhat ambiguous clause with reference to the Little Bear as given above. In this Bertelli concurs, but dissents from D'Avezac's rendering of "suppositam" as if it were "superpositam" and consequent translation of "acum jaculo suppositam" as "a needle mounted on a pivot." It is thought that Bertelli is right, on the principle that no physical discovery ought to be ante-dated merely by a possible change in the signification of words. The burden of proof is on D'Avezac not only to demonstrate that his rendering is reasonable, but also from other sources to show that a pivoted compass was known at or about Neckam's time; and this he fails to do.

formation afforded by the early writers, whose works appeared shortly after Neckam's treatises.

The needle was not in constant use, as it is now, on a ship at sea, nor was it even employed to indicate any particular course. The roughness of the construction of the first compasses made them wholly unsuitable for such purpose. When the Pole-star could be seen at night, the pilot steered by it, as usual, and by day he kept along shore. If, however, the sky at night became cloudy, so that the stars were obscured, the needle was brought out and rubbed with the lodestone. This rubbing was repeated every time the needle was used, and is what is meant by (acum super magnetem ponunt) the placing of the needle over the magnet. This operation was called "inunction." The needle was thrust through a reed or short piece of wood (jaculo suppositam), which supported it floating in a vessel of water. If the needle was left in this receptacle, naturally it would move against the side, and thus be held, by contact, in a position not at all coinciding necessarily with the earth's lines of force. Consequently its magnetic quality would become more or less impaired, and that was apparently one reason for the remagnetization prior to every observation. Another reason probably was the removal, by the rubbing process, of the rust which would accumulate on the iron needle, the effect of the oxidation being to enlarge the needle and to roughen it, and so to impede its free movement on the water.

After the needle had been magnetized and carefully floated it was given an oscillating and circular movement, in order to carry it clear of the sides of the vessel and to overcome its own inertia, and also the normal resistance of the water to its motion. This was done by moving the magnet in the vicinity of the needle in a circular direction, the magnet attracting the needle and causing it to follow. After this motion was established the lodestone was withdrawn and the needle allowed to come to rest, and the point on the horizon noted which the north end designated.

It is plain that this operation must have required considerable thought for its invention. It was necessary to discover, *first*, that a lodestone bar would, when free to turn, place itself longitudinally in a north and south direction : *second*, that an artificial lodestone could be made by rubbing a needle with the natural lodestone: *third*, that such a needle would place itself north and south in the same way as a lodestone : *fourth*, that such a needle would be free to turn if floated on water: *fifth*, that a certain end of the needle would always point to the star, and that the indication of that end must be followed : *sixth*, that, in order to make this extremity always north-indicating, the needle must be rubbed with a definite part of the magnet and in a definite way, that is to say, the needle must be rubbed from south-pointing end to north-pointing end by a certain part of the magnet, or from north-pointing end to south-pointing end by an opposite part of the same magnet; any departure from the foregoing would make the end of the needle regarded as north-pointing turn to the south, and so destroy the utility of the apparatus : *seventh*, that the floating needle would not only follow a lodestone bodily, as the iron moved over the silver dish, as told by St. Augustine, but by suitably moving the lodestone it could be made to rotate : *eighth*, that the inertia of the needle and the resistance of the water could be first overcome artificially and the needle set in motion, so that afterwards the directive force of the earth, tending to set the needle in a particular position, north and south, would act jointly with the inertia and the liquid resistance as a force tending to stop the needle, instead of as a force tending to set the needle in motion in opposition to both of these resistances.

This is a most extraordinary category of discoveries for any period of the world's history, let alone a time when physical research was impeded in every direction, and the human brain supposed competent to evolve all human knowledge. It includes the perception of a difference between the effects produced by different parts of the lodestone, in order that the magnetizing operations above described might be done; the directive tendency of the magnet; the making of artificial magnets by rubbing iron needles with the stone, such magnets also showing different properties at opposite parts; the supporting of the compass needle on liquid, as a rotary armature; the prevention of the disturbing effects of inertia and fluid resistance, and the use of the instrument to reveal the position of the hidden Pole star. It is contrary to the teaching of the history of human invention since the beginning of the world, to the principle which underlies all human progress, to assume that all these discoveries were made simultaneously, and therefore that the compass of Neckam, crude as it is, was the product of a single inventive act. On the contrary, such a chain of phenomena is of necessity the result of evolution, and of slow evolution because occurring at a period when the current of all exact thought moved most sluggishly.

Observe that Neckam has linked together all the electric knowledge of his time. In the same treatise he discusses amber and lodestone attraction, the repelling effect of the magnet, the polarity of it, and as the fruit and flower of all, exhibits the mariner's compass. Klaproth, as I have stated, says that the Chinese had no knowledge of the instrument until during the 13th century, and hence after Neckam's day. The claim made by another authority, that a single Chinese writing asserts that the compass was used on a voyage in 1122 A. D., furnishes no proof that afterwards, and between that date and the period which Klaproth takes as the earliest, there was any continued marine employment of the magnetic needle by the Chinese. It is hardly reasonable to assume that the intelligence of this isolated use in 1122 could have reached England in the very depths of the Dark Ages, at a time more than 150 years before Marco Polo made his famous voyage, when practically no communication existed between China and Western Europe and when no channel can be recognized by which such news could have come by way of the Arabs. Nevertheless, it is impossible, as already stated, to conceive that the mariner's compass had not been slowly evolving somewhere before Neckam described it. Yet, where? We have examined in vain the knowledge of all nations which at various times and by various authorities have been credited with its invention. How came it to be known and in use in Northern Europe before Neckam's day?

Does the intellectual rise in electricity include a lost art regained? True, traditions as to the antiquity of the compass in Europe have never been wholly wanting. The Emperor Charlemagne is said to have given to the cardinal points (which, as we have seen, were established and so termed by the Etruscans) the Teutonic names, North, South, East and West, which they still bear; and to have also named the four intermediate rhumbs, North-East, North-West, South-East and South-West. The sailors of Bruges in Flanders, moreover, have always been reputed to be the inventors of the remaining eight points, completing the thirty-two, to which they gave the present Teutonic designations during the 12th century.[1]

The venerable Dr. Wallis, writing in 1702, at the age of eighty-six, gives it as his opinion that the mariner's compass was originally an English invention—"for the word 'compass' is an ancient English word for what we otherwise call by a French name a 'circle.' And I am sure that within my memory, in the place where I was born and bred, it was wont to be so called, though the word 'circle' is more in use."[2]

---

[1] Anderson: Origin of Commerce, London, 1787, v. 1, 61. Quoting Goropius apud Morisotus, and Verstegan.

[2] Phil. Trans., xxii., 276; xxiii., 278.

"My green bed embroidered with a *compas*," is mentioned in the will of Edward, Duke of York, dec'd 1415. Nicolas: Testamenta Vetusta, London, 1826.

There is material for conjecture, however, perhaps more persuasive than any based on such traditions and inferences as the foregoing. The pursuit of it leads us to the far north, to the sea on the shores of which the amber was first gathered, and to the great island city once grand in marble and brass, but of which now even the ruins are forgotten.

In the Baltic, about equidistant from Sweden, Russia, and Germany, lies the island of Gottland, by some identified as the Kungla of the national epic[1] of the Esthonians, where it is always described as a fairy land of adventure and untold wealth. Hither came the maritime commerce of the Wendic people after their capital city, Veneta, had been destroyed in 1043. Originally occupied by Goths, and later jointly by Goths and Germans, these tribes maintained incessant contests, which ultimately led to the downfall of the place. During the period of its supremacy, the island became a rendezvous for the vessels of all trading nations, and its principal settlement, Wisby or Wisbuy, despite the constant internal strife, grew into a city of large extent, the ruins of which have revealed many works of art and luxury. Olaus Magnus,[2] the great historian of the North, writing in 1555, speaks of it as a noble town, possessing a strongly-defended citadel. He says that it was the emporium of many regions, and that nowhere else in Europe was there such trade: that flocking thither came the Goths, the Gauls, the Swedes, the Russians, the Danes, the Angles, the Scots, the Flemings, the Vandals, the Saxons, the Spaniards and the Finns; these different people freely mingling with one another and filling the streets, the town hospitably welcoming all; that, in his time, there still remained

[1] The Kalevipoeg. See Kirby: The Hero of Esthonia. London, 1895.
[2] Olaus Magnus: Hist. de Gent. Septen. Rome, 1555, lib. 2, cxxiv.

marble ruins, vaulted halls and iron gates, windows decorated with copper and brass, afterwards gilded—all showing the grandeur of a bygone age. By 1288 the city seems to have become dilapidated through the continual feuds; but, in that year, Magnus, King of Sweden, allowed the citizens to rebuild their walls and fortifications—a circumstance which has led some historians into the erroneous belief that the place was then, for the first time, established.

It naturally followed that amid such a vast concourse of foreigners, all seafaring men, disputes constantly arose, based on controversies peculiar to the mariner's calling—the relative rights of masters and seamen, of owners and shippers, the adjustment of marine losses, contracts governing the chartering and maintenance of ships and crews, and so on through the great body of that branch of jurisprudence now known as admiralty.

There is probably no one more stubbornly conservative of his rights than the sailor, or more ready to assert them; and as this has always been found true of his species since time whereof the memory of man runneth not to the contrary, there is no reason to believe that the mariners who took their liberty in the streets of Wisbuy differed materially in modes of thought and action from those who congregate to-day in the great maritime ports of the world. Jack came ashore, and probably spent his hard-earned wages and fought the "beach combers" and the "rock scorpions" and became the prey of the crimps of Wisbuy and the terror of its police, just as he does now at Gibraltar, or Liverpool, or Hong Kong; while the owners and the masters and the average adjusters and the sea-lawyers wrangled over questions of jettison and demurrage and collision with the same fervor that brings them nowadays into the Admiralty Courts. The consequence was that two sets of locally-devised laws came into existence, administered by the consulate courts or authorities of the city—the one known as the Ordinances of Wisbuy, controlling all matters pertaining to the harbor, docks and to vessels in port; and the other, known as the Laws of Wisbuy, governing rights on the high seas. To these statutes merchants and sailors submitted by general custom and consent,[1] and they submit to them still, for they are imbedded in modern codes of marine law. Whether the famous laws of Oleron, supposed to have been framed by Queen Elinor, who died in 1202, or Richard I., who died in 1199, preceded or followed the Wisbuy laws, which they closely resemble, is a mooted point; but apparently the latter are the older.[2]

It is certain that, early in the 13th century, the Wisbuy laws were commonly observed in the eastern ports of the Baltic, which, of course, could not have been the case had these statutes not come into existence, as some suppose, until after the rebuilding of the walls of the city in 1288. Furthermore, recent research has made it plain that the Wisbuy code was a composite structure built up gradually over a long period, during which not only additions but omissions were made; many features, at one time in full force and regarded as wise and proper, becoming obsolete or out of harmony with changed customs or more moderate notions of wrongs and remedies.[3] The code as it appears to-day is extremely brief,[4] and thus bears on its face the evidence that it is probably merely a residuum, and by no means inclusive of all the precepts which at various times formed parts of it.

[1] Olaus Magnus, cit. sup., says: "The laws for sea affairs and the decisions of all controversies severally, far and wide, as far as the pillars of Hercules and the utmost Scythian Sea, are fetched from thence, and are observed ; being given, that all things may be done in a due tranquillity that may be fit and agreeing to peaceable commerce."

[2] Beckmann: Hist. of Inventions, London, 1817, i., 387. Parsons: Treatise on Maritime Law, Boston, 1859, 10, inclines to the opposite view.

[3] The Black Book of the Admiralty, London, 1876. (Monumenta Juridica.) Introduction.

[4] Ibid. Also Appendix to Peter's Admiralty Reports.

Here then was a great central mart or exchange, whither came the ships and mariners of all nations, save only the Saracens; for the infidel vessels would have found scant welcome at the hands of the newly-converted Northmen. Here was a source of sea law observed by all Christian sea-faring peoples. And here, if anywhere, was the focal point from which it may be presumed would be radiated any new item of knowledge, of interest and importance to the maritime world.

Among the ships which came to Wisbuy were those of the Finns and Lapps; and among the northern tribes, the Finns and Lapps differed from all the others in character and customs. Unlike their neighbors, they belong to that great Ugric nomad race which includes the Mongolians, Etrurians and Magyars. Their early history is exceedingly obscure. While the Lapps are commonly regarded as members of the Finnic branch of the Turanian family, some ethnologists consider them to be the original inhabitants of the country now known as Finland, and to have occupied it before the irruption into Europe of the Asiatic hordes which destroyed the Roman Empire. The Finns, on this theory, starting from the foot of the Ural Mountains, came to Bulgaria and Hungary, and being driven thence in the 7th century, made their way to the Baltic provinces, whence they drove the Lapps to the extreme north. Other hypotheses deny the close connection thus predicated between the Finns and Magyars, and place the migration of the former northward at a far earlier date, while extending the area of their settlement over a large part of Sweden and Norway, whence they were expelled by the Scandinavian Teutons and forced into the confines of present Finland.

In the 12th century, at the instigation of the Pope, Eric IX., King of Sweden, undertook to introduce Christianity among them; a series of crusades followed during the next two hundred years, with the result of subduing the Finns, though not of conquering them, and with the further consequence of making their peculiar customs and national life far better known to their northern neighbors.[1]

During the Middle Ages the territory about the Baltic occupied by the Finns, the Esthonians and the Lapps, was regarded as the peculiar home and nursery of sorcerers, whither people from every land, even from distant Greece and Spain, resorted for instruction or for special aid. The Esthonians looked upon the Finns as greater sorcerers than themselves, and the Finns in turn considered the Lapps their superiors in magic skill. But the old writers always single out the Finns by name, as the typical wizards. The mediæval Finns were a gloomy, earnest people, showing on their faces the marks of their Tartar relationship, and retaining in their families the same distinctive appellations as the far-distant Chinese. In their wanderings from the cradle of the human race in Asia, perhaps, they brought with them the Runic characters in which are written the ancient inscriptions found both in the north of Europe and on the Tartar steppes; but in common with the other northern nations, their traditions came down by word of mouth and in the songs of the Skalds and minstrels. They were the earliest iron workers in Northern Europe, and the Finnish swords anciently had a reputation equal to that which the famous blades of Toledo long afterwards acquired. Their great epic, the Kalevala, a composite structure of no definite date, shows them also to have been skilled as ship-builders, and in its descriptions of battles and forays it is not unlike the Anglo-Saxon poem of Beowulf, or the Norse Eddas and Sagas; but its chief characteristic is its wild and gloomy legends of sorcery and magic.

In all forms of witchcraft the Finns were regarded as masters. They devised the magic runes and spells which overcame the enemy while protecting the wearer, the impenetrable garments, the charmed weapons, and raised

[1] Vincent: Norsk, Lapp and Finn. N. Y., 1881. Peschel: The Races of Man. N. Y., 1876. Simcox: Primitive Civilization, cit. sup.

the ghosts of the drowned.[1] They practiced soothsaying as a means of profit. Their traffic in charms was chiefly with the sailor. To him they sold weather, good and bad, and bags of wind ("as Lapland witches pottled air") which would waft his ship to the desired haven, or send that of his enemy to disaster.[2]

The Finn country, with its many inlets and sounds, had an extended sea-coast, so that the early inhabitants became navigators from the beginning of their settlement. Therein they differed from the Mongols, who, as I have stated, remained for a long period dwellers inland. If we may conjecture knowledge of magnetic polarity and of the guidance of the lodestone, existing in the ancient people of Central Asia, whence both the Finns and the Mongols sprang, it is as reasonable to infer persistence of the same knowledge among the Finns as among the Chinese; although, as I have also remarked, the unchanging nature of Chinese customs would render the conditions for its preservation more favorable in the Middle Kingdom than in the Northern land. In such a country as Finland, however, the need for the land compass would quickly disappear; for there long land journeys were both unneces-

[1] Olaus Magnus: Hist. de Gent. Sept., Rome, 1555, lib. iii., c. xvi. See Lea: History of the Inquisition, N. Y., 1888, iii.; Peschel: The Races of Man, N. Y., 1876.

[2] The nautical superstition as to the weather-controlling power of the Finns is still alive (see Bassett: Phantoms of the Sea, Chicago, 1892). Dana, in his Two Years before the Mast, tells of the crew ascribing persistent headwinds to the presence of a Finn on board, whom the captain proceeded to imprison for his refusal to provide good weather. "The Finn held out for a day and a-half, when he could not stand it any longer, and did something or other which brought the wind round again, and they let him up." The "Rooshian Finn" is a frequent character in the forecastle yarns of the United States navy; and that he can alter the wind by sticking his knife into the mast is firmly believed by the old man-of-war's man. Whether any possible connection exists between the insertion of the knife for this purpose, and the savage Norse punishment also involving driving the knife into the mast, noted hereafter as a penalty for tampering with the compass, may interest those curious in investigating such matters.

sary and arduous, the sea when open furnishing a far easier road, while no elaborate buildings or engineering works called for employment of the needle in establishing sites or in determining alignments. On the other hand, adaptation of the guiding needle (assuming it to be known) to marine use would be not at all unlikely. While, therefore, on the one hand, and for ethnological reasons, it may be possible to assume knowledge of magnetic polarity to have existed among the Finns and Mongols, and that both may have availed themselves of it in migrating, the one people northward and the other eastward; on the other hand and for geographical reasons, the probabilities point more strongly to the Finns, seamen and dwellers by the sea, having discovered the sea use of the magnetic needle rather than the agricultural, inland-living, sea-dreading Chinese.

Bringing together now the conclusions which have been thus far suggested, we have found *first*, that the circumstances attending the appearance of the compass among the European sailors all indicate a radiation, so to speak, of intelligence concerning it from some central point or focus : *second*, that at about this time the city of Wisbuy, on the Island of Gottland, in the Baltic sea, was the great gathering-place and mart for all sea-faring men, and that thither came Goths, Swedes, Russians, Danes, Angles, Scots, Flemings, Vandals, Saxons, Spaniards and Finns: and *third*, that a knowledge of magnetic polarity may be more reasonably conjectured to have existed among the Finns rather than among any of the other peoples named, because of the race affiliation of the Finns and their peculiar skill in sea-sorcery. It may readily be imagined that if they possessed in the needle or stone a charm which would guide a ship from haven to haven, even in the narrow seas, how mysterious such a talisman would seem to the ever superstitious mariner, and how eagerly he would seek to obtain it and how quickly the tidings of it would spread throughout all the fleets of the western world. Nor is

AN ANCIENT FINNISH COMPASS. 141

direct proof of such possession wholly wanting. A single Finnish compass has been discovered for which the people claim great antiquity, the card or scale of which is marked for a latitude where the sunrise and sunset at the summer and winter solstices differ by sixty degrees:[1] this condition, curiously enough, being found along parallel 49° 20' N., which crosses Asia at the region which was the cradle of primitive civilizations, and from which began the wanderings of the great family to which both Finns and Mongols belong.[2]

A source from which knowledge of the mariner's compass may have come to Wisbuy, is thus found in its possible Finnish origin. How the Finns, if they had the secret, came to part with it—whether it was forced from them by their Swedish masters, or whether they yielded it up for the benefit of mankind in general, under the exhortations of good St. Henry, the English bishop, who entered their country in the train of Eric, or whether they bartered it with other mariners at Wisbuy, until all the world came to know of it—is a matter of surmise with which we are less concerned than we are with finding corroboration of the conjecture that from the great maritime exchange in the Baltic came the intelligence which Neckam first recorded.

The ancient sea laws of Wisbuy—as I have said—regulated rights and duties on the high sea, and therefore dealt with nautical crimes and offenses. Of these none is more heinous than to falsify the compass, for as every one knows, upon the accuracy of that instrument the safety of

---

[1] Nouvelles Annales des Voyages. Paris, 1823, vol. xvii., 414. The card instead of being divided into quadrants N. S. E. and W. has its four cardinal points 60° to the east and west of North, and two 60° to the east and west of South: the first two marking sunrise and sunset at the summer solstice, and the last two the same at the winter solstice.

[2] This region coincides closely with that in which Bailly conceived a prehistoric people of high civilization to have arisen and from which it migrated. (Lettres sur l'Origine des Sciences. Paris, 1777.) See also Ency. Brit., 7th ed.

the ship, and of all the lives she carries, directly depends. Even a slight error in its indication may lead the vessel far out of her course or into fatal perils. That this must have been perceived by the first sailors who used it, is altogether probable; indeed, we can easily imagine their terror and apprehension, when they found themselves out of sight of land and the familiar Pole star obscured by clouds, relying solely upon the pointing of the little needle, quivering in its bowl of water, to show them the way. Wreckers and pirates in mediæval times were common, and when the sea-villains learned how implicit the reliance was upon the compass, and how by slightly falsifying it they might bring a richly-laden craft upon the rocks and so into their toils, opportunities for this mode of plying their trade quickly revealed themselves. Obviously, it was much easier to conspire with the crew, or to send one of their own stripe on board in the guise of an honest seaman, to tamper with the needle, and so bring the ship to wreck in some previously-determined region, than to seek to capture her in the open fight for which all vessels then sailed prepared. Therefore, as by common consent, all sea-faring men regarded falsification of the compass as an offense, worthy of the severest punishment.

Tampering with the compass, moreover, in those days was supposed to be very easy, and no doubt many an unfortunate sailor lost his life under the charge of so doing when in fact he was innocent. The lodestone (with which the needle was rubbed) was then supposed to be affected by influences which are really destitute of the slightest effect upon it. Among the superstitions relating to it none, for example, was more common than the belief that its attractive power could be destroyed or weakened by the touch, or even the odor, of onions or of garlic. As Neckam notes, the proximity of the diamond was supposed to have a like effect; but diamonds were not ordinarily in the possession of mariners, while the odorous vegetables were, and so much was their effect feared that

THE GARLIC MYTH. 143

Baptista Porta[1] expressly ridicules the delusion prevailing even in his time which caused mariners, when in charge of the lodestone, to avoid eating onions or garlic, which not only may "deprive the stone of its virtue, but, by weakening it, prevent them from perceiving their correct course." So potent was this garlic myth that it was repeated steadily for fifteen hundred years. "I cannot think," observes one philosopher of the 17th century,[2] "that the ancient sages would write so confidently of that which they had no experience of, being a thing so obvious and easy to try: therefore I suppose they had a stronger kind of Garlick than with us." It began with Pliny, and came down by way of Solinus, Ptolemy, Plutarch, Albertus Magnus, Matthiolus, Rueus, Langius, Marbodæus, and the Arabian physicians and philosophers. True, Pietro of Abano first contradicted it before 1316 and Cardan[3] followed in 1550; nevertheless, the vitality of the notion[4] not only survived these attacks, but attained such vigor that when Philip Melanchthon, the great theologian of the Reformation, undertook to write a book on Physics,[5] in 1575, this same delusion is the only phenomenon concerning the magnet which he mentions; and he introduces it as an illustration of an accidental effect. It got its quietus in 1646 at the hands of that genial and witty iconoclast, Sir Thomas Browne,[6] who says "for an iron wire heated red hot and quenched in the juice of the

---

[1] Magia Naturalia, 1589, Lib. vii., c. 48.

[2] Ross: Arcana, 192.   [3] De Subtilitate, lib. vii., 474.

[4] Numerous theories have been evolved to explain the origin of this fiction. The most ingenious is that noted by Bertelli in his Memoirs of Peregrinus (Mem. ii., p. 39). He says that the passage in Pliny's Nat. History, "Ferrum ad se trahente magnete lapide et *alio* (theamede) rursus abigente a sese," is given in some codices so that "alio" reads "allio," thus transforming "other" into "garlic." This hypothesis relieves Pliny of responsibility for the error, and places it upon some unknown transcriber.

[5] Initia Doctrinæ Physicæ. Wittenberg, 1575, 221.

[6] Pseudodoxia Epidemica, ii., iii.

garlick doth notwithstanding contract a verticity from the earth and attracteth the southern point of the needle. If also the tooth of lodestone be covered and stuck in garlick, it will notwithstanding attract: and needles excited and fixed in garlick until they begin to rust do yet retain their attractive and polary respects." And Sir Thomas well knew whereof he spoke, for he had tried actual experiments with lodestone and garlic, and wrote down what he saw: wherein he differed from his learned predecessors who merely commented, with more or less profundity, upon one another's speculations.

To return now to the Laws of Wisbuy: From what has been said concerning the dangers attending the falsification of the compass, it may easily be inferred that in any code prescribing penalties for maritime offences, would appear a prohibition of the crime, and provision for punishment of the criminal. If in the old Wisbuy statutes such a law appears, then the existence and use of the compass is of course established as of an earlier date than their compilation. Now, if we may credit Olaus Magnus, writing before 1555, there was in the ancient code just such a provision, and he gives it because it is still in force in his own day. It is as follows:

"Whoever, being moved by sedition, shall menace the master or pilot of a ship with the sword, *or shall presume to interfere with the nautical gnomon or compass, and, especially, shall falsify the part of the lodestone upon which the guidance of all may depend*, or shall commit like abominable crimes in the ship or elsewhere, shall, if his life be spared, be punished by having the hand which he most uses fastened, by a dagger or knife thrust through it, to the mast or principal timber of the ship, to be withdrawn only by tearing it free."

The savagely cruel character of the penalty tends to show its antiquity, and affords abundant reason for its abandonment as people became more civilized. But beyond this the language used seems to draw a distinction

THE PENALTY FOR FALSIFYING THE COMPASS. 145

between the compass needle and the lodestone; a distinction which, as I have explained, obtained in the early compass, but which had long since ceased to exist in the time of Magnus. Observe also that it is the lodestone of which falsification is especially feared, because it was supposed that if the stone were wrong, then the needle rubbed by it would also be wrong. And this accords with the prevalent idea before mentioned, that the lodestone power could be annulled, as by garlic. Thus, the Wisbuy statute was undoubtedly framed under the common

THE PUNISHMENT OF THE FALSIFIER OF THE COMPASS.[1]

belief that the falsification could be very easily accomplished; and this was true, for the perpetrator, for example, might rub the needle with the lodestone so as to reverse its polarity, or so as greatly to diminish its directive tendency. In whatever way the result was actually produced there was the garlic or diamond theory which would suffice to account for it.

The facts which point to the European invention of the mariner's compass, may now be recapitulated as follows:

[1] From Olaus Magnus' History of the Northern Nations, Ed. of 1555. The old engraving, besides showing the compass-falsifier with the knife thrust through his hand and into the mast, illustrates the punishments of "keel-hauling" and throwing the criminal overboard, which were inflicted for mutiny and treason.

10

A description of the instrument appears for the first time in Neckam's treatise, written toward the end of the 12th century. The nature of this description is such as to make it clear that the writer is not referring to something of his own devising, but, on the contrary, to a contrivance which has then been known to sailors for some indefinite period. So many discoveries concerning the magnet are necessarily involved in it, moreover, as to justify the presumption that it is the product of evolution and of many minds. But, neither in the writings of William Appulus, nor in the Bestiary of Phillippe de Thaun, is there any evidence of similar knowledge; although it is hardly supposable that de Thaun, especially, would have failed to mention it somewhere in his long categories, had he possessed any such information. This places the probable time of the appearance of the compass in Europe at about the middle of the 12th century.

After this description of the compass appeared in an English work, descriptions of it in the literature of other nations followed so rapidly as leave their true chronological sequence in doubt, and under conditions which not only preclude the idea that the writers got their information from Neckam, but also that of the transmission of such knowledge *seriatim* from people to people. This suggests the radiation of the intelligence to the world from some central focal point. Such a point is found in Wisbuy on the Island of Gottland, then a great trading place for sea-faring people. Hither the knowledge may have been brought by the wonder-working Finns. Finally, the ancient sea laws of Wisbuy, dating from the time of the first appearance of the compass before noted, contain a direct provision against tampering with the instrument, and impose a terrible penalty for so doing.

We may imagine that the lodestone fell at once into its proper place in nautical employment. It belonged to the category of appliances used by the pilots to make their crude observations. It was not especially exposed to the landsman's gaze, being used only at sea, and then as occasion required. Thus it might perhaps escape chronicle, until some one, like Neckam, intending to write an encyclopædia, instituted an inquisition into things maritime sufficiently minute to bring the device to light. If Neckam's description be re-read in the light of this hypothesis, it seems to be framed on just such broad lines as would naturally be chosen by any one setting forth, for the first time, a (to him) new and extraordinary appliance. He tells simply what the contrivance does, but he is totally ignorant why it so acts, and of the long series of discoveries which separates it from the magnetic knowledge of Isidore; as ignorant as were the sailors in the English ships who came into the English harbors, and who probably told him just what they knew themselves, and no more

CHAPTER VI.

THE thirteenth century found ecclesiastical authority sovereign in every department of thought. It was an offence against religion, as well as against reason, to reject the truth; and the truth, it was insisted, was in the dogmas which the Church in its wisdom had arbitrarily defined. The members of a university, who had developed a spirit of investigation, found it sternly repressed, with an admonition "to be content with the landmarks of science already fixed by the fathers, to have due fear of the curse pronounced against him who removeth his neighbor's landmark, and not to incur the blame of innovation or presumption." In vain did the Italians, especially, show an intrepid desire to pursue the truth, or reveal prophetic visions of discovery. "Who shall say," asks Ranke, "whither this tendency would have led? But the Church marked out a line which they were not to overstep; woe to him who ventured to pass it!"[1]

The century had not far advanced, however, when the first faint signs of emancipation of the intellect from theological fetters began to show themselves, although the completion of the enlargement was still many a score of years distant. The work of scholasticism as the "solvent of theology" became manifest, while scholasticism itself commenced to pass into mysticism. As the military and clerical power started upon its decline, so the industrial and scientific forces of the world began once more an upward course.

The works of Aristotle and the Alexandrians had now

---

[1] Whewell: Hist. Induc. Sciences, ii.; Tennemann: Geschichte der Philosophie, viii.; Ranke: History of the Popes, i.

been given new life through the commentaries of the Greeks and the Arabs, and were being eagerly restudied by those who had hitherto denounced them as the ravings of pagans and infidels. The gathering of physical facts was gradually becoming regarded as an objective proceeding, and philosophy began its movement away from the subjective methods of theology.

While the philosophers and the theologians were pursuing endless disputations resulting from these changing conditions, the imaginative spirit of Christendom burst forth almost unchecked. The new language of the Normans yielded the new romance, and chivalry and love replaced piracy and murder or the dull category of saintly virtues, as the burden of the poems which the jongleurs recited, or the songs which the trouvères sang.

Among these new singers was one little known to fame, but still the most prolific of all. He wrote one of the Romances of the Round Table, but, like some few others, his muse favored subjects of a religious and moral character rather than those of a sprightly or amatory turn. He called himself William the Clerk,[1] and he was a vassal of Sire Rauf or Raul, who fought in the wars of Frederick I. in Italy (1159 to 1177). Robert Wace, the most eminent of the trouvères, vouches for the multiplicity of William the Clerk's writings; but if, as seems to be the case, they were generally of the stripe of the rhymed natural history interspersed with moral lessons (Li Bestiare Divins), which he composed by order of Rauf, whom he eulogizes in a fulsome manner through thirty verses, we need waste no regrets over their loss. In fact, William has spared us that trouble by himself deploring that he ever wrote them.

[1] Sur un MS. du Commencement du XIVme Siécle, etc. Bulletin du Bibliophile. Paris, Sept., 1836. D'Avezac: Anciens Témoignages historiques relatifs à la Boussole. Bull. de la Soc. Geog., 10 Feb., 1858. Jal: Archéologie Navale. Paris, 1840, 208. De la Rue: Essais Hist. sur les Bardes, les Jongleurs et les Trouvères. Caen, 1834. Wright: Biog. Brit. London, 1842, vol. ii., 426.

After he had become a monk he made atonement by informing mankind that

> "William, a Norman clerk who verses strung
> In flowing numbers of the Romance tongue,
> Too oft, alas, indulgent his refrain
> In fable foolish and in legend vain—
> Too oft he sinned—and him may God forgive
> Who loved the world, and in it loved to live."[1]

Among the poetic effusions which their author thus lamented is one discovered by M. Paulin Paris, a distinguished French antiquarian, in a MS. of 1329, which he attributes unquestionably to William the Clerk. It is entitled Love's Complaint (Complainte d'Amour), and in it the poet, after comparing his inamorata to the Pole star or Tramontane, gives the following description of the compass.[2]

> "Such of Tramontane the guise
> Shining blazing in the skies.
> Who, to far Venetia's strand
> Greece or Acre, Frisian land,
> Wandering sees its friendly ray
> Pointing out the hidden way.
> Knows it faithful guide to be
> O'er the bosom of the sea.
> Whether storm vext or at rest,
> Blow the north wind or the west.

> "When before the northern gale
> Flies through raging waves, the sail,
> That pure beam serene and clear,
> Saves the bark from danger near.
> When the blackness of the night
> Cloud-enshrouded veils its light,
> Still it doth a virtue own
> Drawing iron to the stone.
> Guiding safely those who roam,
> To the sweet delights of home.

[1] The free translation is the author's. Wright, T.: Biog. Brit., cit. sup.
[2] Author's translation. Bulletin du Bibliophile, cit. sup.

"Who would of his course be sure,
When the clouds the sky obscure,
He an iron needle must
In the cork wood firmly thrust.
Lest the iron virtue lack
Rub it with the lodestone black,
In a cup with flowing brim,
Let the cork on water swim.
When at length the tremor ends,
Note the way the needle tends;
Though its place no eye can see—
There the polar star will be."

This is apparently the first attempt to account for the north and south pointing of the needle, and represents probably the generally-accepted notion of the time; for we can hardly imagine the poet as the originator of it. The reasoning seems to have been that the needle points to the star because it has been rubbed by the stone. Therefore it receives a virtue from the stone. Whence does the stone get its virtue? Clearly from the Pole star, else why should the needle point to that star in preference to any other object in the universe—say the moon.

This is a long stride ahead in scientific reasoning, in that it seeks to explain a natural phenomenon by natural causes, and not by the intervention of supernatural machinery, or by an appeal to faith, or by the exercise of dialectic ingenuity. Whether the hypothesis be right or wrong is therefore of no consequence; it was an effort at straight rectilinear thought, made at a time when minds ran around in small circles; and as such it denoted progress. It was, moreover, encouraging to the intellects who had begun to feel the influence of the new centrifugal force, of which they could not understand the meaning, pulling them out of their little orbits.

While William the Clerk was bewailing the shortcomings of the world which he had left, the world in turn—even the Church itself—was scourging the iniquities of the clergy.[1]

[1] The Lateran Council of 1215. See Lea: History of the Inquisition, cit. infra.

"Thy ministers rob here and murder there,
And o'er thy sheep a wolf has shepherd's care,"

sang Walther von der Vogelweide, the Minnesinger of Germany: and the troubadours in France echo the same strain in even fiercer invective under the Arabian influence from across the Pyrenees, couching their denunciations in the new and flowing rhythms learned from the same source.

Among the troubadours was Guyot de Provins,[1] a minstrel, who, like the great Minnesinger, wandered from court to court singing his lays, and who had followed the Templars in the Crusades. Becoming tired of the world, or, perhaps, his world tiring of him, he entered the Cistercian novitiate, but abandoned that order in favor of the Cluniacenses, and then repented his choice and sought to return to his first association. The result was an increase in the jealousy between the two orders, and finally a triangular contest in which Guyot stood aloof and poured out upon both of them the vials of his wrathful sarcasm in gall and wormwood, none the less biting because of his intimate knowledge of monastic secrets.

The principal satire written by Guyot is entitled "Bible," as common a name for productions of the sort as De Natura Rerum was for encyclopædic treatises. It is a long poem of some 2,700 lines, written between the years 1203 and 1208, and it brings all sorts and conditions of men under the lash, beginning with monarchs and ending with "theologues, priests and physicians."

The second book is devoted to the clergy, and opens with a criticism of the Pope himself. It might well be supposed that such startling audacity would have brought the earthly pilgrimage of the writer to an abrupt conclusion; but Guyot was speaking only the popular thought, and other troubadours—Pierre Cardinal,[2] for example— of

---

[1] Wolfart, J. F.: Des Guiot von Provins bis jetzt bekannten Dichtungen, etc. Halle, 1861.
[2] Lea H. C.: History of the Inquisition, N. Y., 1887, i., 55.

far higher rank and consequence than himself, were attacking Innocent with even greater rancor and openness. The Pope settled most of these scores to his own satisfaction, during the Albigensian Crusade.

Guyot's onslaught on the papacy is mildness itself compared with his vituperations against the hierarchy generally, or even as contrasted with the poem of Pierre Cardinal, who openly accused the Pope of betraying his sacred trust and "vending his pardon briefs from cot to hall." He merely holds up the Pole star as an example of constancy and rectitude for papal emulation, but, in thus doing, so closely copies the verses of William the Clerk that before we know it we are laughing at the grotesque substitution of the supreme pontiff for the fair unknown of the subsequently remorseful monk.

Guyot begins by wishing that the Pope resembled the Pole star, whereby the sailors guide their course, and which, unlike other stars, is fixed and immovable; which, of course, is entirely inoffensive, except, as a schoolman of the time might remark, in so far as it inferentially suggests that the successor of St. Peter has not that "Petreity" which is the rock of his foundation. Still some change had to be made in language originally designed to celebrate the young woman whose brilliancy and attractive allurements William intended the Pole star to typify. But Guyot tamely follows the Clerk of Normandy, dragging in identically the same description of the compass, with the slight addition that in dark weather the needle can be illuminated. After which he returns to the Pope, and wishes him to be beautiful and clear like the star; but as he leaves out the whole of the ingenious theory whereby William connects the star with the lodestone, the precise relation of the Pope to the compass is left as obscure as Darwin's famous linkage of cats and red clover would have been had the great naturalist never explained it.[1]

[1] Guyot's poem has been so frequently published during the last century that its bibliography is now quite voluminous. A carefully edited

Although it is representative of the temper and mode of thought of the times, the Bible of Guyot would scarcely merit the notice here given it were it not constantly referred to in modern literature as the earliest known writing on the compass. It has frequently also been made the basis for the claim to the original invention of that instrument by the French. Both the treatise of Neckam and the poem ascribed to William the Clerk are in all probability of much earlier date, while the signs of the copyist are certainly more apparent in the imperfect work of the troubadour than in the logically complete structure of the trouvère monk.

The theory to which William the Clerk alludes as explaining the action of the needle soon begins to assume definite form, and align itself with the general hypothesis of magnetic virtue laid down by Galen. Thus the two lines of magnetic discovery, attractive power of the stone and its directive tendency, hitherto merely linked by Neckam, now begin to coalesce. "The magnet is found in India, and draws the iron to it by a certain occult nature. An iron needle, after it has touched the stone, always turns to the northern star, which does not move around the axis of the heavens as do the other stars; whence it is very necessary to those who navigate the sea,"[1] writes Cardinal de Vitry in 1218, thus bringing the statement of both phenomena side by side in a single paragraph.

Still more suggestive are the lines of Guido Guinicelli, the first of Italian poets who embodied in verse the subtleties of philosophy, and whose fame Dante has recorded:

---

text appears in Wolfart's work (cit. sup.), and in Fabliaux et Contes des Poètes François des xi., xii., xiii., xiv. and xvme siècles. Nouv. ed. Paris, 1808, pp. 327-8. Bertelli, in his Memoria sopra P. Peregrinus, 59, gives the poem, and a partial bibliography in a foot-note. An English translation of it appears in Lorimer's Essay on Magnetism. London, 1795.

[1] Historiæ Hierosolimitanæ, cap. 89.

> "Kindles in noble heart the fire of love
>   As hidden virtue in the precious stone;
> This virtue comes not from the stars above
>   Till round it the ennobling sun has shone;
> But when his powerful blaze
>   Has drawn forth what is vile, the stars impart
> Strange virtue in their rays;
>   And thus when nature doth create the heart
> Noble and pure and high,
>   Like virtue from the star, love comes from woman's eye.[1]

Even more closely knit are the facts in the following stanza by the same poet, for here the traditional magnetic mountains once more come to light—

> In what strange regions 'neath the polar star
> May the great hills of massy lodestone rise,
> Virtue imparting to the ambient air
> To draw the stubborn iron; while afar
> From that same stone, the hidden virtue flies
> To turn the quivering needle to the Bear,
> In splendor blazing in the northern skies.[2]

This adds another step to William the Clerk's original theory. The Pole star communicates its virtue to the magnetic mountains, and from the magnetic mountains comes the lodestone wherewith the needle is rubbed. But, for another reason, this stanza is very curious, in that it shows an early form of the hypothesis of the field of force surrounding the lodestone, in which field the power or strength or virtue of the stone is exerted. Note that the virtue is imparted "to the ambient air to draw the stubborn iron." The idea of action at a distance—of the magnet influencing its armature through no material bond—was not so thinkable to the poets and commentators of the twelfth and thirteenth, as it afterwards became to the natural philosophers of the seventeenth and eighteenth centuries. Not long after Guinicelli's poem

---

[1] Longfellow's translation: Poets and Poetry of Europe, 511.
[2] Author's translation. Ginguené: Hist. Litt. de l'Italie, i. 413.

appeared, Guido delle Colonne[1] of Messina expressed the same thought—

> " It is a secret of the lodestone,
> That to itself the iron 'twill not draw
> Unless the mediate air consent ;
> Although it hath the nature of a stone,
> Yet of its nature stones do not partake
> For lack of this same strange capacity."

This idea that the intervening medium takes part in the phenomenon of magnetic attraction was one which did not replace, but supplemented the prevailing doctrine that the stone operated solely by reason of its occult virtue. It was suggested in the poem of Lucretius and favored by the greatest of the Arabian philosophers, Averrhoës,[2] who also explained the attraction of rubbed amber for chaff by the same conception. A century later it was adopted by St. Thomas Aquinas.[3] Certainly a physical hypothesis which enlisted the concurring advocacy of the most eminent of Christian and Pagan commentators, and which appeared, ostensibly at least, to rest upon the principles of Aristotle, whom the world then regarded as the fountain-head of philosophy, could have had no stronger support.

The references to the lodestone and to the compass now begin to multiply rapidly in all classes of literature. Gautier d'Epinois,[4] in 1245, writes amatory verses comparing the object of his affection, not to the Pole star, as William the Clerk had done, but to the magnet, and the whole world to the needle which turns in response to such transcendent attractions. Matthew Paris,[5] perhaps also

[1] Author's translation. Nannucci: Man. della Lett. Florence, 1856, 81. Bertelli: Mem. sopra Peregrinus, 35.

[2] Colliget., V.

[3] In Phys., VII., lect. 3. See, also, Albertus Magnus : Phys., lib. VIII., tract. 2.

[4] D'Avezac: Aperçus Hist. sur la Boussole. Bull. Soc. Geog., 20 Apr., 1860.

[5] McPherson: Annals of Commerce. London, 1805, i.

not without some poetic license, though of a different kind, tells us that the first papal legate sent to Scotland in 1247 "drew the money out of the Scots to himself as strongly as the adamant does iron." Hugo de Bercy,[1] in 1248, speaks of the compass as in common use, and notes a change in its construction, the needle now being supported by two floats and arranged in a glass cup. The Norwegians,[2] by the middle of the century, not only had the instrument in constant employment, but were using it as an especial reward of merit and as the device of an order of knighthood.

Meanwhile the influence of the philosophy of Aristotle had greatly augmented, and his writings were the subjects of commentaries innumerable. But the world was indebted to the Arabs for the Aristotelian text, and it had come down from copyist to copyist, gathering errors as the rolling snowball gathers snow on its way; for the transcribers of the East were not the patient and accurate writers of the monasteries, and they had little compunction about adding paragraphs here and there drawn from their own imaginations. But worse even than this, there also appeared works attributed to Aristotle which are now generally conceded to be entirely spurious and of purely Arabic origin. Such, for example, is the Arabic translation of a Book of Stones, of which, if it ever existed, no trace remains, nor can any reference to it be found in any classic author.

The Arabic treatise does not purport to be even a complete translation of the alleged work of Aristotle, but merely a collection of excerpts. Nevertheless it seems to have been received with the same respect accorded to the philosopher's genuine writings, and this despite the fact that the manuscripts of it must have materially differed among themselves. In certain of these codices, though evidently not in all, for the passage is wholly absent in

---

[1] Riccioli: Geograph. and Hydrograph., lib. x., cap. 18.
[2] Torfaeus: Hist. Norweg., lib. iv., 345.

that possessed by the great National Library of France,[1] there appeared an account of the magnet which did much to retard the progress of the science.

It was unearthed by both Albertus Magnus[2] and Vincent de Beauvais,[3] who refer to it in their works, so that there is still the further hypothesis that it was originally invented by one of them, and hence not chargeable even to the Arabs. It sets forth that the point of the magnet which attracts iron is to the north, and the point which repels it is to the south; and it asserts that if the iron be held to the point which respects the north the iron will turn to the north; which is untrue, for the pole of the magnet which is directed to the north is the south pole, and it will induce a north pole in the iron, and that north pole will turn to the south, and not to the north. But the substance of what is said, whether right or wrong, is of much less moment than the historical fact that here probably began that complex tangle of relations between the poles of the lodestone, the poles of the needle magnetized from it by induction, the poles of the heavens, and, later, the poles of the earth,[4] in which the philosophers of the 16th century were even more hopelessly enmeshed than those of the 13th, and which is not clearly unraveled yet in our own terminology; for we still persist in calling that

---

[1] MS. Arab., No. 402, St Germ., quoted by Klaproth, cit. sup., 52.

[2] De Mineralibus, lib. ii., tract iii., c. v.: Opera. Leyden, 1651.

[3] Vincenti Bellovacensis: Speculi Naturales, etc., tom. ii., lib. ix., c. 19.

[4] The modern confusion arises from referring to the magnet needle as having a north and south polarity. The end which points to the north magnetic pole of the earth is, of course, south in polarity, although it is often marked N, and spoken of as the north pole of the needle. French writers frequently omit the inversion, and designate by north end of the needle that which in fact points southerly. Maxwell proposed the terms "positive" or "austral" magnetism to indicate that of the north end of the magnet, and "negative" or "boreal" magnetism that of the south end. So also it has been suggested to speak of the poles alternately as "red" and "blue." It is gradually becoming common to call the extremity of the needle which turns to the north the "north seeking" or "marked" end.

end of the needle which points to the north the north pole, when, as a matter of fact, its inherent polarity of course is south.

After this follow a series of falsehoods which we shall find afterwards cropping up everywhere. We are told that the magnet attracts lead because it is the softest of metals, and that the magnetic ardor penetrates and corrodes stones and tarnishes their brilliancy. That some magnets attract gold, others silver, and others iron; and that, if the gold be in a fine powder and mixed with sand, the magnet will separate out every particle of the metal.

This last is the first suggestion of the process of magnetic separation of metals from other substances mixed with them. The removal of iron in this way from an admixture with sand, etc., is elaborately described, as we shall see, by Porta and others, in the sixteenth century; so that the same idea of late years applied to the magnetic extraction of the same metal from its crushed ores, is of much antiquity.

Lastly, there is described the "creagus" or "flesh magnet," a stone "which, when once attached to the body, cannot be removed without tearing with it the flesh, although, in the latter, not a drop of blood will be found." This was probably nothing more than pumice, which adheres slightly to the lips or other moist surface of the body; but, none the less, the delusion lasted well; for, three centuries later, the wonder books told of "a kind of adamant which draweth unto it fleshe, and the same so strongly that it hath power to knit and tie together two mouthes of contrary persons and drawe the heart of a man out of his body without offending any part of him."[1]

[1] Fenton: Certaine Secrete Wonders of Nature. 1569. The Rev. Henry N. Hudson, in his excellent edition of Shakespeare, cites this passage in apparent explanation of Hermia's speech: "You draw me, you hard-hearted adamant," etc. (Midsummer Night's Dream, Act II, Sc. 1). There will be some, I fancy, who will be unwilling to take the poet in quite so literal a way, or to accord to him less play of imagination in the premises than was shown by Gautier d'Epinois.

In about the year 1250, Bartholomew de Glanvil, or, as he was commonly termed, Bartholomæus Anglicus, an English monk of the Minories order, wrote an encyclopædic work,[1] as usual on the lines of that of St. Isidore. His chapter on the magnet is of no intrinsic importance, for it is partly copied from the Etymologies and partly taken from the same source from which Albertus Magnus and Vincent de Beauvais drew their information—the false treatise of Aristotle. But Glanvil's work fell into the hands of the man who was easily first among the philosophers of his time, and whose genius towered over that of his contemporaries like a mountain peak above mole-hills.

For forty years, Roger Bacon studied science through the medium of experiment, which extended chiefly over the fields of alchemy and optics. Meanwhile he found time to learn Greek, Arabic, Hebrew, Chaldaic, and to master all that was known of mathematics. In an evil hour, he joined the order of Franciscan monks, and then found that he had literally thrown himself, body and mind, into chains. His writings were forbidden. If he attempted to instruct others, punishment awaited him. He was denied books, and because, despite all the obstacles cast in his way, it was evident, even to the dull minds of those who harassed him, that his knowledge of nature was far beyond that of the world in general, he was accused of sorcery. When he was not treated like a disobedient school-boy, he was dealt with as a suspected heretic.[2]

At length there came a pope—Clement IV.—whose leaning toward scientific inquiry caused a desire to know what Bacon could teach him; so he ordered the monk to disobey his superiors, hastily and secretly, and to write out his treatises and send them to Rome. Bacon had already exhausted his pecuniary resources, for he had expended some 2000 livres on his experiments; and how was he, a mendicant friar and penniless, to find the sum necessary

[1] Lib. de Proprietatibus.
[2] Lewes: Hist. of Philosophy. London, 1867, ii. 77.

to pay the scribes for transcribing his works? Further than this, how was such a task to be done in the monastery, where he met hostility at every hand? The Pope sent him no money, nor even dared to interfere in his behalf with the ruling powers of his order.

Nevertheless, he undertook the task single-handed, and in eighteen months, by dint of labor which, in the face of the difficulties encountered, seems almost superhuman, he had composed and written out and dispatched his *Opus Majus*, *Opus Minus* and *Opus Tertium*. Almost immediately after receiving these, the pope died. For ten years thereafter Bacon was allowed to prosecute his studies in peace. Then, in 1278, in his 64th year, a council of Franciscans condemned his works, and he was sentenced to solitary confinement in his cell, and it is generally believed that he died while thus immured.

Such, in brief, was the career of the first great apostle of experimental science who, in an age the whole temper of which was against scientific and philosophical studies, conceived of the essential connection between all sciences and their dependence upon the fixed and universal laws of nature; who brought grammar, philology, geography, chronology, arithmetic and music into scientific form; who laid the foundations of optics; who discovered the explosive force of gunpowder, and probably invented the telescope; and whose "Greater Work" was at once "the Encyclopædia and the Novum Organum of the 13th century."

Through the treatise of Glanvil the attention of Bacon seems to have been directed to the magnet, which he calls the "miracle of nature." He says that the iron which is touched by the lodestone follows the part of the latter which excites it, and flies from the other part; and that it turns to the part of the heavens to which the part of the magnet wherewith it was rubbed conforms. He says that it is not the Pole star which influences the magnet, for, if such were the case, the iron would always be attracted toward the star. On the contrary, the rubbed portion of the iron will follow the rubbing part of the magnet in any direction, backwards or forwards, or to the right or left; and if the iron be floated in a vessel of water and the magnet placed beneath, the same part of the iron will submerge itself to meet the magnet, while, if the magnet be placed above, it will rise upward. On the other hand, if the opposite portion of the magnet be presented, the iron (rubbed part as before) will always fly from it, "as the lamb from the wolf." Consequently he concludes that the magnet is influenced by the four parts of the heavens, and not by the one part in which the Pole star is located.

If Bacon did not actually discover the law of magnetic action (like poles mutually repel, unlike poles attract), it is manifest, from the foregoing, that he came very close to doing so. At all events, he brought the condition of general knowledge on the subject to a point where the very next step resulted in discoveries of the highest importance.

Bacon was pre-eminently a teacher, and seems to have freely communicated his knowledge to others whenever he was not restrained from doing so. To Brunetto Latini, the celebrated Florentine grammarian and preceptor of Dante, he not only told what he knew about the magnet, but repeated his experiments in his presence. Latini, at that time, was in exile, and visited Oxford, where Bacon resided, in about the year 1260. He died in 1294. He describes the compass in his Li Livres dou Tresor,[1] and, in certain letters written during his sojourn in England, he tells how Bacon showed him the "ugly and black stone to which the iron voluntarily joined itself," and the needle which, when rubbed by the stone, turned to the star and guided the mariners.

In the *Opus tertium* Bacon says that there are but two perfect mathematicians, Master John of London and "Master Petrus de Maharn, curia, a Picard." John of

[1] Li Livres dou Tresor. Paris, 1863, p. 3. Mainly a collection of excerpts from earlier authors.

London was his own disciple, "nurtured and instructed for the love of God," and the trusted bearer of his completed works to Rome in 1267. For John, Bacon predicted a glorious future "if he live to grow old and goes on as he has begun." But upon Peter—this Picard from Maricourt—he lavishes all his praise, all his enthusiasm. And Master Peter had well deserved it. From the trenches before Lucera he had written an epistle, which later came to be known as the "Letter of Peter Peregrinus"—a missive, little remembered now, often misunderstood, often plagiarized centuries ago, more often misinterpreted, but none the less a great epoch-making deliverance—an imperishable landmark in the path of physical discovery.

Before entering upon the examination of this work, a brief reference to some features of the generally accepted cosmical philosophy of the Middle Ages is here necessary. That our globe was the centre of the universe, and thus fixed and immovable, was undisputed. Encompassing it were supposed to exist ten heavens, successively enveloping one another; all except the outermost being in constant rotation about their common centre. The highest or external heaven formed the boundary between creation and space, and here abode the Deity, forever hearing the harmony of the spheres which lay below Him, in an endless hymn of glory and praise. Beneath the Empyrean came the crystal heaven, or *primum mobile*, then the heaven of subtle elements without weight, constituting the fixed stars, while the successive inner shells were respectively the heavens of Saturn, Jupiter, Mars, the Sun, Venus, Mercury, and, finally, of the Moon; the Earth and its atmosphere being sublunary things. All motion of these heavens was the direct work of angels or intelligences, and the laws of Nature were merely divine precepts which they carried into execution.

Ages, however, before these notions were conceived, the

apparent rotation of the heavens had been observed ; and not only this, but also that this revolving motion was seemingly about an axis, the intersection of which with the celestial vault marked the places of the poles of the universe. The conception of such poles was of still more ancient date. The story of Creation, deciphered from the broken and scattered remains of Assyrian and Babylonian tablets, recounts how "Maiduk embellished the heavens, prepared places for the great gods, made the stars, set the Zodiac * * * and fixed the poles." This carries the idea of these points back fully to 3,000 B. C.; but it probably had its rise very much earlier in prehistoric times. The Kushite-Semite race, who were the first imperial rulers of the primeval world, called themselves "sons of the pole," and substituted, for the reckoning of time by the Pleiades, one founded on so purely a physical motion of the heavenly pole, that they conceived the heavens to move about it with friction ; a fact which they deemed proved by the apparent movements of the fixed stars. They even believed the pole to be an ever twirling fire-drill, the heat of which influenced the stars. The race of Yakotas, the sons of Jokshan, or Joktan, in Genesis, likewise believed that the pole in its revolutions produced the burning heat of summer.[1]

This material idea of the poles, of course, has no place in the mediæval conception. They were simply the points about which the concentric heavens revolved, and that one which was visible to Europeans was marked by the presence of the Pole star. The progress of electrical knowledge owes much to this mediæval cosmic philosophy. It was because of the belief in the rotary heavens that the great discoveries now to be recounted were made, and, as I shall show hereafter, it was because of a disbelief that the earth stood still, that the even greater work which immediately ushered in the present science was undertaken.

[1] Davis: Genesis and Semite Tradition, New York, 1894. Hewitt: The Ruling Races of Prehistoric Times, London, 1894.

CHAPTER VII.

THE town of Lucera or Nocera, situated in the province of Apulia in southern Italy, was founded early in the thirteenth century by Frederick II., Emperor of Germany, as a place of free refuge and dwelling for the Saracens. In 1266, Charles of Anjou, who had been crowned king of the two Sicilies by Pope Urban IV., captured the town. Subsequently it rebelled and he besieged it a second time. The defense was obstinate and the town was finally reduced, in 1269, only because of starvation and after a year's siege.

Among the partisans of Charles who were encamped under the walls of Lucera during this long investment was the Magister Petrus de Maharne-Curia (or Master Peter de Maricourt), of whom Roger Bacon speaks in glowing terms. The surname "de Maricourt" is derived from a little village in Picardy, whence he came, and is classed among the territorial designations of the French nobility. The title "Magister" indicates the academic grade of "Doctor," showing that the bearer had studied and attained scholastic honors. The eulogiums of Bacon are so unstinted that there is reason to believe that Peter was already a man of wide celebrity for his learning and skill. Bacon[1] calls him "a master of experiment" seeing in full brilliancy the things which others grope for in darkness, like bats in the twilight, and says that through experiment he had become "versed in all natural science, whether medicinal, or alchemical, or relating to matters celestial or terrestrial." He is skilled, the monk tells us, in minerals and metal working—in arms, whether military or pertaining to the hunt, in agriculture and geodesy and magic;

---

[1] Brewer, V.: Fr. Rogeri Bacon, Opera. Lond., 1859. Op. Tertium, c. xi., p. 46.

and that he pursued learning for its own sake, neglecting all rewards, although his wisdom was sufficient to have enabled him to accumulate immense wealth had he so willed. That his experiments were continued over a considerable period of time is shown by Bacon's statement that he worked for three years upon burning glasses—evidently following in the footsteps of Archimedes. But, of all his achievements, that which most excites the admiration of the Friar, is his invention of a perpetual motion : the first recorded contrivance of the kind which came into the world, and probably the only one which in the end served a good purpose.

Here again the influence of Archimedes is apparent. There had always been a tradition that that philosopher constructed a sphere which reproduced the motions of the heavenly bodies. Cicero[1] refers to it in a general way, and Kircher[2] devotes a chapter to speculation on its possible construction; but probably it was nothing more than an orrery, showing the supposed relative positions and movements of the planets, but destitute, of course, of any automatic mechanism.

The circumstances which led to Master Peter's presence at the siege of Lucera are not difficult to conjecture. He probably belonged to one of the semi-military religious orders which, like the Templars, took an active part in the Crusades. The name of "Peregrinus" or Pilgrim, which later writers substitute for the surname "de Maricourt," shows that he had made the pilgrimage to the Holy Land—for this was a common honorary title accorded to persons who had taken part in the efforts to rescue the Holy Sepulchre; and, as Charles of Anjou, under whom we now find him serving, had joined the first crusade

[1] De Nat. Deorum, ii, 35. Tusc. Disp., i. 25.
[2] De Arte Magnetica. Rome, 1654, lib. ii., part iv, p. 245. See also, Claud. Ep. xxi. In Sphaerum Archim., Sext. Empiric. adv. Math. ix. 15. Lactantius: Div. Inst., ii. 5. Ov.: Fast vi. 277. Smith: Dict. of Gr. and Rom. Biog. and Myth. i. 2711.

of his brother Louis IX., of France, Peter or Peregrinus—as for the sake of uniformity with the old writers we shall hereafter term him—very probably went to the Orient in Charles' train. Friar Bacon indicates plainly enough what his functions were. He was skilled in arms and magic, and as pretty much all mechanical and physical knowledge, in those days, over and above what Archimedes had taught, was included broadly under the last-named term, Peregrinus was, in brief, an engineer. He probably devised engines for throwing stones and fire-balls, or for breaching walls; while his knowledge of geodesy came into play in building fortifications and digging mines.

During this employment, Peregrinus seems to have conceived the idea of converting the sphere of Archimedes into a self-moving magnetic motor, and then to have gone a step further and evolved a magnetic perpetual motion on an entirely different principle. It is a most singular fact that he reached these delusions through a series of brilliant discoveries, in which he not only overthrew most of the old notions concerning magnetism, but established, for the first time, the great fundamental laws of the science. Yet he cannot well be condemned for thus landing in an impossibility. No one knew that such a thing as a self-moving machine was impossible. The force of such a conception, especially when attained through the medium of experimentation which was correct in itself, and upon an intellect educated perhaps to as high a degree as was attainable in those days to the appreciation of the magnitude of it, may well have been overwhelming. A machine moved by the virtue which God had put into the lodestone and requiring no human aid—such was the initial idea which, running on to other conclusions, must have developed itself into speculation concerning the stupendous results which many such machines could accomplish, the possible accumulation of their powers, and the vast aggregated mights—and that was an age when might made right—which should be at the disposal of whoever controlled them. And beyond all this, conceive of the tremendous influence upon this soldier-monk, imbued with the superstitions of his creed, of the conviction that he might be the chosen of the Almighty to remove the curse of Eden, and to relieve man from the earning of his bread by the sweat of his brow.

He does not say this in the letter which he wrote on the 12th day of August, 1269, from the trenches in front of Lucera. The stake would, no doubt, have claimed him in short order, had he dared even to breathe a word of such a doctrine. But no one can read that missive without seeing how deeply the writer's soul was stirred within him. The person to whom he sent it was not a philosopher like himself, not even a scholar, but a knight, one Sigerus of Foucaucourt, and his next-door neighbor at home. "Amicorum intime"—"nearest of friends"—is the form of address, and the story is told as if in answer to some question put by Sigerus concerning the occult virtue of the magnet. But it all leads up to the machine which its inventor thought would run forever, and which is described in his last chapter; and what precedes is introduction, evidently intended simply to educate the recipient to a comprehension of the great result which the writer believed he had attained. It was the beginning of the arch-delusion in mechanics which ran for centuries parallel with the arch-delusion in chemistry, and with consequences very similar. For, as the search for the philosopher's stone and the elixir of youth brought to light many of the basic truths of the one science, so the equally vain quest for the perpetual motion has resulted in the discovery of many of the underlying principles of the other.

But let us examine the letter itself. It begins with a brief table of contents designed to show the orderly plan on which it is arranged. There are two parts—the first divided into ten chapters and relating to general principles; the second, into three chapters, which set forth the apparatus in which these principles are embodied.

After stating that he proposes to describe the occult nature of the lodestone in simple language, Peregrinus lays down the principles of experimental research. While he admits the value of general reasoning, he warns the reader against relying upon speculation and theory alone.

In the abstract, he says, many things appear true and correct which cannot be done by hands. The student must exhibit the wonderful effects by his work; for, by actually doing things, he can remedy errors which he never can correct by mathematics. This may seem curious counsel from the inventor of a perpetual motion, and lead to the query whether he practiced what he himself preached. The answer is suggested further on, when Peregrinus describes the first of his self-moving contrivances. If it does not work, that fact, he says, is to be ascribed to the lack of mechanical skill in the maker, rather than to inherent difficulties of the mechanism. This, of course, is one way of avoiding a troublesome issue; but it must be remembered that Peregrinus is writing from the seat of war, where he probably has had no means of obtaining accurate workmanship. He is sure of the conclusions which he has deduced from experiment; and, having tested some probably rude form of his machine and finding that it refuses to work, he considers this due, not to erroneous deductions, but to imperfections in the making. Hence this warning at the outset.

He next tells how to select a good magnet. In color it must be iron-like—slightly bluish and pale. The best comes from the northern regions, and is used by sailors who travel between the ports of the northern seas, notably those of Normandy and Flanders. This preference for the northern magnet is noteworthy, not only as showing that the best lodestone existed in the part of Europe where the compass found its first employment, but also because it is in direct variance with all the earlier writers who invariably give first place to the Indian stone. The heavier and more compact the magnet, the better, although such stones are the most costly. A mode of testing the lodestone is now for the first time announced. The best magnet, we are told, is that which will attract the greatest weight of iron, and draw it most strongly. In other words, Peregrinus considers not only the lifting power of the stone, but the magnetic strength, and apparently recognizes the difference between these effects. It is difficult to believe that a thirteenth century mind is evolving these concepts. Not until three hundred years later did Baptista Porta and Cardan and other philosophers of the time begin to measure the attractive force by causing the magnet to draw iron suspended on a scale arm.

All of the foregoing is prefatory to his announcement of greater discoveries. The ancient notions, as we have seen, were that the Pole star governed the magnet; then, that the Pole star influenced the magnetic mountains, which, in turn, governed the magnet; then, that the magnet was controlled, not by the Pole star, but by all parts of the celestial sphere, and, by their resultant action, the needle was brought to the north and south position. Peregrinus takes the next step forward, and reveals the poles in the lodestone itself. There are two points in the heavens, he says, of greater note than the rest, "because the celestial sphere revolves about them as if it were on pivots, one of which is called the Arctic or north pole, and the other, the Antarctic or south pole." So in the stone which he looks upon as an image of the celestial sphere, "you must understand there are two points, the one north and the other south."

Bacon knew that different parts of the same magnet would affect iron (as he supposed) differently, one attracting, the other repelling; but he had no notion that these parts had any definite position. Peregrinus not only tells us that they have precise places—as precise as the poles around which the celestial sphere apparently revolves— but now proceeds to explain how they may be found.

FINDING THE MAGNET POLES.   171

The stone is to be made in globular form and polished in the same way as are crystals and other stones. Thus it is caused to conform in shape to the celestial sphere. Now place upon it a needle or elongated piece of iron, and draw a line in the direction of the length of the needle, dividing the stone in two. Then put the needle in another place on the stone, and draw another line in the same way. This may be repeated with the needle in other positions. All of the lines thus drawn "will run together in two points, *just as all the meridian circles of the world run together in two opposite poles of the world.*"

Here was a magnet made in spherical form, the poles of it recognized and named, and the magnetic meridians found. More than this, although the lodestone sphere was regarded as an image of the celestial sphere, a certain analogy between it and the terrestrial globe was also plainly seen. Yet, again, more than three centuries were to intervene before William Gilbert should perceive in the globular magnet of Peregrinus a miniature earth, or, in the world itself, only a great magnet—a colossal reproduction of the Pilgrim's lodestone ball.

Peregrinus probably first found the poles in the way that is above described. Then afterwards he remarked that, at the points so determined, the needle was more strongly attracted than elsewhere. Consequently, he sees that the poles can be detected without marking the meridians, by simply noting the places on the stone where the needle is most frequently and powerfully drawn. If, however, he continues, you wish to be precise, break the needle so as to get a short piece, about two nails in length. Place this on the supposed polar point. If the needle stands perpendicularly to the surface of the stone, such point is the true pole; if not, then move the needle about until the place is found where it does thus stand erect. If these points are accurately ascertained and the stone is homogeneous and well chosen, he adds, "they will be drawn diametrically opposite one another like the poles of the sphere."

Here was still another advance. The idea that the lodestone somehow influenced the space between itself and the iron had been in existence from the time of Lucretius. But this in its latest form implied simply that the intervening air "consented" to the passage of the magnetic virtue, and thus the lodestone became, as it were, permitted to draw the iron. But Peregrinus goes far beyond that or any other earlier theory of the magnetic influence. He sees for the first time that the lodestone not only attracts the iron over the intervening space between them, *but compels the iron to take a definite position in that space.* In other words, he perceives that when his little needles are placed at the poles of the stone they stand erect, while elsewhere they stand more or less inclined. That was the first definite recognition of the directive action of the magnetic field of force: the first revelation of the direction in which the strains and stresses therein are exerted, shown by the turning of the little bits of iron in response thereto, as an anchored boat swings to the tide, or a weathercock to the wind.

Having found the position of the poles, the next step is to distinguish one from the other. "Take," says Peregrinus, "a wooden vessel, round, like a dish or platter, and put the stone in it so that the two points of the stone may be equidistant from the edge; then put this in a larger vessel containing water, so that the stone may float like a sailor in a boat." There must be plenty of room in the large vessel, so that the one containing the stone may not meet the side and so have the free motion impeded. Then "the stone so placed will turn in its little vessel until the north pole of the stone will stand in the direction of the north pole of the heavens, and the south pole in that of the south pole of the heavens;" and if it be removed from this position, it will return thereto "by the will of God." "Since the north and south parts of the heavens are known, so will they be known in the stone; because each part of the stone will turn itself to its corresponding part of the heavens." Here the naming of the magnet poles leads to confusion because Peregrinus gives to each magnet pole the same name as that of the quarter toward which the end of the free needle pointed; an example ever since followed.

Having thus both located and identified the poles, the next step was to determine their action upon one another; and then fell all of the old theory which began with the "theamedes," and ended with the supposed power of the magnet to repel as well as to attract iron. Two stones he says, are to be prepared, and the poles determined and marked by cuts. One stone is to be placed in a cup, and floated as before. The other stone is to be held in the hand. Then, "if the north part of the stone, which you hold, be brought to the south part of the stone floating in the vessel, the floating stone will follow the stone you hold, as if wishing to adhere to it;" and, if the south part of the held stone be brought to the north part of the floating stone, the same thing will happen. "Know it therefore as a law," he says, "that *the north part of one stone attracts the south part of another stone, and the south, the north.*" But, if the reverse be done, if the north part of the stone in the hand, be brought to the north part of the floating stone, the latter will flee; and the same will happen if south be joined to south. Thus was found the fundamental law that unlike magnetic poles mutually attract.

Peregrinus does not lay down the further law, that like magnetic poles repel; for, singularly enough, he does not recognize any actual repulsion occurring between these poles of like name, but assumes that the stone merely turns itself around so that the law already stated may come into play—that is, so that unlike poles may attract one another. Finally, he attacks the theory that the iron is the natural affinity of the magnet, and that the magnet will attract iron rather than another magnet. Here he finds further support in the doctrine of similitudes, which, as we have seen, was so generally prevalent. The magnet, he thought, attracted the magnet more powerfully than the iron because the magnet was like the magnet; and he uses the same illustration given by Neckam—the attraction of scammony for bile.

The more one reads of this remarkable letter, the more evident becomes the conflict in the mind of Peregrinus between the conclusions drawn from experiment, and those deduced from existing theories and speculations. It is also curious to note how much further he had extricated himself from the prevailing atmosphere of delusions and false conceptions than Neckam had done seventy years earlier. In Neckam's treatise, the wholly speculative ideas predominate; in that of Peregrinus, those which rest purely upon experiments obviously control. Neckam endeavors to reconcile the teachings of experiment with the prevailing theories, evidently through some sort of impression that he must do so, even if the results of investigation are out of harmony with those evolved from speculation. Peregrinus, on the contrary, does just the reverse, and tries not only to harmonize his experimental conclusions with one another, but to adapt the existing theories to them.

Yet, in one instance, he seems to fail completely, and to allow theory to lead him entirely astray. It has already been stated that in giving names to the poles of the magnet, Peregrinus calls that pole "north," which points to the north when the needle is supported so as to be freely moved. He says, "You will infer what part of the iron is attracted to each part of the heavens from knowing that the part of the iron which has touched the southern part of the magnet is turned to the northern part of the sky. The contrary will happen with respect to that end of the iron which has touched the north part of the stone, namely, it will direct itself towards the south." It is difficult to see how he could have made this error in the face of his experiments; for, as a matter of course, the end

THE NAMING OF THE MAGNET POLES. 175

of the needle which touched the south part of the lodestone must have acquired north polarity, and, therefore, have pointed to the south, which is exactly the reverse of what he states. True, the doctrine of similitudes would lead him to infer that the north pole of the magnet would point to the north pole of the heavens; but why should he allow that theory to control his ideas in the face of this particular demonstrated fact, when he has no hesitation in stating conclusions drawn from other facts in the same series of experiments, which were directly in the teeth of that theory? Two reasons may be given to account for this. The first is that the error was not due to Peregrinus, but to a transposition of terms by some copyist.

The second and stronger reason becomes clear when it is remembered that the doctrine of similitudes was more commonly applied with reference to the magnet and needle than with reference to needle and Pole star. The end of the needle in the compass was always rubbed by one and the same end of the magnet, and thereafter it turned to the north. Therefore it was concluded *a priori* that the pole of the magnetizing lodestone must also be north. Peregrinus undoubtedly, as others had done, rubbed the north end of his magnet to the needle and saw the latter point to the north, and thus, as he supposed, he established the principle, not by theory, but by actual experiment. And that the prevailing theory harmonized with the experiment tended, of course, still further to support the latter.

If he had presented to the supposed north pole of the needle the south pole of the magnet, he would have seen repulsion instead of attraction, and possibly have been led to question his hypothesis; but that is asking altogether too much of an investigator of the thirteenth century. In that he experimented on the subject at all connotes important progress. To suggest that he might have experimented to test the apparently plain conclusions of observation, is simply to impute to him a capacity for inductive reasoning far in advance of his age.

The next great discovery which Peregrinus notes is the possibility of changing the magnetic poles. "If," he says, "the south part of the iron which has been rubbed by the north part of the stone be *forced* to meet the south part of the stone; or the north part of the iron, which has been rubbed by the south part of the stone, be *forced* to meet the north part of the latter, then the virtue of the iron will be altered; and if it were north, it will be made south, and *vice versa. And the cause of this is the last impression acting, confounding, or counteracting and altering the original virtue.*" Unless Peregrinus drew some occult distinction, in his own mind, between the influence of the heavenly sphere upon the magnet and upon the needle receiving its virtue therefrom, it is difficult to perceive why this remarkable revelation of the possibility of destroying or reversing magnetic polarity did not suggest to him that such influence must be of a strange and inconsistent kind if it could be thus neutralized or inverted. But this again is a nineteenth century criticism.

Peregrinus does see, however, that the poles of the lodestone are apparently unstable, in a curious sort of way: and he announces that the unlike poles of two magnets come together not only to assimilate, but to unite and make one. Then, to prove this, he cuts a magnet in two and shows that each part has two different poles. And yet, when the parts are brought together to reconstitute the magnet in its original form, the polarity is the same as before the cutting, and two of the four poles which the two fragments possessed have seemingly vanished. That is the first announcement of the persistence of polarity in the separated parts of a lodestone, and it was a refutation in advance of the later theory of two magnetic fluids residing only in opposite ends of the stone. The experiments[1] are stated in some detail, but, as they amount

[1] In the printed copy of Peregrinus' letter which the British Museum possesses, Dr. John Dee, Queen Elizabeth's favorite astrologer, has covered the pages relating to them with underscorings and diagrams, as if he regarded that part of the work as the most important of all.

THE SOURCE OF MAGNETIC VIRTUE. 177

merely to transpositions of the pieces of the divided stone, it is not necessary to trace them minutely here. The conclusion is that the unlike poles attract because naturally they desire to unite and make one; whereas the like poles, also because of their nature, have no such desire.

Peregrinus next remarks that some unlearned people have supposed that the virtue by which the magnet attracts iron is already existing in the mineral veins in which the magnet is found; "whence they say that the iron is moved to the poles of the earth because of the mines of the stone there existing." But, he declares, the mines of the stone are found in various places in the earth, and hence the needle influenced by them should stand irregularly in different positions; which is not the fact. Now, he concludes, "wherever a man may be he may see with his eyes this motion of the stone, according to the place of its meridian circle. But all meridian circles meet at the poles: wherefore *from the poles of the world the poles of the magnet receive their virtue.*"

He evidently regards the poles of the earth, and those of the heavens, as in the same axial line, and attributes no especial directive faculty to those of the earth. For, he adds that the needle does not point to the Pole star, which varies in place, but to the heavenly poles, thus showing that he knew, possibly by means of astronomical observations, that the common opinion of his contemporaries, that the position of the Pole star coincided with that of the pole of the heavens, was erroneous.[1]

The first part of Peregrinus' letter, which I have now reviewed, ends with the description of his first form of perpetual motion, and this, as I have already stated, is apparently based on the Archimedean sphere. He introduces it as a means of showing how all parts of the

---

[1] This opinion, however, was not universal in the Middle Ages, as is shown by a celestial globe (Cufic-Arabic) in the National Museum at Naples which dates from 1225, and in which the Pole star is indicated $5\frac{1}{2}°$ distant from the pole.

heavens, and not the poles only, influence the magnet; but he is very cautious, and throws the burden of success or failure upon the maker. Make a globular magnet, he says, and find the poles. Then affix two pivots, on which the globe may turn. See that it is equally balanced and turns easily on the pivots, and try this repeatedly for many days and at different times in the day. Now place the stone with its axis in the meridian of the place, and dispose its poles so as to correspond to the elevation or depression of the heavenly poles in the region where you may be. And then—

But Peregrinus here drops his affirmative style and takes refuge in "ifs."

"*If* the stone is moved according to the motion of the heavens, you will delight in having found so wonderful a secret; but if not, impute the failure rather to your own unskillfulness than to nature."

What he believed would happen—he had never tried the experiment—was that the globular magnet, being on the (then considered) motionless earth, and being influenced by the heavenly vault revolving about it, would follow the motion of the sky and so rotate.[1] He thought it might even serve as a timepiece. But this delusion nevertheless bears good fruit—for, he adds, "in this position (*i. e.*, in the magnetic meridian) I believe the virtues of the stone to be best preserved," which is true.

In order to appreciate the remarkable improvements which were embodied in the instruments which Peregrinus now proceeds to describe, it may be recalled that the existing compass was nothing but a needle supported on a reed so as to float in a vessel of water. It simply showed,

---

[1] "It was the opinion of Pet. Peregrinus, and there is an example pretended for it in Beltinus (Apiar. g. Progym., 5, pro., 11) that a magnetical globe or terella being rightly placed upon its poles, would of itself have a constant rotation like the diurnal motion of the earth; but this is commonly exploded, as being against all experience." Wilkins: Mathematical Magick, London, 1707, 5th ed., Chap. XIII.

during cloudy weather, the position of the Pole star. It was not combined with any scale, nor was any means provided whereby a vessel could be steered on a given course by the direct aid of the compass itself. In other words, the compass had no "lubber's point" or fiducial line, and the angle of the course to a true north and south line was only guessed at. The nautical astrolabe, however, was fairly well known, and was used for measuring the altitude of the sun. It was a ring of metal divided into quadrants and graduated in degrees. It had cross-pieces, so that there could be pivoted at its center a bar with sight notches at opposite ends. The user held the ring in suspension by his left hand, so that its vertical diameter would be plumb. Then, with his right hand he manipulated the sight-bar before his eye, glancing first along it at the horizon line, and then elevating it to the position of the sun, thus rudely measuring the angle of altitude of the sun above the horizon. Peregrinus now combined the nautical astrolabe and the compass, and then, for the first time, he produced a compass having a graduated scale and a fiducial line or "lubber's point," which not only could be steered by, but which could be used for taking the azimuth of any heavenly body. This was very ingeniously done. He makes his magnet in ovoid form and puts it in a bowl in symmetrical position. Then, on the upper circular edge of that bowl he places marks, so that a diametral line will coincide with a line passing through the poles of the magnet, which last he has already determined. Then he marks another line at right angles to this, and finally divides the four quadrants into ninety parts each, so that each division is of course one degree of the circle. Now he places the bowl in a large vessel (probably glass) of water, in which the bowl floats, and the magnet, of course, places itself with its poles in the magnetic meridian. Thus he can recognize all points of the horizon by the marks which he has put on the edge of the bowl. Lastly, he rests upon the bowl edge a light bar of wood. This has, at each end, an upright pin. Normally it is placed in the north and south line. Hence, if he wishes to take the angular bearing of the sun from the north and south line, he moves this bar until the shadows of the pins coincide with the longitudinal axis of the bar; and simply notes the angle between the final direction of the bar and the north and south line marked on the edge of his bowl. To do this he has to hold the bowl steady with one hand, as he describes, and move the bar around its center with the other. Then, of course, he has only to read on his scale the angle between the bar and the north and south line, and he has the solar azimuth. In a similar way he can find the direction of the

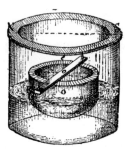

PEREGRINUS' FLOATING COMPASS.[1]

wind, by turning his bar until it is in line with the direction in which the wind blows. Or, he can find the azimuth of the moon, or a star, by placing his bar in the direction of the heavenly body. Of course it is only a step beyond this, and one that was probably fully known to him, to place the same bar in a line fore-and-aft the ship, and then his instrument would show the course of the vessel.

The evolution of this instrument from the astrolabe and the old floating compass is obvious; but Peregrinus is not contented with it, and now he proceeds further, and, for

[1] From Bertelli, cit. sup.

the first time, produces the pivoted compass. The floating bowl and the large vessel of water are abolished, and in place of them there is the ordinary circular compass-box of to-day. Its edges are marked as those of the bowl were —with the degrees of the circle. It is covered with a plate of glass. In the centre of the instrument, and stepped in the glass cover and in the bottom of the box, is a pivot, through which passes the compass needle, now no longer an ovoid lodestone, but a true needle of steel or iron. Then, at right angles to this needle is another needle, which, curiously enough, he says is to be made of silver or copper. Pivoted above the glass cover is an azimuth bar, as before, with sight pins at the ends. Now, he says, you are to magnetize the needle by means of the lodestone

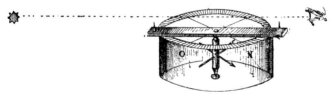

PEREGRINUS' PIVOTED COMPASS.[1]

in the usual way, so that it will point north and south; and then the azimuth bar is to be turned on its centre so as to be directed toward the sun or heavenly bodies, and in this way, of course, the azimuth is easily measured. In fact, the device is the azimuth compass of the present time. "By means of this instrument," says Peregrinus, "you may direct your course towards cities and islands and all other parts of the world, either on land or at sea, provided you are acquainted with the longitudes and latitudes of those places." Or, in other words, find the position in latitude and longitude of the place whither you wish to proceed, which is obviously the first thing necessary; note the direction of that place from the place

[1] From Bertelli, cit. sup.

where you are, and by means of the compass as he describes it you have simply to follow the course you have plotted. There would not be any difficulty in steering a modern ocean steamer by means of Peregrinus's compass, and in exactly the same way.

After the time of Peregrinus, as we shall see, the compass card was invented, and all of the thirty-two points of the compass (beginning with north, and thence pass-

PEREGRINUS' COMPASS.[1]

ing in order to north by east, north northeast, northeast by north, northeast, northeast by east, and so on) were named.

The presence of the little needle of copper or silver which Peregrinus thrusts through the pivot at right angles to the iron needle is a matter of curious interest. Peregrinus does not say why he uses it, nor what purpose he expected it to serve. Probably it was intended merely to indicate the east and west points and made of non-mag-

[1] From the Vatican Codex of Peregrinus' Letter. This is apparently intended as a plain view of the floating compass.

netic metal so as not to interfere with the magnetic needle. Yet the currents generated in a non-magnetic conductor moved in a field of force cause it seemingly to meet resistance as if the field contained some retarding medium, so that a copper bar or disk has been applied to the needle of a modern galvanometer to utilize this retarding effect to prevent undue vibration of that needle. The non-magnetic needle in Peregrinus' compass may have had the same effect. Its retarding influence might not be sufficient to interfere with the impressed force of the earth's magnetism upon the iron needle, and yet enough to check the vibrations of the latter due to inertia; so that Peregrinus may thus have unwittingly stumbled upon a phenomenon the discovery of which belongs to recent years.

The last chapter of this famous letter relates to the supposed perpetual motion for the understanding of which, by his friend Sigerus, all of these discoveries have been made and described. We have no contemporary record of any earlier attempt to construct a self-moving machine, although Peregrinus in the very beginning says that others have vainly tried to make them. The description which he presents of his own conception is incomprehensible: and in this respect it is the prototype and exemplar of all the subsequent so-called elucidations of the mysterious and power-generating "motors" which have been devised since his day. It had a ring of silver, which he rendered light by perforating it in various places. This he supported in some way so that it would rotate on its center. In the ring he arranged a series of iron teeth, the sides of which were at different inclinations, something after the fashion of the teeth of a ratchet wheel. The magnet was placed at the extremity of a radial arm disposed within the ring, with its end close to the teeth. The magnet was fixed. The description of the operation is unintelligible, but presumably Peregrinus expected that the magnet would draw the prominent portion of each tooth to itself, and then the momentum of the wheel would carry the tooth beyond the magnet, and then the magnet would attract the protruding portion of the next successive tooth, which he probably imagined would be brought nearer to it than would the rapidly-retreating face of the tooth which had just passed. He states that the tooth came alternately to the north and the south portion of the magnet, and there was alternately attracted and repelled; but the pictures of the machine which appear in the various old manuscript copies of the letter do not accord with this explanation. He also adds a small ball, which falls from one tooth to the other as the wheel rotates; and possibly he supposed that the movement of this ball would add something to the momentum of the wheel. Of course the contrivance never could have worked. But then, paper inventions often have that failing, and a large and goodly company of imitators, walking in Peregrinus' footsteps, are even now constantly finding this out. The law which asserts that two and two make four, and no more, at all times, and in all places, and that it is not given to man to create anything whatever, has never been suspended in favor of any mechanism, no matter how expensive or ingenious,—not even when it meets the approval of persons of superior consequence, financial and otherwise, in the community. Not having fully realized this fact ourselves at the end of the nineteenth century, we may perhaps look with some lenience upon the similar misapprehension of Peregrinus six hundred years ago.

Let me now recapitulate the foregoing remarkable achievements. Peregrinus discovered and differentiated the poles of the magnet. He revealed the law that unlike magnetic poles mutually attract. He showed how to detect the magnetic poles and demonstrated that in every part or fragment of a divided magnet the two poles persist. He proved that not only is the iron needle attracted by the lodestone, but that it will assume definite inclined or angular positions when brought into proximity thereto. Thus he, for the first time, disclosed the state of strain and stress existing in the medium surrounding the magnet, which, acting upon the light needle, compelled it to set itself in the direction of lines of force proceeding from the stone. Thus he first exhibited the condition of the magnetic field. He saw, though dimly, that the directive quality of the freely-suspended lodestone depended, not alone upon some inherent virtue of the stone, but upon an external influence acting upon it—an influence which he regarded as emanating from the celestial sphere. He found the position of the poles on a globular magnet, and recognized the magnetic meridians upon its surface. He first perceived the correct way of measuring magnetic strength. He discovered the mutability of the magnetic poles, and that the poles of a weaker magnet could be reversed or obliterated by the inductive action upon them of a stronger magnet.

He invented the first mariner's compass which could be constantly used to steer by as we steer by it now, instead of being employed merely to indicate the direction of the Pole star; the first compass having a fiducial or "lubber's" point and a graduated scale : the first compass capable of being used to measure azimuth or bearing : the first compass having a pivoted needle—the prototype of all electrical measuring instruments in which such an indicator is employed : and, if he did not actually recognize the retarding effect of a magnetic field upon a non-magnetic body (such as silver or copper), and combine such non-magnetic metal with the needle of his instrument in order to dampen or check its natural vibrations and so to bring it quickly to rest at its indication, he at least perceived that the magnetic field had no directive force upon such a body, and that therefore it could be employed as an additional index under the control of the magnetized needle. Finally, he first suggested the conversion of magnetic (electric) energy into mechanical energy in an organized machine and to do useful work ; and thus he proposed the first magnetic (electric) motor.

That even such remarkable discoveries as these should have remained unknown or have been forgotten for so long a period is easily accounted for by the intellectual condition of the times. Education was restricted to the few, and mainly to members of the religious orders, who, knowing little or nothing concerning maritime matters, would be unlikely to appreciate improvements in the compass or experiments upon it. The contents of the learned treatises of the time are not to be regarded as common knowledge, for manuscripts were costly and rare, and the masses of the people could not read them in the vernacular, much less in Latin. Frequently all that was known relative to a certain subject was confined to one author or group of authors, or to some one town or region. This sort of isolation constantly occurred in scientific matters not taught in the universities. Intercommunication by letter was difficult, instruction in the schools was mainly by lecture, and the universal reverence for Aristotle caused all novelties which were not accounted for by his teachings to be slightingly considered if not ignored. Undoubtedly, also, Sigerus regarded Peregrinus' information as a secret confided to his care, and thus general knowledge of it might have been delayed indefinitely. And, finally, when it came to light, people who tested the perpetual motion apparatus and found it a delusion naturally would discredit all other statements.

Of the later history of this extraordinary man nothing is known: not even the lavish encomiums of Roger Bacon availed to save him from oblivion, for such was the fate of Bacon himself.[1] The few manuscript copies which had been made of the famous Letter lay buried in the monasteries for nearly three hundred years.

---

[1] The Opus Majus was not published until 1733, nor the Opus Minus and Opus Tertium until 1859, and not a single doctor of the thirteenth and fourteenth centuries mentions Bacon either for blame or praise. Charles, E.: Roger Bacon, sa vie, etc. Paris, 1861, p. 31.

There are two well known features of the modern compass which Peregrinus' instrument lacks; and these are of importance. The needle passes through a vertical pivot shaft, so that the pivot and needle turn together, while the modern compass needle rotates on a fixed point; and, besides the scale marked in degrees around the circle, the modern instrument has also the so-called card on which appears a species of star of 32 points, each having its appropriate name, as N. E., N. N. E., S. W., W. S. W., and so on. Almost the first piece of nautical learning acquired by the young sailor is the learning of the names of these points of the compass in their order, or, as it is commonly termed, "boxing the compass." The older seamen of to-day still steer by the same points; but the modern fashion is to go back to the idea of Peregrinus and to lay a course from north so many degrees east, for example, instead of designating it by the arbitrary name of the point toward which the vessel is steered. The star itself has been known for centuries as the Rose of the Winds, and it is likewise inscribed on very old charts to show the direction of the various points. The earliest maps on which it has been found are Genoese, and date from 1318, and hence it has been supposed that both the star and the pivoted needle were invented by Mediterranean rather than by northern mariners, at some period between the time of Peregrinus' letter and the above-named year.

The suggestion that any part of the compass is of Italian origin recalls at once the man whom the world, for scores of years, believed to be the inventor of the entire instrument, and whom many modern encyclopædists, and Italian writers generally, still delight as such to honor. Flavio Gioja,[1] or Giri, or Gira (for the name is in doubt), lived

[1] See Nuova Enciclopædia Italiana. Boccardi, Turin, 1880. The most elaborate arguments in favor of Gioja's claims are given by G. Grimaldi in Memorie dell' Acad. Etrusc. di Cortona. See also Brechmann: Hist. Pandectarum Amalphi, Diss. 1, No. 22. Inter Scriptores Rerum Neapo-

at Pasitano near Amalfi at the beginning of the fourteenth century. His fame rests on the line of Anthony of Bologna, who lived later in the same century:

> Prima dedit nautis usum magnetis Amalphis.
> (Amalfi first gave to seamen the use of the magnet.)

Of Gioja's life nothing definite appears to be known, and even the line quoted above does not ascribe to him the invention of the compass, but only its introduction. Flavius Blondus speaks merely of the "rumor" of the Amalfitans being "entitled to the credit of the magnet by the assistance of which navigators are directed to the North Pole."[1] But, long before Gioja's day, the Italian vessels from Venice, Genoa and other ports had been transporting Crusaders by thousands to the Holy Land, and making trips with a regularity which, since the compass was fully known, leaves little doubt that they depended upon that instrument to some extent; although the pilots of that time, after the fashion of their ancestors, kept close to shore. Peregrinus, as we have seen, had fully pointed out that vessels could be steered from one place to another by means of his pivoted needle when the latitude and longitude of the objective point was known; but that description, it must be remembered, was buried in a private letter. Gioja seems to have re-discovered this and hence to have taught the adventurous sailors of the Mediterranean that the compass could be used, not only to find the Pole star, but directly to steer by. In so doing he earned a title to fame but little inferior to that which he would have merited had he been the original inventor of the apparatus. It was probably Gioja also who first added to the instrument the compass card or Rose of the Winds, of which the Etruscans, ages before, had designed the pattern—one doubtless repeated over and over again in their

litarum, Napoli, 1735, p. 935. Also McPherson: Annals of Commerce, Lond., 1805, Vol. I., 365.

[1] Italia Illustrata. Basle, 1559, 420, g.

ornaments, and bearing a resemblance to the Rose of the Winds, which, as I have already pointed out, seems too close in detail to make denial of some relationship between the two designs altogether reasonable. The invention of the needle turning on a fixed pivot seems to follow that of the card as a matter of course; for, by that means, the needle could be brought much nearer to the surface of the card below it than if it were on the long pivot shaft which Peregrinus employed to bring it near to the graduated edge of the bowl which was above it.

I have already stated that the first authentic description of the Chinese marine compass is of later date than the appearance of the instrument in Northern Europe. It is found in a work[1] known to have been written in 1297 under the reign of the Mongol Emperor Timour Khan, and is therefore after the letter of Peregrinus. In this the sailing directions for ships are indicated by the rhumbs or diagonal lines of the compass card.

It is undoubtedly true that, at this time, the Chinese were making voyages of great extent and duration. That famous traveler, Marco Polo, (whom Gilbert,[2] and other writers on the magnet in the 17th and 18th centuries, erroneously insist first brought the knowledge of the compass from China to Europe) describes a great expedition from the Pei-ho river, which occupied three months in making the journey to Java, and afterwards wandered for eighteen months in Indian seas before reaching "the place of their destination in the territory of King Arghun." Polo also records the enterprises of Kubla Khan against Madagascar.

[1] Chrestomathie Chinoise. Paris, 1833, p. 21.

[2] De Magnete, 1600, p. 4. There is not a word in Polo's narrative which describes the compass, and no evidence that he imparted otherwise any information on the subject. Furthermore his travels occurred between 1271 and 1295, and hence had not begun in 1260, the date when Gilbert says he learned of the compass from the Chinese. Gilbert also speaks of Polo as Paulus Venetus, which is an error, the latter being the name commonly given, not to Polo (Messer Million), but to Fra Paolo Sarpi.

But just what sort of compass the Chinese then had, even so strong an advocate in their behalf as Klaproth fails to discover. He quotes from a work of the 16th century a description of the floating compass in common use before the time of Peregrinus, and goes to Dr. Barrow for the details of their pivoted needle compass; so that, while we may infer that, with characteristic conservatism, they may have passed down these instruments unaltered from some far distant period, there is still that fatal absence of direct proof which renders all early Chinese invention open to more or less suspicion. That the deviation of the needle from the astronomical meridian—or, in other words, its variation—was well known to the Chinese long before that phenomenon had been remarked in Europe is sufficiently well established; and therefore I shall not devote space to the long discussions based on the assumption of its European invention which fill the treatises. A spurious addition to a Leyden codex of Peregrinus' letter, in which the variation is mentioned, has led many writers to credit Peregrinus with its discovery. But he knew nothing of it, and if, as Bertelli concludes, the variation in Europe was in fact zero at his time, there was nothing to direct his attention to it. The first practical knowledge among European people of the fact that the needle is not strictly true to the earth's geographic pole belongs to a later period than that now under review.

During the following century little was added to the magnetic discoveries of Peregrinus, nor was the compass as he and Gioja left it materially improved. The mention of the magnet and of the needle became more frequent, philosophers, poets, and theologians dealing with the subject with the same catholicity as in the past, and finding in it an unfailing source of supply for simile and metaphor.

Raymond Lully,[1] metaphysician and monk, entangles

---

[1] De Contemplatione. Capmany: Memorias Historias Sobre la Marina. Madrid, 1792, 1, 73.

the purely physical facts of magnetic attraction with his occult teachings, and does this at so early a date (1297) that books have been written to advocate his right to the credit for Peregrinus' achievements. Dante in the Paradiso[1] speaks of

> "a voice
> That made me seem like needle to the star
> In turning to its whereabouts,"

which, if recording no discovery, at least led to the first mention of the pivoted compass card itself carrying the needle; which is the form now used, wherever recourse is not had to the still older notion of the floating magnet. Da Buti,[2] the commentator on the great Florentine, writing in 1380, tells us that "the navigators have a compass in the middle of which is pivoted a wheel of light paper which turns on its pivot, and that on this wheel the needle is fixed and the star (Rose of the Winds) painted."

In the north, Barbour, writing in 1375, says that in 1306, King Robert, of Scotland, in crossing from Arran to Carrick, steered by a fire on the shore; for he "na nedill had na stone;" and the adoption of the Mediterranean compass seems to have been long delayed, for not until 1391 does Chaucer[3] mention the substitution of the horizon circle divided into thirty-two points in place of twenty-four.

NOTE.—The text which I have followed in the foregoing epitome of Peregrinus' researches is the one which Bertelli Barnabita has prepared from a careful collation of all of the existing manuscripts of the letter. (Sopra Pietro Peregrino di Maricourt e la sua Epistola de Magnete. P. D. Timoteo Bertelli Barnabita, Mem. Prima. Rome, 1868. Sulla Epistola di Pietro Peregrino de Maricourt e Sopra Alcuni Trovati, etc. Mem. Seconda: Bull. di Bib. e di Storia delle Scienze. Math. e Fisiche. Vol. I., Jan., Mar. and April, 1868.)

The first printed edition edited by Gasser (Petri Peregrini Maricurtensis de Magnete, seu rota perpetui motus libellus. * * Per Achillem P.

[1] Canto XII., v. 28.

[2] Da Buti, Francesco: Comment. Sopra la Div. Commedia. Pisa, 1862.

[3] Treatise on the Astrolabe. Ed. Skeat. Early Eng. Hist. Soc. London, 1872.

Gasserum, L. nunc primum promulgatus. Augsburgi in Suevis,) appeared in 1558, at which time manuscript copies of the letter were regarded as very rare. At the present time, several codices are known to exist—there being two in the Vatican, and six in the Bodleian Library of different dates, besides others elsewhere. In 1562 the material portion of the work was stolen by John Taisnier (Opusculum Perp. Mem. Digniss De Natura Magnetis et ejus effectibus. * * Authore Joanne Taisnierio Hannonio, etc. Coloniae, 1562), and published as his own in a treatise on the Nature of the Magnet and its Effects. The only English version of Peregrinus' letter is a translation of Taisnier's work by one Richard Eden, which seems to have been originally printed without date, and then reprinted in 1579 by Richard Jugge, London.

The perpetual motion of Peregrinus was also copied by a writer of the 16th century—Antonio De Fantis, of Treviso—and to him the invention of the apparatus is most commonly ascribed by authors subsequent to Jerome Cardan. The rotary magnetic sphere of Peregrinus was also plagiarized by Cornelius van Drebbel, who, in his letter to James I., of England, his protector, solemnly avers his ability to construct the apparatus so that it will automatically operate. Cardan: De Varietate Rerum, 1553 lib. 9, c. 48; Vuecher: Les Secrets et Merveilles de Nature. Lyon, 1596, 912. Cornelii Drebbeli Belgae Epistola ad Sapient. Bret. Monarchi. Jacobum, De Perp. Mobiles Inventione. Hamburg, 1628, p. 66. The possibility of Peregrinus' apparatus is doubted by Gilbert (De Magnete, 1600, lib. vi., c. iv.), and denied altogether by Galileo. (Opera, Florence, 1842, 443-9.)

In the beginning of the present century a mythical person was invented, one Peter Adsiger, and to him a few facts, which some one had exhumed from the old manuscripts or the Augsburg edition of the Letter, were duly credited; so that, for a long time, the names of Peter Peregrinus and Peter Adsiger were found in the text-books and histories, and so appear even up to to-day. But the name "Adsiger" was simply a translator's blunder, and is a part of the Latin dedication of the letter which Peter writes to Sigerus (Ad Sigerum). On such small errors as this, fame too often depends.

Some question has been raised as to whether certain of Peregrinus' discoveries were not earlier made by Dr. Jean de St. Amand, who was a celebrated physician and a canon of the cathedral church of Tournay. He lived "after the year 1261," but just when is not known. He seems to have been merely a copyist who restates Peregrinus' conclusions in an obscure way.

CHAPTER VIII.

THE revival of literature throughout Europe was everywhere manifest as the 14th century drew to its end. Giotto, Dante, Petrarch, Boccacio, Chaucer, Froissart, Wicliffe —such were the men whose great works both mark this period and serve as indices of the directions which the newly-aroused intellectual forces were taking. Yet the rise of positive science was none the less steadily continuing; before it the dogmas of authority, and especially those of Aristotle, were as steadily weakening. Meanwhile the commercial rivalry between Venice and Genoa, the great centers of Mediterranean trade, had brought the spirit of maritime adventure to the highest pitch. In the war between the republics, Genoa had been worsted; and the Venetians, by advantageous treaties with the oriental rulers, had established trading stations in the East, which gave them advantages unattainable to their competitors. The narrative of Marco Polo of the prodigious wealth of the far distant India, had inflamed the cupidity of his countrymen. However much the fathers of the church might assert the flatness of the earth, the sailors of Genoa and of Amalfi knew to the contrary, for they had learned that the ship which vanished beneath the brink of the horizon was neither sunk nor lost, and that, in all the seas wherein they had adventured, the quivering needle was a safe guide. So began, in Italy, the desire to sail, under the safeguard of the compass to the westward, and thus to reach the golden realm of Cathay.

In 1450, the invention of printing from movable types was made, and with this means of communicating and influencing opinion, the extension of knowledge was vast and sudden. Books fell four-fifths in price. The fruits of the new education and of the rapid spread of intelligence soon began to appear.

At this time the Italian trade, principally Venetian, came overland from India by way of the Persian Gulf to the Caspian and Mediterranean, and through the Red Sea to Egypt. The transhipments were many, the delays serious and the expense of carriage enormous; yet the Italian merchants were the most opulent in the world. The adventurers of Spain and Portugal looked upon their commerce with envy, and hungered for its profits. Meanwhile the knowledge of the compass had reached the Spanish and Portuguese mariners, and the latter, from their Atlantic ports, had begun to make, by its aid, voyages upon the Western Ocean longer than ever before. Their audacity, however, was checked by their superstition. They believed in a region of fire about the equator, in a weedy and entangling sea far to the West, and in the certain destruction of vessels that doubled Cape Bojador; currents bewildered them, and the trade-winds suggested only gales always blowing them away from home. Yet, if these men could be got to steer around the African cape, as the Egyptians had done ages before, it was certain that a water-way to India would thus be opened, and the nation to which they belonged might well hope to wrest from the proud Venetians the commercial supremacy which made all Europe their tributary. So thought Prince Henry of Portugal, son of John the First—Henry the Navigator—and, thereupon, he set to work to educate the sailors. He founded a naval college, got together the cosmographers and the mariners and the artificers skilled in instrument making, and having corrected the charts, and improved the astrolabes and the compasses—the last more especially—he provided the money and equipment for great voyages. Ultimately, under this stimulus, the Portuguese doubled Cape Bojador, penetrated to the tropics and found there no deadly heats, explored the African coast to Cape de Verde, and sailed to the Azores, passed with impunity through the weedy terrors of the Sargasso Sea, and learned to lay their courses homeward despite the trade-winds and the currents.

Such was the first great work of the magnet. Henry died in 1473, with the object of his ambition—the opening of the water-route to India—unfulfilled. Yet all Europe knew of his achievements, and the Italians better than all others, for their commercial existence was at stake. In Genoa, the interest was extreme, for she saw the opportunity, through her mariners, not only of surpassing the hardy Portuguese, but of avenging the crushing humiliation which she had received at the hands of Venice. But, in the long and wordy discussions which ensued among her learned men, in their wanderings amid the labyrinths of what Aristotle said, or Cosmos Indicopleustes asserted, or Augustine and Lactantius thought, the golden hour passed by, and from the little harbor of Palos, in Spain, and not from the great port of Genova la Superba, sailed the ships which carried forth the visionary son of the wool-comber and brought back the Admiral of the Indies.

Columbus, as is well known, went to Lisbon in 1470, where he supported himself by chart-making in the intervals of voyages to the Guinea coast. He was well aware of the advances in navigation which Prince Henry's mariners had made, and, in fact, had married the daughter of one of the ablest of the Portuguese sailors. From the Imago Mundi of Cardinal Pedro d'Aliaco, written in 1410 and published in 1490, he culled the opinions of Aristotle, Strabo and Seneca, on the possibility of reaching India by sailing to the westward. D'Aliaco's scientific knowledge came chiefly from the Etymologies of St. Isidore, but the particular part of his work which, as the annotations in Columbus' own hand on the copy now in Seville show, seemingly most influenced the discoverer, was plagiarized from the Opus Majus of Roger Bacon.[1]

[1] Major, R. H., F. S. A.: Select Letters of Columbus. Hakluyt Soc., London, 1870. Introduct., p. xlvii. Humboldt: Ex. Critique. Vol. i., pp. 64, 70.

The knowledge which Columbus had of the compass and of the magnet, therefore, rested on both practical and theoretical grounds. Of the compass, his early voyages had taught him even more than the ordinary use. In a letter to Ferdinand and Isabella, dated 1495, he describes how, having been sent by King René to Tunis to capture a galley, he found, on arriving at the island of San Pedro, in Sardinia, so powerful a force arrayed to meet him that his crew became alarmed and insisted on returning to Marseilles for reinforcements, "upon which," he continues, "being unable to force their inclination, I yielded to their wish, and, having first changed the points of the compass, spread all sail, for it was evening, and at daybreak we were within the Cape of Carthagina, while all believed, for a certainty, they were going to Marseilles."[1]

This is of a piece with his alteration of the reckoning of the ship's progress during his first voyage to the New World. It not only shows his familiarity with the compass, but incidentally furnishes an instance of that very tampering with the instrument against which the severe provision already noted in the Laws of Wisbuy was directed.

If Columbus was not familiar with Roger Bacon's work, he at least had learned, somehow, of the theory of the magnet, in which both Bacon and Peregrinus believed; namely, that the magnet was not controlled by the North star, but by all points of the heavens; for, in the history written by his son, he is expressly credited with this idea.

The figure of the great Admiral is one of especial interest in this research, because of the remarkable magnetic discoveries which he made. As this subject appears to have been confused by certain of his biographers, some detailed consideration of it is necessary.

A geographical meridian of the earth passes, as we know, through any given point of observation and the earth's geographical poles. The compass needle may

---

[1] Major: Select Letters, cit. sup., p. xxxvi.

stand longitudinally in the direction of this meridian. It then points to the geographical north pole, and is said to have no variation. But if the north-pointing end lies to the east or west of the meridian, then it is said to have east or west variation. This it is absolutely necessary to allow for in steering a ship, or in running a line in surveying land, or in laying out a railway. The variation is not the same at all points on the earth, nor is it constant at any one point. Therefore, there is a variation of the variation, which is secular in that it occurs over very long periods, besides being annual and even diurnal. Besides these changes there are irregular variations or perturbations, due to disturbances in the earth's magnetic field, which need not here be considered.

As I have already stated, the Chinese knew, certainly as early as the 11th century and probably before, that the needle did not point to the true north and south, and they constructed their land compasses to allow for the angle of discrepancy. But there appears to be no record showing that any European ever recognized the variation of the needle as a cosmical phenomenon before Columbus did so on his memorable voyage.[1]

Now, briefly, what happened to Columbus was this. On his first voyage to America, on the evening of September

[1] True, much has been written (See Libri: Hist. des Sci. Math. en Italie. Paris, 1830, vol. ii., 71. Formaleoni: Saggio Sulla Nautica Antica dei Veneziani. Venice, 1783, 51-2. Humboldt: Cosmos, v. Irving: Life of Columbus) concerning an old chart made in 1436 by Andrea Blanco, and now in the Library of St. Mark in Venice, upon which appears a figure supposed to represent the points of the compass with a correction for variation. This, however, Bertelli (Sulla Epistola di P. Peregrino. Rome, 1868, mem. iii., 77.), has investigated and finds no suggestion of variation present—the correction simply being that necessary to apply to the courses of a ship sailing on rhumbs of the compass (as N. E., N. W.) to keep clear of the loxodromic curve, or endless spiral due to the curvature angle between the earth and the meridian, which would never lead to any determined point. I shall not take space here to repeat Bertelli's demonstration of this error; while the other anticipations upon which he comments are merely inferences which he easily disproves.

13, 1492, the needle varied to the northwest half a point, and at dawn nearly half a point further. From this he states that he knew that it was not adjusted to the Pole star, but to some other fixed and invisible point, the variation of which no one had, up to that time, observed, and hence, on the third day, having sailed about one hundred leagues further, he wondered because he observed the needle come back to the star. On September 17th the pilots, having measured the sun's amplitude,[1] found the needles were a whole point in error; and then the seamen were greatly terrified, for they believed that their trusted guide had failed them. This is the time when Columbus is said to have invented the fiction of the movement of the Pole star in order to quiet their apprehensions, the personal narrative in Martin's collection stating "the Admiral discovered the cause, and ordered them to take the amplitude the next morning, when they found that the needles were true. The cause was that the star moved from its place, while the needle remained stationary."[2]

As Peregrinus had pointed out this movement of the Pole star more than two hundred years before, and Prince Henry's College had probably sifted all theories of the compass, including the notion that the needle pointed to the pole of the heavens and not to the Pole star, it is probable that Columbus simply stated a fact as he understood it, and in that respect invented nothing. Of course, to the ignorant seamen, any reasonable explanation would

---

[1] The sun's true amplitude is the number of degrees that the sun rises or sets to the northward or southward of the east or west points of the horizon. As the sun has no variation, by means of such an observation the variation between the true north and the magnetic north as indicated by the compass can be determined.

[2] Hist. del S. D. Fernando Colombo * * * dei fatti del l'Ammiraglio D. C. Colombo, suo padre, etc. Venice, 1621, cap. xvii.

Martin: Coleccion de los Viajes y Descubrimientos que hiceron por mar los Españoles desde fines del Siglo XV. Madrid, 1825.

Kettel, S.: Personal Narrative of the First Voyage of Columbus to America. Boston, 1827.

have sufficed. Columbus also believed that the lodestone was influenced by the different parts of the heavens, so that if the needle were touched with one part of the stone it would point east, with another west, and so on; and in fact he says that those who rub the needles cover the stone with a cloth so that the north part only is exposed, and the needle being touched with this possesses the virtue of turning to the north.[1]

Whatever interpretation Columbus may have given to the phenomenon in order to quiet the fears of his men, or whatever his own ideas may have been as to the cause of it, there is certainly no disputing the fact that he did then fully observe and recognize the variation of the compass. Moreover, he saw the needle vary at other times on other voyages, and the net result of his observation is given in his letter to the King and Queen on his third voyage, in his own words, as follows:

"When I sailed from Spain to the West Indies I found that as soon as I had passed 100 leagues west of the Azores, there was a very great change in the sky and the stars, in the temperature of the air and in the water of the sea: and I remarked that from North to South in traversing these hundred leagues from the said islands, the needle of the compass, which hitherto had turned toward the northeast, turned a full quarter of the wind to the northwest, and this took place from the time when we reached that line."[2]

He even drew a deduction from his observations which is curious, and characteristic of both the man and the time:

"I have come," he says, "to the conclusion that the earth is not round, but of the form of a pear, or of a ball with a protrusion—being highest and nearest the sky situated under the equinoctial line. . . . In confirmation

[1] Hist. del Almirante, C. 66. Muñoz.: Hist. N. Mundo, lib., vi., § 32. Also authorities before cited.

[2] Major: Select Letters, cit. sup.

of my opinion, I revert to the arguments which I have detailed respecting the line which passes from north to south one hundred leagues west of the Azores; for, in sailing thence, westward, the ships went on rising smoothly towards the sky, and then the weather was felt to be milder, on account of which mildness the needle shifted one point of the compass; the further we went the more the needle moved to the northwest, this elevation producing the variation of the circle which the North star described with its satellites."

The denial of credit to Columbus for the actual discovery of variation depends chiefly upon the supposed indications of the 1436 chart of Andrea Blanco or Bianco (see note, page 197) and in a general way upon inferences that earlier navigators may have observed the same behavior of the needle.

Humboldt[1] states that three places in the Atlantic line of no variation for September 13th, 1492, May 21, 1496, and August 16th, 1498, can be certainly determined, and that the line at that time ran from northeast to southwest, touching the South American coast a little east of Cape Cordova. That distinguished scientist, however, summed up the achievement of Columbus in the words: "The rediscoverer of the New World found a line of no variation 3° west of the meridian of the Island of Flores, one of the Azores,"[2] and elsewhere explicitly says that he has no right to the title of discoverer of the variation itself. But then, Humboldt, who as the foregoing quotation shows, was equally averse to according to Columbus the greater honors which the world's opinion now freely bestows upon him, appears to have based his conclusion upon the show-

---

[1] Examen. Critique de l'Hist. de la Géog., vol. iii., p 44-48. Cosmos, i., 169-197; v., 49-60.

[2] He also says (Cosmos, v. 54) that Columbus "had the great merit of determining astronomically the position of a line of no variation 2½° east of the Island of Corvo in the Azores on the 13th of September, 1492."

ing of the Blanco chart, which modern research has since proved to have been misinterpreted. The fact that a similar dictum to that of Humboldt is advanced by Washington Irving in his fascinating life of the Admiral has done much to place the matter apparently beyond dispute; but an impartial study of the history of the rise and progress of magnetic knowledge up to the time of Columbus, and of the condition of it during his life, and a recognition of the fact that much important data underlying such history has been made known since both Humboldt and Irving wrote, indicate the need for a revision of their verdict.

Little weight can be given to the argument that the first freely suspended magnetic needle certainly showed variation, as did all later ones when influenced by the earth's field, and that therefore the phenomenon was always open to observation. Unfortunately many a physical effect has thus presented itself for ages to the perception of mankind—nay, forced itself under the very eyes of the keenest investigators—without gaining recognition, or adding in the slightest to the world's stock of knowledge, until suddenly hailed as a great discovery. Moreover there were cogent reasons why, even if navigators had noted an aberration of the needle, they would have been likely to ascribe it to other causes than the true one, and so have failed to recognize the real variation at all.

Thus, in May, 1496, when the Genoese and the Flemish compasses on the ships of Columbus were found to disagree, one varying to the northwest and the other indicating the star, Columbus himself concludes the reason to be the difference in the magnets with which the needles were rubbed. In such rudely constructed instruments as then existed, it was equally possible to have assigned the errors to difference in shape of the needles, or weakness of magnetization, while it is not at all unlikely that both their form and treatment resulted in the production of consequent poles, which imported into them still further error. There was much better reason, therefore, for the European pilots before Columbus to have regarded any deviation of the needle as due to faulty construction or faulty magnetization, than to have assumed that it varied because of some external influence.

Because Columbus laid most stress upon his observation of the variation of the variation and of a line of no variation, is no more reason for disputing his right to be known as the discoverer of the variation itself, than is his notion that he had visited a part of India one for denying him his title as the discoverer of the New World. Mankind has long since decided that the forgotten voyages of the 9th century Icelanders detract nothing from his renown: equally immaterial is the hidden knowledge of the Chinese. The planets moved in accordance with definite law before the eyes of millions before Newton or Kepler lived; but the originality of the conceptions of these men is unimpaired. Moving planets and moving compass needles merely produced images on retinas: it was inconsequent whether of men or of sea-gulls. But to discover meant the establishment of connection between retina and a thinking, intelligent brain, and the application of the result of thought to the world's benefit. That is what Newton and Kepler did—and Columbus did likewise. He was the first discoverer of the New World who made his discovery known to the Old World. He was the first discoverer of the variation of the compass needle who made that fact known to the rest of mankind. And the true discoverer is not only he who has eyes to see and ears to hear, but he who has a tongue and uses it to tell to others what his keener senses have told to him.

The variation of the compass needle having been discovered, the importance of it was soon perceived by the sailors, for such a vagary of the needle would lead ships far astray if not known and allowed for in laying the course. But the philosophers, who cared little about nautical matters, and knew less, were more interested in speculating upon it and evolving new causes. Despite the light shed upon the problems of navigation by Prince Henry and his wise men, the myth of the magnetic rocks still survived among sailors the world over. With the discovery of variation, this assumed new vigor. Here was an explanation of the aberration of the needle ready at hand, and it was promptly and universally adopted. When the chart of the New Continent was added to the Edition of 1508 of Ptolemy's geography, the magnetic rocks, having traversed the Indian Ocean and the Mediterranean, at last came to final anchorage north of Greenland, which was depicted as the eastern part of Asia, and the earth was given a magnetic pole in the shape of an insular mountain.[1]

The idea of a magnetic pole of the earth governing the compass, to which Peregrinus alludes but dismisses because of the wide distribution of mines of magnetic ore, had never been forgotten. Cecco d'Ascoli, satirist and astrologer, fifty years later expressly affirmed in his bitter poem, l'Acerba[2]—a most heterogeneous gathering of learning of all sorts, and hence appropriately termed "the Heap"—that both poles of the earth were magnetic and exercised attraction, and perhaps he would have gone further and found out the magnetic character of our globe, and made who knows what other discoveries, if he had not fallen foul of the Inquisition, which burned both him and his books.[3] Between 1324 and 1508, however, great intellectual changes had taken place. Where people before assigned physical phenomena to causes entirely evolved from their inner consciousness, and hence without any foundation at all, they now explained them by physical facts wrongly selected; which, on the whole, was in the direction of progress.

[1] Humboldt: Cosmos. Lond. 1872, vol. v., 56.
[2] Poeti del Primo Secolo della lingua Ital. Florence, 1816.
[3] Lea: Hist. of Inquisition, vol. iii., 444.

In 1546, Jerome Fracastorio[1] questioned the existence of the polar mountains, and says that Bishop Oviedo, having made diligent inquiry "about that part of Sarmatia now called Moscovia," could find no such elevations. Olaus Magnus,[2] who lived in the North, however, not only affirms the existence of the mountains, but directly avers that the "compass follows them in direction." He also finds an island near the Arctic circle where the needle becomes demagnetized.

If the magnetic mountains of the North governed the needle, the variation was still to be accounted for. That, however, presented little difficulty. The mountains, it was explained, were not at the north pole, but at a distance from it. "The needle does not point to the true North," remarks Francesco Maurolycus[3] in 1567, "but by nature to a certain island which Olaus Magnus calls the magnetic island." Martin Cortez[4] had evolved this and more twenty years before, for he had not only to account for variation, but for secular variation—the needle, in his time, not pointing to the rocks, as located in the 1508 edition of Ptolemy. But mountains which had already traveled from Cochin China were abundantly movable, so Cortez merely shifted them sufficiently to the south to meet the changed conditions—a course evidently approved by Livio Sanuto, who did the same thing years afterward.

Meanwhile the world began to speculate as to what sort of a place this magnetic pole—*el calamitico*—might be, and what would happen to those who went there. Thus the sailors, urged by love of adventure and curiosity, began those journeys to the far north of which the end is not yet, and thus the quest of the north pole has its rise in the desire to attain the great island of lodestone to

[1] De Sympathia ; Opera, Venice, 1555, 103.
[2] Hist. de Gent. Sept. Rome, 1555, lib., ii., c. xxxvi.
[3] Op. Mathematica. Venice, 1575.
[4] Breve Compendio de la Sphera: The Arts of Navigation, trans. by R. Eden. London, 1561.

which, it was supposed three hundred years ago, all the compass needles turned themselves.[1]

The discovery of the line of no variation by Columbus (which was substantially a part of his discovery of the variation itself) became at once of great political moment. Immediately upon his return to Spain, in March, 1493, the King and Queen despatched an embassy to Pope Alexander VI., with a prayer for the securing to them of their rights in the newly-discovered lands. Martin V. had already given to Portugal all the territory which her mariners might discover between Cape Bojador and the East Indies. Alexander now made over to Spain all lands west and south of a line drawn from the Arctic to the Antartic Pole, one hundred leagues west of the Azores; or in other words, all of the world yet to be discovered was partitioned between these two nations with the line of no variation to separate their respective possessions.

The Portuguese lost little time in cultivating their hemisphere. The great dream of Henry the Navigator remained still unrealized, although the three years' voyage of the sailors of the Egyptian Pharaoh, centuries before, showed that the doubling of the Cape of Good Hope was not impossible. Bartholomew Diaz had confirmed this in 1486, by reaching the cape with a couple of fifty-ton pinnaces. The India, which Columbus had not found by sailing westward might still be open to discovery through an eastward voyage. The Jewish physicians said so. John of Portugal, who had seen the prize of the New World slip through his grasp, burned to retrieve his error. Again the compass led on a great adventure, and in 1498 the ships of Vasco da Gama, having sailed around the African continent, came to anchor on the Malabar coast.

The maritime supremacy of the Italians was now vanishing, and the rivalry lay between the nations of the Iberian peninsula. It was not long before it dawned on them that the earth, being globular, an imaginary line on

[1] Humboldt, cit. sup.

only one side of it would not divide it into hemispheres, and that serious dispute might easily arise as to the ownership of territory at the antipodes, depending upon whether ships sailed thereto in an easterly or westerly direction. The Portuguese had found the Molucca or Spice Islands by traveling constantly eastward, and had established a fine trade with them in cloves and nutmegs. Upon this trade the Spaniards looked with longing eyes, so that, when Ferdinand Magellan suggested that it might be entirely practicable to go to the same place by sailing constantly westward, his persuasions found a ready and favorable reception. For, obviously, although the islands had been reached by traveling constantly to the eastward of the line of no variation, and hence claimed by the Portuguese, if they could be attained by traveling constantly to the westward of the same line, the claim of the Spaniards to them, under the provisions of Pope Alexander's bull, would be just as good. Such was the inception of the magnificent voyage of the good ship San Vittoria in 1520–22, from the port of Seville to the port of Seville. Her intrepid commander died before the task was completed, but through his indomitable perseverance and faith, the world was circumnavigated.

Thus the first practical application of electricity to human use—for of electricity it must be remembered magnetism is but one form—had resulted in the greatest of human achievements. And the consequences—who shall measure even the most immediate of them? The whole commercial condition of civilization profoundly changed; new political questions engendered, to precipitate new conflicts amid the clashing interests of the nations; new interminglings of races; new issues of religion in its relation to the heathen; old theological tenets, so far as they depended on assumed flatness of the earth, overthrown, and the Scriptural interpretation of physical phenomena discredited; the unsettlement and moving of great bodies of people; a new thirst for adventure and a new spirit of enterprise; new distribution of wealth; new thought: in a word, the world, which had halted for a dozen centuries, now moved onward, not doubtingly and feebly as the invalid regaining health, but with the might and majesty of its new and irresistible energy.

Granted that it took great acts to do this, and that nothing less than the discovery of the new continent, the opening of the water-way to India and the circumnavigation of the globe would have sufficed; beyond them all, making them all possible, lay the slender bit of magnetized iron, quivering on its pivot yet always looking to the far north.

Of the three methods of finding the magnetic poles which Peregrinus describes, two, it will be remembered, are based upon the position assumed by the needle or short bit of iron when placed upon the surface of the spherical magnet. In one instance, we are told to draw several lines upon the globe corresponding in longitudinal direction to the needle, the latter being placed at different points and permitted freely to direct itself. These lines are found to be meridians, and the poles of the stone are at their intersection. In another method, the needle is moved about on the sphere, until a point is observed where it becomes inclined and stands perpendicular.

Plainly, both of these methods reveal, as I have already suggested, not a force drawing the iron to the stone, nor yet anything happening in either stone or iron, but a peculiar condition in the space immediately around the stone, by reason of which the needle is moved both in a horizontal and in a vertical plane, into a determinate position.

The circumstances here are in all respects remarkable. The compass needle was then supposed to point to the Pole star under the influence of virtue from that star; or, as Peregrinus believed, under the effect of virtue from all parts of the heavens. But the bit of iron which Peregrinus placed on his globe was not a compass needle, nor did it point to the Pole star; nor does he attach to it any theory of control by anything celestial or terrestrial, other than the lodestone itself. It turned to the pole of the round stone, and not only did that, but adjusted itself with such accuracy that its very action was the best means of finding the pole.

How the discoverer reasoned over this, we can only conjecture. The bit of iron being already in contact with the stone, he must certainly have remarked that here was a force which did not act to draw the metal to the magnet, but simply to turn it into a new position in a horizontal plane, and that, one always pointing to the pole.

This, however, was not all. In Peregrinus' third method the needle is turned, tilted, in a vertical plane. If nothing but attraction were involved, it ought simply to be drawn to the stone in any position, sidewise or endwise. But here it is turned endwise and then inclined until perpendicular. Here then was another force acting to make one end of the needle move downward and so to point to the pole beneath.

Of course the needle or bit of iron which Peregrinus used was itself a magnet; or became one immediately by induction from the lodestone globe.

Now what had he shown? First, the existence of the field of force around the stone, in which he had seen the needle deflect both laterally and vertically in order to point directly at the pole—second, he had marked out lines of force and determined their direction and that they ended in the poles—for these were the meridians which he traced on his globe: and he had also seen that they existed at considerable distances from the pole—for, in order to discover the place where his needle would stand perpendicular he must have moved it to other places where it was simply more or less inclined.

Thus Peregrinus revealed the presence of the magnetic field—a discovery which lies at the very foundation of all electrical development. We may look upon him as beginning a cycle which ended five hundred and fifty years later, when Oersted saw his needle turn and place itself anew in the field of force surrounding, not a lodestone, but a wire through which an electrical current was passing.

In the middle of the sixteenth century this seed which Peregrinus planted reached the end of its long period of germination. Or, perhaps, there had come into the world people capable of reading more from the pages of his manuscript, than was there in words. At all events, by direct inspiration from his writing came the discovery of the dip or inclination of the compass needle, and the still more definite recognition of a magnetic field of force. How this happened I have now to tell.

In March, 1544, Dr. George Hartmann, a native of Eckholtsheim, and a mathematician and astromoner of eminence, wrote to Duke Albert of Prussia an account of magnetic discoveries made during the preceding year, which he had already explained to King Ferdinand of Bohemia. He says:

"In the second place I find also this in the magnet: that not only does it decline from the north, and turn to the east for nine degrees more or less as I have said, but it also shows a *downward inclination* which may be demonstrated as follows: Take a compass needle about the length of a finger and place it on a point in a position exactly horizontal (or on the water-level) so that neither end inclines to the earth and both sides are in exact equilibrium. Now, if I rub either end of the needle once with a magnet, the needle does not stand any longer balanced, but inclines downwardly about nine degrees more or less. I have not been able to demonstrate to his Majesty the cause of this phenomenon." This was the first announcement, after

Peregrinus, of the dip or inclination[1] of the magnetic needle.

In that same letter, Hartmann describes the precise mode of identifying the north and south poles of a magnet, by placing it in a wooden bowl, floating on water, which Peregrinus gives; and revealing the source of his information beyond peradventure, he says—

"I have received an old parchment book of the time of the wars of the Contadini, in which I have found mentioned the force of a magnet, and the mode of constructing by means of a magnet, an instrument which moves automatically in equal form, time and manner as does the heavens: so that, as the sky, every 24 hours, makes its revolution around the terrestrial circle, so also this instrument in the same period completes its revolution. I have not been able to believe it."

It is easy to identify, from the foregoing, the first of Peregrinus' perpetual motions, and to recognize in the "parchment book" a manuscript of Peregrinus' letter; for no printed edition of it had at that time appeared. The error which Hartmann makes as to the extent of the dip, is easily accounted for by the fact that his needle was arranged on a vertical, instead of a horizontal pivot; and hence was impeded in inclining.[2]

[1] The term "Variation" expresses the action of the earth's magnetic force in a horizontal plane, but that force has another action upon a freely suspended needle. Only along the line of the magnetic equator (which varies but little from the earth's true equator), does the needle lie in a horizontal plane; proceeding northward, the north pole of the needle is drawn downward at an increasing angle called the "dip" or "inclination," until it reaches a value of 90° at the magnetic pole; but proceeding southward the north end of the needle is tilted upward. In some modern compasses sliding weights on the frame which carries the needle are used to counteract this inclining tendency.

[2] See Bertelli: Memoria Sopra P. Peregrinus, p. 111, quoting from Hartmann's letter. Dove: Repertorium der Physik, Berlin, 1838, Band ii. Also, Volpicelli: Intorno alle prime scoperte della proprietà che appartengono al magnete, Atti dell. Accad. Pontif. de' Nuovi Lincei, Vol. XXX., 8 March, 1866.

DELUSIONS CONCERNING THE LODESTONE. 219

of the magnet to induce its property in another, and another, and another iron nail, and so on indefinitely, until he bethinks him of musk, which, "having a sweet savour or smell itself imparteth the same to another thing, as to a pair of Gloves : and those Gloves give out savour and perfume a whole Chest of Cloaths," and so concludes that "the Vertue of the stone is distributive." Note the physical character of all this, when contrasted with the older notions of the affection of the magnet and iron, or the hunger of one for the other, or the doctrine of sympathy and similitudes.

It is necessary, in order to appreciate how singular the position which Norman assumes, and how completely he adopted the inductive method, to recall some of the general ideas concerning the magnet which were then in vogue. I mean the beliefs of the great mass of the people—the conceptions which infiltrated through all sorts of literature, and which made up the sum total of the world's knowledge on the subject.

There was not a single myth which had come down from antiquity which was not in full vigor. That garlic would destroy magnetism, that the lodestone had no attractive power in the presence of the diamond, that it was a useful medicament when administered internally—even the ancient superstitions of Samothrace—all were preserved and implicitly accepted. They had persisted unimpaired by the dialectics of the schoolmen or the physical discoveries of the philosophers; and they had become folk-lore and chimney-corner gossip. A few examples will suffice:

A magnet (it was believed) carried on the person will cure cramp and gout, draw poison from wounds, prevent baldness, cure headache, obtund pain and facilitate parturition. It will draw gold from wells, speak when sprinkled on water with a voice like that of an infant, and when mixed with nettle juice and serpent fat, make a man " mad and drive him from his kindred, habitation and country." It is both a test of connubial fidelity and a potent means of effecting marital reconciliations. It takes away fears and jealousies and renders a person "gracious, persuasive and elegant in his conversation." It is the especial friend of burglars; because, "if burned in the corners of a house, it causes the inmates to believe that the building is falling; and so terrified are they with fancies, that they fly out, leaving everything behind them, and, by this artifice, thieves seize on goods." Such were the typical absurdities which filled people's minds at the beginning of the sixteenth century. At about the middle of the same period, they received a vast accession of others, which, if not more absurd, were far more pestilent—for then began many of the delusions and deceptions which still prevail under the generic name of "animal magnetism."

To the arch impostor Bombast of Hohenheim, or, as he is commonly called, Paracelsus, the conception of these last is chiefly due; and as we shall encounter more or less of his influence in tracing further developments, some brief consideration may be given to his magnetic theories. He had learned the rudiments of medicine from his father; but he became by choice a professed astrologer, alchemist and magician, traveled widely "to observe the secrets of nature and the famous mountain of lodestone,"[1] and eventually imbibed, from the East, a rude sort of theosophy. This, mingled with the mysticism of Europe, produced a new and complicated form of quackery which, being less comprehensible than any which had preceded it, found especial favor with the fanatics, demonologists and philosopher's-stone hunters. Thus encouraged, he denounced Galen, Hippocrates and Averrhoes and all their adherents as imbeciles, claimed the discovery of the elixir of life, asserted communication with spirits, and entered upon a career of the grossest dissipation. Finally, he lost the support of the more intelligent portion of his followers,

[1] Biographie Univ., Paris, 1822.

and sank into a mere strolling charlatan, wandering from town to town telling fortunes, casting nativities, selling alleged receipts for producing the philosopher's stone, and preying generally upon the most ignorant classes. In 1541 he ended his career in abject poverty, an inmate of the public hospital at Salzburg.

The modern tendency toward psychical research, the ever-present inclination of the credulous to accept old delusions if revamped in novel guises, and probably Mr. Robert Browning's poem[1] have given to Paracelsus and his cult a new and wholly factitious importance. His absurdities have been dignified by the name of a "philosophy," and the man himself converted into a martyr. He knew human nature, and played upon its foibles with consummate skill. In that he proclaimed the doctrine of free thought in medicine and developed the therapeutic value of opium and mercury, the world is indebted to him. But, as a teacher of physical science, the best that can be said of him is that he preached reliance upon phenomena rather than faith, and practised exactly the opposite.

His notions concerning the magnet are of moment, because of their retarding effect upon scientific progress. Yet they were merely amplifications of myths as old as the race itself, re-told in manner suited to that era, as they are still rehearsed in terms suited to the present time. Lies, in which the people wish to believe, rival the truth in immortality. The finger rings from Samothrace, the Martial Amulets of Paracelsus, and the magnetic cure-alls of to-day are all accounted for in the persistence of human gullibility and ignorance through all civilizations and all ages. Exposures seem powerless to destroy them. Human imagination and chance recoveries afford ample sustenance.

According to Paracelsus, every human being is a mag-

---

[1] To this it is but fair to say that Mr. Browning himself provides an antidote in his appended notes. There is no one work, I apprehend, which shows more fully the influence of Paracelsus on the thought of his time, than Burton's Anatomy of Melancholy.

net "possessing a magnetic power by which he may attract certain effluvia of a good or evil quality in the same manner as a magnet will attract iron." This is merely an attribution, in fact, to man of the attractive quality which the ancient writers were so fond of expressing as pertaining to the Deity, and as metaphorically illustrated by the drawing power of the magnet or amber. The old poets and philosophers as I have shown constantly personified the magnet and spoke of its love, or its appetite, for the iron. Paracelsus simply reverses the figure, and, instead of attributing to the magnet the capacities of the man, gives to the man the capacities of the magnet. The metaphors of the poets became more literal after that. The difference is apparent between Gautier d'Epinois' asseveration that all the world turns to his lady because of her beauty, even as the needle to the lodestone, and the complaint

> You draw me, you hard-hearted adamant,
> But yet you draw not iron, for my heart
> Is true as steel—

of Helena following Demetrius. It is a reflection of the Paracelsan notion—and not a gross allusion to the mediæval "flesh-magnet"—which is thus found in the only extended reference to the lodestone which Shakespeare makes.

"A magnet," continues the astrologer elaborating his theories, "may be prepared from iron that will attract iron; and a magnet may be prepared out of some vital substance that will attract vitality." Such a substance, he holds, is one which has remained for a time in the human body; and it may serve to allay inflammation because "it will attract the superabundance of magnetism carried to that place by the rush of the blood." Diseases can be transplanted from the human frame into the earth by similar means. As to the lodestone itself, Paracelsus asserts that it "attracts all martial humors that are in the human system;" and that "martial diseases are caused by *auras* coming and expanding from a centre outwards and at the same time holding on to their centres." The front (north pole) of the magnet attracts, and the back (south pole) repels; and, in cases of nervous "epilepsy where there is a great determination of nervous fluid towards the brain, the repulsing (negative) pole of a magnet is applied to the spine and to the head, and the attracting (positive) pole of other magnets upon the abdominal region."[1]

All of this reads very like an extract from a "paper" by some modern "hypnotist" or magnetizer. We shall encounter more of it as we approach the period of Van Helmont, and Charleton, and Digby, so that it is not necessary, for present purposes, to dwell longer on the subject. The traces of Paracelsus' fancies, either original or revamped, constantly appear in the scientific works of the sixteenth, seventeenth and eighteenth centuries, like barnacles fouling and delaying the ship. Even Norman turns aside to contradict his statement, that a magnet, when "made red-hot, and quenched in the oil of Crocus Martis," will become so increased in strength as to be competent to pull a nail out of the wall. "But I suppose he meant not that the nail should be fast," adds Norman drily, "for then it were a miraculous matter;" which being applied to a miracle-monger of singular flamboyancy, savors of the sarcastic.

[1] The foregoing extracts are from the Paramirum of Paracelsus. See his Life by Mr. Franz Hartmann (London, 1887, pp. 138-431).

CHAPTER IX.

AMONG the men of the past, whose true greatness the world is only now tardily appreciating, stands Pietro Sarpi,[1] better known by his monastic name of Fra Paolo, for he was a friar of the Servite order. He was born in 1552, and died in 1623. The erection of his statue—the highest honor which the Republic of Venice could bestow upon a citizen—was decreed three weeks after his death, and carried into effect two hundred and seventy years later.

It is not my province to recount the strange history of Fra Paolo's political career; wherein, by sheer force of ability, he successfully opposed the Pope in the plenitude of his power, and became the chief consulter, guide and *de facto* ruler of the proudest state in Europe. The greatest of the Venetians was equally, in his day, the greatest of Italian scientists. A history of any branch of physical science, known in his time, must of necessity deal with some part of his work.

"What he did," says Macaulay, "he did better than anybody;" and, perhaps, it will suffice to recall Galileo's reverent address to him, as "my father and my master," to show that the encomium of the historian applies not alone to his achievements as a statesman. His private secretary and intimate friend, Fra Fulgenzio Micanzio, in a list of subjects in which he declares Fra Paolo to have been profoundly versed, mentions, besides the Hebrew and Greek languages, and mathematics, "history, astronomy, the nutrition of life in animals, geometry, including conic sections, magnetism, botany, mineralogy,

---

[1] Robertson: Fra Paolo Sarpi, London, 1894. Griseleni: Vita de F. P. Sarpi, 1760. Giovini: Vita, etc., Brussels, 1836. Micanzio: Vita, etc., Verona, 1750. Fabronio: Vitæ Italorum, Pisa, 1798, xvii.

hydraulics, acoustics, animal statics, atmospheric pressure, the rising and falling of objects in air and water, the reflection of light from curved surfaces, spheres, mechanics, civil and military architecture, medicine, herbs" and "anatomy." And, in almost every one of these great fields, Sarpi made discoveries of the highest importance. He first observed the dilatation and contraction of the uvea of the eye; first found the valves in the human veins, and first discovered the circulation of the blood (Harvey experimentally demonstrated this afterwards), and invented artificial respiration. He made the first maps of the moon, anticipated Kepler in his observations on the reflection of light from curved surfaces, first recognized the effects of refraction, and declared that the sun is fed, and that stars are suns. He announced that heat is motion, and exemplified its generation by heating iron with a hammer; that light is motion, and that it comes to us in waves or pulsations through a medium less material than the atmosphere; that sound is motion, but not (as he thought) motion of the atmosphere, for it travels against the wind and through water, moving like light in waves or pulsations; that color is caused by the atmosphere and by the reflection of different rays of light; and then he identifies sound, color, heat and light together, thus correlating these physical phenomena. The desire is strong to dwell upon Sarpi's researches in these fields, but it must be foregone to turn to his discoveries in magnetism. Unfortunately, here the actual records are meagre. He wrote a treatise on the magnet, which, after his death, remained, with his other manuscripts, in the Servite Monastery, where he spent his life. As late as 1740 his literary remains were minutely examined and arranged in order by the learned Fra Giuseppe Bergantini. Twenty-six years afterwards they, with the buildings in which they were stored, were completely destroyed by fire.

While Sarpi's original treatise on the magnet was thus lost, a brief record of its contents is contained in his biography, written by Griselini, and published about 1760. It appears to have contained, first, a mass of scattered data (probably lecture notes), followed by 140 propositions, based on magnetic phenomena. They relate to the discovery of the two points or poles of greatest attraction on the magnet by means of the inclined magnetized needle and to the "new generation of the same;" to magnetic attraction and repulsion, and the communication of magnetism, both from the lodestone and from magnetized iron; to the increase of magnetism in magnetic bodies; to the action of one magnet upon another; to the various effects produced "in the sphere of the horologe through different positions of magnetized bodies with respect to it;"[1] to the irreparable loss of magnetism which happens in the lodestone and in magnetized bodies when submitted to fire; and, finally, to the magnetization of iron, by means other than by rubbing it with a lodestone.

Another volume of Sarpi's writings, the original of which was also destroyed in the same fire, contained 674 propositions or "Pensieri" on all kinds of subjects, pertaining to every branch of natural science. Fortunately, a copy[2] was made of this before its destruction, which is now in the Library of St. Mark in Venice. Accompanying the manuscript are notes, made during the last century, in which Sarpi's discoveries are compared with those then claimed by Peter Van Musschenbroeck of Leyden. This gives a little clearer idea of Sarpi's investigations, in that it states that he determined the reciprocal relation of one magnet upon another, but did not measure or determine the magnetic force: also the action of the magnet on iron: also the manifestation of magnetic activity around the poles as an atmosphere—or in other words the field of force: also the maximum and minimum of attractive force of the magnet on the iron according to the magnitude of

---

[1] This may possibly relate to the supposed rotary sphere of Peregrinus.
[2] Class II., No. cxxix., cited by Bertelli, Mem. Sopra Peregrinus, p. 88.

the mass of the latter: also the inversion of polarity which may take place during the magnetization of the needle, although he seems to have known nothing of consequent poles: also magnetic variation (but not the variation of the variation) and magnetic inclination: also the magnetic properties acquired by iron "freely exposed to the air."

Robert Norman's book, to which I have referred in the preceding chapter, was published a few years before Sarpi is believed to have made his principal magnetical investigations; and it is altogether unlikely that it escaped the friar's attention. The Letter of Peregrinus had been in print for more than two decades. Moreover, a manuscript of it existed, and was at Sarpi's disposal in the Castellan Library of Venice. We are therefore justified in eliminating from the two categories, before given, all matters anticipated by Norman and Peregrinus, so far as these can be recognized. This done, the net result is to leave the destruction of magnetism by fire, the magnetization of iron by means other than induction from a lodestone— afterwards alluded to as the acquirement of magnetic properties by iron freely exposed to air—and the existence of the field of force *around the magnetic poles*, now directly made known for the first time.

That a lodestone could be deprived of its attractive quality by heating it to a high temperature was a new discovery, which may well have excited the incredulity of those who believed with Norman that the virtue in the stone was implanted by Providence, and hence was presumably ineradicable. The revelation that iron could be magnetized without the aid of the stone at all was not original with Sarpi, but was the result of an accidental observation made by one Giulio Cesare, a surgeon of Rimini, early in 1586, and not long before Sarpi wrote concerning it. An iron rod which supported a terra-cotta ornament upon the tower of a church in the before-named town had become bent by the force of the wind, and had remained thus distorted for about ten years. It was taken down and the iron sent to a blacksmith to be straightened, and while it lay in the smithy, Cesare, by chance, noticed that it possessed attractive properties. By an odd coincidence the church was dedicated to St. Augustine; so that one might almost fancy that the influence of the Saint whose discoveries concerning the magnet have already been noted, was somehow still potent to lead others in the same path. The circumstance puzzled the philosophers greatly; for how, they asked, could iron which is a metal be thus converted into lodestone which is a stone? For the time being the old doctrine of sympathies and similitudes—the great likeness and sympathy between iron and magnet—furnished a sufficient answer: but after a few years the true explanation appeared in a great work, to which the orderly progress of this narrative forbids further reference at present.[1]

Although the limiting of Sarpi's magnetic discoveries to the destruction of magnetism by heat, and the apparent concentration at the poles of the atmosphere or virtue which Norman thought to be spherical seems to be the consequence of the process of exclusion followed, it would be unjust to the great Consultore to assume that there are here defined the actual metes and bounds of his accomplishments in magnetic research. The evidence so far adduced concerning them is at best imperfect; while it must be remembered that to depreciate their importance or to obliterate them wholly, powerful forces have acted for centuries.

Still, to have conceived the first clear idea of the field of force about the poles of a magnet is sufficient to give the discoverer an undoubted pre-eminence, and that Sarpi did this is not only indicated by a comparison of his reputed achievements with what was already known, but is strongly substantiated by the efforts which have been made to deprive him of all credit for them. Sarpi had no worse enemies than the Jesuits, whom he caused to be

---

[1] Aldrovandus: Musaeum Metallicum. Milan, 1648, lib. 1, 134.

driven from Venice after they had refused to comply with the statutes passed by the state in contravention and defiance of the Pope's interdict. The rancor against him, which resulted in an attempt to assassinate him, and the removal of his remains nine times from place to place before they found safe and permanent sepulture, had not undergone the slightest abatement when the Jesuit Cabæus,[1] six years after Sarpi's death, wrote his book on the magnet, and with ingenious indirection, proceeded to ascribe to Leonardo Garzoni, another Jesuit who died in 1592, the discoveries of which, as I shall shortly show, John Baptista Porta obtained knowledge directly from Sarpi.

Garzoni seems to have written, at some indefinite time (but very close to and possibly even after the periods when Sarpi made his researches), a treatise on the magnet which he left uncompleted. His brother, after his death, announced an intention of publishing it, but if he did so, Bertelli[2] (despite a thorough search through all the principal libraries of Italy and especially through those in which Cabæus found his literary material), has been unable to discover any trace of it. He unearths, however, a book, published in 1642, which says that Garzoni's magnetic discoveries were well known, and on no better basis than this, permits himself to accept, without question, the assertion of Cabæus that the whole idea of the field of force originated with Garzoni, and hence, by necessary implication, not with Sarpi. But against the tacit opinion of even so learned a scholar as Father Bertelli, stands the total lack of evidence in favor of Garzoni, and the intense antagonism to Fra Paolo characteristic of the Jesuits, in which Cabæus evidently shares.

While, however, as I have stated, proof of Sarpi's discoveries, based on his own writings, is now meagre, it

---

[1] Philosophia Magnetica. Ferrara, 1629, lib. i., c. xvi.
[2] Mem. sopra Peregrinus, 24.

is probable that, through the intervention of Baptista Porta, we have always been in full possession of a record of his work: in fact, we may not unreasonably assume, of a better one than such as may have been contained in the brief treatise which was burned: for Porta has told *in extenso* matters which Sarpi (who seems to have been the prototype of Faraday in his predilection for reducing his results to brief paragraphs and numbering them) would have reported in the most concise and abbreviated form.

John Baptista Porta was a prodigy. He died in 1615 at the reputed age of seventy, the author of many discoveries and many books—the last not all philosophical and scientific, for he is said to have written "fourteen comedies, two tragedies and one tragi-comedy." Despite some confusion and discrepancies in dates, it appears to be the fact that he produced his work on Natural Magic, in four books,[1] when only sixteen years of age. It bears the date of 1558, and deals learnedly with astrology, abounds in the wildest vagaries on the generation of animals, discusses agriculture and horticulture in similar manner, ends with an *omnium gatherum* on domestic economy, and has not, from first to last, a word about the lodestone. Despite his tender age, Porta seems to have been the moving spirit in what was probably a *ridotto* or club of persons interested in one another and in some especial subject, which met for purposes of discussion and mutual entertainment. Whatever the precise nature of the assemblage originally may have been, it developed finally into the first of all learned societies, the Academia Secretorum Naturæ—abbreviated ordinarily into "the Segreti," in which it was an essential condition of membership that the

[1] Magiæ Naturalis, sive De Miraculis Rerum Naturalium, lib. iv. Io. Baptista Porta Neapolitano Auctore. Neapoli, 1558. Io. Baptista Portæ Neapolitani: Magiæ Naturalis, libri xx. Neapoli, 1589. There have been many editions in translations.

applicant should have successfully prosecuted an original research in medicine or philosophy. It is said that, because the participators called themselves the "Otiosi" (idle, lazy), the people became aroused and denounced the organ-

ization to the Pope as a gang of sorcerers; but it is a much more reasonable assumption that the astonishing statements in Porta's book, on such subjects, for example, as the production of birds and frogs from decaying matter, had more to do with the indictment than any name which the members might have chosen to assume. In fact, scores of such societies—mainly literary—were organized in Italy before the close of the sixteenth century. Tiraboschi gives a list of one hundred and seventy-one of them, and among their designations were such singular names as "Inflammable," "Pensive," "Intrepid," "Unripe," "Drowsy," "Rough," "Dispirited," "Solitary," "Fiery," "Sympathetic," "Grieved," "Re-ignited" and "Drunken."

At all events, Porta, as head and front of the offending, was summoned to Rome, whence he escaped with no worse penalty than the dissolution of his society and some fatherly advice, which indicated the extreme imprudence of ever starting it again. While at Rome he gained the favor of Cardinal Luigi D'Este, who gave him means of traveling through France and Spain, and who afterwards called him to Venice to build for him a parabolic mirror.

By this time Porta's attainments had gained him considerable celebrity. He had studied optics closely, and, although he did not invent the camera obscura (which was the work of Leon Baptista Alberti nearly a century earlier),[1] he first pointed out and taught the analogy between that apparatus and the human eye. He probably originated the magic lantern, however, and had some notion of the telescope, although his reference thereto is by no means unambiguous.

That the most eminent natural philosopher of Naples should have encountered the most eminent natural philosopher of Venice, and that the two should find in one another mutual attraction, seems to have inevitably followed. Porta instantly assumed the role of pupil, as most men did who came in contact with Sarpi, whatever their callings or attainments might be; and Sarpi, who delighted in teaching, found in Porta a congenial and tireless disciple. Nor did this relation cease even after great honors

[1] Tiraboschi: Storia della Lett. Ital. Firenze, 1810, vol. vii., 495.

had come to the Venetian; for when, as Procurator of his Order, he made an official visitation to Naples, it was with Porta that he sojourned, and into Porta's eager ears poured the story of the magnetic researches which he had then just completed. The Neapolitan had reached middle age, and for thirty-five years had been collecting material to add to the work on Natural Magic. Perhaps he deemed this latest teaching of Sarpi the cap-sheaf of all, and an indication that the auspicious time for publication had come. He had been admitted to the great Academy of the Lyncei, and, shrewdly considering that nothing issued under such sanction would be taken as savoring of the black art, he induced that society to give the work, now extended to twenty books, its official approval. This completed edition appeared in 1589. The seventh book is devoted wholly to the magnet, and is one of the longest in the volume. Cabæus intimates that it is merely an epitome of knowledge gained by Porta from Garzoni; and Bertelli, perceiving the necessity of bringing the two men at least into geographical proximity, thinks that Porta may have met the Jesuit when he went to Venice to make the Cardinal's mirror.

But this is disposed of by Porta himself, who, in his preface to his book on the magnet, says "We knew among the Venetians Paulus Venetus, vigilant in this study. He was of the order of Serviti, then a Provincial, now most worthy Procurator, from whom we not only do not blush to have copied, but we rejoice therein, since we know no one than he more learned or more subtle among those that we have seen. He was born to universal knowledge, and is an ornament not merely to the city of Venice and to Italy, but to the world. If we begin from his fundamental ideas and proceed to his completed studies of transcendent sublimity and accurate labor, we shall never be disappointed." Sarpi was Provincial of his order from 1579 to 1582, so that, at the time Porta's work appeared, the two men must have been in communication for more than seven years.

Porta's writing bears all the ear-marks of the compiler. It is exceedingly diffuse, often self-contradictory, and the same fact is repeated over and over again in different guises, as if the change in form were regarded as involving a material change in substance. Much that is set down may be laid out of sight at once, since it is merely a re-statement of the discoveries of Peregrinus. The phenomena of attraction and repulsion, the mode of determining the poles and the persistence of the poles in a divided magnet, are all described as by Peregrinus with no material variation, except that in the last-named instance Porta carries the separation further, and finds the poles "in the smallest fragments as well as in the great magnetic rock."

It requires but little critical study of this treatise to reach the conclusion that the notion of the field of force was regarded as by far the most important subject within the knowledge of its writer, or more correctly, of the individual from whom that knowledge was acquired. It is recurred to over and over again, examined from many points of view and tested in many different ways; so that we may almost see the conception grow as experiment made it clearer. This growth I shall now briefly trace.

Porta, having pointed out that the magnetic virtue and polarity remains, even when the stone is divided into minute grains, avers that when these grains are brought together, the strength of all will become unitary. Then he says, "But what is more wonderful, although the strength may be received in the middle of the stone, *it is not diffused at the middle but at the extremities of the polar lines and . . . comes forth openly.*" Not long afterwards, we encounter an experiment which consists in grinding a magnet into the minutest grains and mixing it with some inert white substance. Then, for the mystification of the bystanders—and Porta delights in that sort of thing—a magnet, hidden by a cloth, is brought up to the mass. At once the magnet grains rush to the stone, packing themselves densely "like hairs" or "like the down on a man's chin;" or in other words, behaving just as do the well-known iron filings. (Porta sees at once the possibility of drawing iron out of sand mixed with the ore by this means, and mentions it.) "This shock of hairs," he says, "adheres to the stone so persistently that they can hardly be detached: even when the stone is struck by a hammer, or two stones covered with them are rubbed together, still they stick on. They stand erect like spurs, and the more the stone is rubbed by another stone, the more they congregate." So much Lucretius had also seen in the Samothracian marvels.

But now follows the direct recognition of the field—"It is to be noted that the point diffuses virtue in its sphere as from centre to circumference, and just like the light of a candle, which is diffused everywhere and illuminates a chamber, and the further it recedes the more languidly it glows; and after a little further movement, it is lost; and then, as much as it approaches nearer, the more vividly it shines. In the same way, this force emanates from the point; and the nearer the latter, the more strongly it draws, while the greater the distance the more it is remiss: so that, if it recedes much, it vanishes and does nothing; therefore, in place of any other term we will call the extent of its power 'the sphere of its virtue.'"[1]

What more was there left in substance to discover as to the law of magnetic attraction and repulsion? Every particle of matter in the universe, Newton tells us, "attracts

[1] Literal translation—The English edition of Porta's Natural Magick (London, 1658,) renders the passage as follows: "Giving you to understand that the Pole sends its force to the Circumference. And as the light of a Candle is spread every way, and enlightens the Chamber; and the farther it is off from it, the weaker it shines, and at too great a distance is lost; and the neerer it is, the more cleerly it illuminates: so the force flies forth at that point; and the neerer it is, the more forcibly it attracts; and the further off, the more faintly: and if it be set too far off, it vanisheth quite, and doth nothing. Wherefore for that we shall say of it, and mark it for, we shall call the length of its force the compass of its Virtues." Note *passim* the old use of the English word "compass" for "sphere."

every other particle with a force whose direction is that of a line joining the two and whose magnitude . . . is inversely as the square of their distance from each other." Light radiation equally moves in a straight line, and its intensity varies as the square of the distance between the surface on which it falls and the source of radiation. Those are the laws of the phenomena stated in the foregoing quotation. Put the two together, and the result is Michell's sixth law, first expressed in 1750 : "The Attraction and Repulsion of Magnets decrease as the squares of the distances from the respective Poles increase."[1] Neither Sarpi, who had so narrowly studied the human eye and had dived into the deeps of all existing optical knowledge, nor even Porta who had followed him, could perhaps have expressed that law as Michell did a couple of centuries later; but they knew how the light of a candle increased and diminished, and they saw the resemblance between this and the increase and diminution of the virtue around the poles of a magnet. How better could they state their knowledge than by making the comparison and pointing out the resemblance?

But let us go a little further. Whenever Porta becomes grave and states an important fact, he shortly afterwards relaxes and tells about "jokes of the magnet with which we often exhilarate friends," and such parts of the book are undoubtedly his own. He is particularly fond of the iron-filings experiment, and, after stating the similarity of the sphere of magnetic attraction to the candle radiations, he recurs to it. This time he places on a table two masses of magnet fragments, and, holding a lodestone in each hand under the table, he makes the particles move, as he fancied, to represent contending armies. The individual grains rise up like erected spears, and advance and retreat and enter into "deadly struggles, now conquering, now conquered, now with arms raised, now lowered;" and then, in the midst of the play, he remarks, "the nearer

[1] Michell, J.: A Treatise of Artificial Magnets, London, 1750, 19.

the magnet approaches the more strongly the force extends its sphere," which to him is confirmatory of the light analogy.

The ingenuity which could evolve this conception was not slow to perceive the consequences. Iron is visibly attracted when placed in that sphere of virtue, but does anything else happen to it? Here is the answer, "Not alone by adhesion does the magnet diffuse its virtue to the iron, but, what is more wonderful, within the radii of its own virtue it causes virtue in the iron. For, if you approach a magnet to iron, so that the latter may be in the sphere of virtue, this iron will attract another iron, and the one so attracted will draw another, and thus you may see a chain of needles or rings in the air hanging." (The Samothracian rings, with the upper one not in contact with the lodestone but magnetized therefrom by induction.) "But while the chain exists, if you gradually remove the magnet for a short distance, *the last ring falls*, and then the next, and so on in succession all fall, and thus you see the stone is able to cause its virtue in the iron without contact."

It is curious to note how this led Porta into a false conclusion. He supposed that the sphere of virtue around the magnet had a definite outer limit whereat the radii of virtue ended; and that, so long as it overlapped a certain number of rings, all would remain suspended; but, if its centre were retracted, so that the last ring were left out of the sphere, then that ring would fall. Consequently, he says, if you try to magnetize a bar three feet long with a stone having a sphere of only two feet radius, you cannot do it except over two feet of the bar; the protruding one foot will be inert. Neither Porta nor Sarpi (seen through him) appears to have had any idea of the virtue extending from pole to pole, or to have made any deductions from the positions assumed by the inclined needles of Peregrinus or Norman. But their knowledge of a field of force and of magnetic induction due thereto was certainly well defined.

The many other discoveries which Porta records may

now briefly be stated. The destruction of magnetization by high temperature seems to have been found by burning the lodestone, which, when surrounded by heaped-up coals, emits a "blue, sulphurous and iron flame; and, at the same time, with the dissipation of this, the soul of the magnet departs and it loses its attractive power." The suggestion by Peregrinus of measuring the force of the magnet is carried into practical effect, and the strength determined is that required to resist separation of the armature from the magnet, which is arranged in one pan of a balance. The fact is noted that, while the magnetic virtue passes freely through brass, etc., iron acts as a screen to cut it off. The guarding of compass needles after they are magnetized from proximity with lodestones is strongly advised, lest "they be inebriated;" and clean iron is said to receive the virtue much more tenaciously than metal containing rust or earthy matters. Here occurs the first mention of the fact that "sailors prefer steel" for compass needles— "which will keep its value for a hundred years." Iron filings, we are told, wrapped in paper, receive virtue like a solid magnet, but, if they are shaken, they lose it.

Porta's recital abounds in absurdities, many of which are due to experimental errors. His theory of the cause of the attraction of the lodestone is correlated to the observed behavior of the iron filings; for he considers the magnetic forces to be due to minute particles of the stone springing from friction and concentrated into hairs, which, becoming attached to iron, impart thereto magnetic virtue. Yet, on the other hand, he believed with Alexander of Aphrodiseus that the lodestone actually fed on iron, and therefore buried a stone surrounded by filings and occasionally exhumed it to note without success the amount devoured. He imagined that iron rubbed by a diamond would become magnetic, and so avers; and, he believed that a magnet has east and west, as well as north and south, poles. A closer analysis of his work will show many more such delusions, and, to counterbalance them, suggestions which perhaps proved the germs of later useful discoveries.

The rise which we have been tracing has been followed mainly through effects which, although of like nature, are commonly defined as magnetic rather than electric; and all roads have led us to the mariner's compass as a *ne plus ultra* of invention. But now Porta begins an advance movement. The doctrine of sympathies and similitudes is still in force, and it is common belief that nowhere is sympathy stronger or likeness closer than between magnet and magnet. Meanwhile, there has arisen the conception of the sphere of virtue surrounding the lodestone. Conjectures as to the extent of that sphere have become confused with speculations as to the potency of the sympathetic influence, and out of all this has grown a curious notion that distance is no bar to the mutual effects of magnets, that they will even copy one another's positions, so that, if one magnet point in a certain direction, a second and sympathetic magnet will indicate the same direction, even if they be situated far asunder. A step further and Porta's thought thus leaps ahead :

"To a friend, that is at a far distance from us, fast shut up in prison, we may relate our minds; which I do not doubt may be done by two Mariner's compasses having the alphabet writ about them."

So came into the world the fancy which finds its modern embodiment in the great wire cobweb which envelops the earth and brings all people into converse, as it were face to face. Yet this initial notion of the telegraph is of less historical significance than the fact that Porta is here, for the first time, seeking to put the pivoted magnet needle to a new use. In other words, he is trying to invent beyond the compass; and he is taking from it as his instrumentality the pivoted needle moving in and controlled by a surrounding virtue. A few more years and it will be this same instrument in another hand, which will usher in electricity as a distinct manifestation of natural force and as the world now knows it.

In tracing the history of magnetic discovery, and especially that of the conception of the field of force, it has been necessary, in order to avoid complication, to lose sight, for a time, of the progress which the world was making toward a better recognition of the phenomenon of the amber. At the middle of the fifteenth century the identity of the attractive force exercised by magnet and amber was generally accepted as certain. No one thought of seriously disputing the matter, no reason for investigating an occurrence so manifest obtruded itself, and no practical employment of amber in any wise akin to that of the magnet invited research to discover what further occult capabilities the resin might possess. By the end of the century, however, the use of the compass had brought people into greater familiarity with the lodestone, and, as the knowledge of it increased, the time approached when differences between amber attraction and magnet attraction began to excite remark. But this was the period when the Greek influence, which attended the revival of learning after the fall of Constantinople, was making itself felt throughout all Europe. The new school of Platonists, under the leadership of Marsilio Ficino[1] the Florentine, challenged the supremacy of the Aristotelian philosophy, and precipitated new discussions which divided the learned into opposing camps, wherein the wordy warfare raged and the experimental study of nature was forgotten. Yet it was Ficino himself who virtually repeated the question asked centuries before by St. Augustine, by suggesting a difference between the amber and the magnet; not, be it observed, by describing the respective phenomena and comparing the facts—for that would be far below the dignity of any Platonist, new or old—but by promulgating a speculation on the subject, which could have arisen only from some previous knowledge based upon the actual observation of such a difference.

He says that iron is rendered magnetic, and maintained

[1] Born 1433, died 1499.

so, by rays from the Bear—that is, the North star or Arctic pole—and the "lodestone attracts iron because of a superior grade in the properties of the Bear." Following the prevailing notions, he would naturally have accounted for the attractive quality of amber in the same way; for, as I have stated, no one had drawn any distinction between the effects of the stone and the resin. But it is significant to note that Ficino does not do this, because he has clearly found out that, while the magnet attracts iron and points to the North pole and hence is controlled by the latter, amber does not attract iron but chaff, and does not point to the North pole at all. Yet because iron, under a supposed control, attracts, so some control must likewise be assumed for amber, because it also has an attractive quality, although of a different character. Therefore he triumphantly concludes that it is not the Arctic pole, but the *Ant*arctic pole which influences the resin—and the argument stands forth in symmetrical perfection; the lodestone is a thing, which is caused to attract iron by the Arctic pole: the amber is a thing, which is caused to attract chaff by the Antarctic pole.

Many years after Ficino's time, Jerome Fracastorio,[1] poet, physician and philosopher of Verona, reverts to the old doctrine of similitudes to deny its application to the magnet and the amber, and incidentally, for the first time, announces that the amber property exists in another natural body—the diamond; for the gem, he says, when rubbed, will attract hairs and twigs in the same way as the amber.

He is much more concerned, however, in evolving a new theory which will explain why hairs and twigs are thus attracted, when clearly there is no affinity between such substances and the amber or the diamond, than in recording experimental details. Yet he also sees clearly that the attracting bodies are widely different from one another;

---

[1] Born 1483, died 1553. Authorities differ as to the orthography of the name, some giving it as Fracastoro, others as Fracastorio.

so that, even granting hairs to have an affinity for the resin, that fact in itself, to his mind, seems to negative the idea of their possessing any similar affinity for the gem. The discovery of another substance which, while different from either magnet or amber, still possesses the same singular drawing power, is of no importance to him in comparison with getting these stubborn facts within the safe confines of a new theory—a conclusion eminently characteristic of his time. And he is not unsuccessful—at least, to his own satisfaction. The amber and the diamond, he finally announces, do not attract hairs and twigs because hairs and twigs *are* hairs and twigs; but because, in the thing attracted, there is a principle, perhaps in the included air, which is first drawn by the analogous principle existing in the thing attracting. In other words, the reciprocal attraction and repulsion of the magnet, the amber or the diamond, depends upon whether the principles entering into their composition—principles of a spiritual character apparently—are analogous or contrary.[1]

This was published in 1546. Fracastorio had then attained great fame as a physician—a fame which lives yet; for he was the first to assert that contagion is due to "invisible effluvia" and not to occult causes, and to distinguish the exanthematic typhus of the plague, which, up to that time, included all the grave epidemic maladies; while from the hero of his famous poem comes the name of that hideous disease of which the Old World is said to have known nothing until after the discovery of the New. That his simple opinion that Trent was unhealthy should have resulted in the removal of a great council of the Church from that town to Bologna, is sufficient to show the immense influence he exerted.[2] The announcement of the foregoing theory by so high an authority therefore

---

[1] Hier. Fracastorii, Veronensis: Opera Omnia. Venice, 1555. Lib. de Sympathia et Antipathia.

[2] Biographie Universelle, Art. Fracastorio. La Grande Encyclopédie, Paris, 1893, Vol. 17.

JEROME CARDAN.

may well have been considered conclusive as to the identity of magnetic and electric attraction. Yet within a very few years it was challenged.

On the 24th of September, 1501, there was born in Milan the first of that trio of Italian philosophers whose achievements in physical science seem all the more brilliant by contrast with the ignorance and superstition of the period covered by their lives. To two of these men—Fra Paolo Sarpi and John Baptista Porta—some reference has already been made. Girolamo Cardano,[1] or Jerome Car-

dan, as his name is commonly Anglicized, belonged to the generation immediately preceding theirs; but the three lives overlapped, and much of their work was done contemporaneously. There is little resemblance to the mercurial, inquisitive, precocious Porta, still less to the

[1] Morley, H.: The Life of Girolamo Cardano of Milan, Physician. The portrait of Cardan here given is from a contemporary print forming the frontispiece of the 1553 edition of his treatise, De rerum Varietate. The statement of his age as 49 years does not accord with the date of his birth as given by his biographers.

majestic figure of the Venetian Consultore, in the personality of Cardan; yet he far exceeded the former in ingenuity, and probably (statecraft and theology excepted) he equalled the latter in the variety and profundity of his attainments. Cardan's character was a bundle of contradictions—his life, a series of vicissitudes; and hence, as this or that group of traits or events is selected as typical, so he may be made out a martyr and a philosopher, or a charlatan and a magician. He was the natural son of an aged Milanese geometer, who made him a wretched drudge, until, astonished by the learning the boy had managed to acquire under difficulties, the disheartening quality of which he of all the world knew best, he consented to enter him as a medical student in Pavia. Thence Cardan went to the University of Padua, the affairs of which were in great disorder. For years there had been no rector, mainly because no one wanted the place. Cardan offered himself and was elected by one vote. But the honor was empty. The mother, slaving at menial labor in Milan, worked to defray the bare official charges. The symbols of his mock-majesty, if he had them—his robes of scarlet and purple silk and his gold and jeweled badges, his fife-players and his spearmen and all the stately, ceremonial appurtenances of the office—were paid for, if at all, from the proceeds of the gaming table. He called his term of office his "Sardanapalan year;" the University sardonically termed it the last of the ten years in which there was no rector.

In time he became a doctor, and practised in a little village, and wrote books on therapeutics and the plague. His health was wretched—his poverty, extreme. His marriage helped him a little; but an inordinate passion for gambling resulted in chronic destitution. The Milan physicians would not permit him to practice because of his origin; but a lectureship on geography, geometry and astronomy yielded a pittance sufficient to ward off starvation. So he lived, writing more treatises, mainly on the subjects of his lectures, and developing a genius for fancies and dreams which hardened eventually into a superstition as controlling and as uncontrollable as the attraction of the dice-box. But he had a fine taste for music, he loved the melodious words of Petrarch and Pulci, he read Aristotle and Plotinus for pleasure; and even if the scanty contents of his purse were the products of his gambling skill, they went for no grosser pleasures than expensive writing materials and rare books. Add to this that he was a skillful physician—especially for those days—and, though blunt in speech, warm-hearted and charitable almost to extremes, and we may safely leave his condemnation to those inerrant moralists who believe that there are no virtues, however great, which the small vices cannot eclipse.

The dream fancies gradually acquired a stronger hold— astrology, first critically examined, became entangled with his faith—the casting of a horoscope of Christ brought him perilously near to prison for blasphemy, and a book pointing out errors in medical practice called down upon him with renewed vigor that uncompromising odium which the elderly medical tortoise, even to this day, especially reserves for the youthful medical hare. The people said he was mad—made so by poverty; the inordinate number of printers' errors in his book, which he himself says drove him nearly to distraction would have furnished a more probable reason.

Thus he lived until nearly forty-five years of age before the tide of his fortunes began to turn. In 1545 he published his great work on algebra, wherein he laid down rules for all forms and varieties of cubic equations, established the literal notation, applied this form of mathematics to the resolution of geometrical problems, and accomplished other results of great importance, though of too technical a character to be noted here. Up to this time, he had written in all some fifty-three treatises. His success as a physician now began to tell, and resulted in his Milan brethren, after twelve years of denial, giving him the stamp of regularity. The rapidity of his rise was phenomenal. From the half-starved, unpaid, flouted student barely able to keep body and soul together, his advice, in less than ten years, was sought by the Emperor himself, by the King of France for the Queen of Scotland, and by the majesty of England, then embodied in the weakling son of Henry VIII. He journeyed in state throughout Europe—men of rank and learning everywhere eager to obtain his aid or recognition. And finally, after his travels, he returned to Milan, loaded with honors and rewards, the undisputed greatest living authority in the healing art.

The temptation to dwell upon the dramatic episodes of Cardan's later life—ending, as it did, in crushing sorrow—is strong, but must be resisted, to proceed at once to the remarkable work which will always hold a prominent place in the history of electricity: for in it, for the first time, the phenomena of the amber are clearly differentiated from those of the magnet.

Cardan's Books on Subtilty occupied, in the writing, three years, and were published at Nuremberg and Paris in 1551.[1] The work attained an enormous popularity, and well it might—for it was calculated to arouse the keenest curiosity, in that it related to "subtle" things or those which are "sensible by the senses or intelligible by the intellect, but with difficulty comprehended." It is hardly possible to figure to one's self a book nowadays claiming to be a treatise on everything not easily understood; but, at that time, such a work was a welcome improvement upon and a distinct advance beyond the old De Natura Rerum treatises, whereof I have noted numerous examples, and which generally undertook to explain not only "things with difficulty comprehended," but, with equal ease and readiness, things not comprehended at all. It is a curious medley, discussing abstruse mathematics and

---

[1] Hier. Cardani, Medici Mediol: De Subtilitate, Lib. xxi. Paris, 1551. There have been numerous later editions. The first French translation is dated 1556, and this I have used.

dreams, hydrostatics and fortune telling, metallurgy and card tricks. It stands squarely on the dividing line between mediæval magic and modern physical science. That the sixteenth century reader might well have regarded the work as be-deviled it is easy to imagine. If he trusted himself to the figments of the author's boiling imagination, he found himself in the end disconcerted with the dry remark that "many things appear admirable until the cause is known; then admiration ceases:"[1] if he pinned his faith only to the statements of fact, again he is laughed at and told that "some things seem more true than they are—others are more true than they seem." The bewildered disciple, especially if imbued with the philosopher's faith in demons and ghosts and apparitions, may well regard this as t. nimbleness of Mephisto, and, recalling Cardan's wonderful cures and vast learning, his strange luck at gambling, his, at times, reckless prodigality and dissolute existence, may see in the Milanese doctor another Faust and the slave of a Satanic compact. But another and final contradiction awaits him on the very last page of the book, where he finds this child of the devil, prostrate as "an humble worm of the earth," acknowledging, in a prayer of singular beauty, that "to Thee I owe all that is here written in truth," that "the errors and faults are of mine own ambition, rashness and haste," and imploring for the Heavenly pardon and "guidance to better things."

The statements in Cardan's treatise which relate to the amber so closely follow those on the same subject in the famous work of George Agricola,[2] which appeared a few years earlier, that the discoveries recorded which are Cardan's own are easily distinguished. Agricola's summary of the uses and properties of amber contains probably all that was then known concerning it. It was utilized in the

[1] Often paraphrased since: *e. g.*, "Science is anything we do not understand: the moment we understand it, it ceases to be science."

[2] Agricola: De la Natura de le Cose Fossili. Venice, 1544. lib. iv.

manufacture of printing-ink and of incense; when burned it was supposed to be a sure preventive of plague—and as that terrible scourge then ravaged Europe almost unchecked, the demand for the resin was great. It was carved into rings, beads for rosaries, and statuettes. But, next to its employment for fumigation, its most extensive use was as a specific for checking hemorrhage, nausea and catarrh. Thus it found its way chiefly into the hands of the physicians, and thus it doubtless came to pass that to the members of the faculty were owing the remarkable discoveries which had their basis in its attractive property.

Agricola, however, perceives no difference between its attraction and that of the magnet. He enumerates down, chaff, hairs, leaves and other small things—including even metal filings—as drawn to it, the last probably fortifying him in his individual belief in the similarity between it and the lodestone. Yet he notes that, when rubbed even with the finger, it becomes hot, and still hotter when the friction is applied with a coarse cloth, or even with a hard substance; but one's faith in his accuracy is somewhat rudely shaken by his culminating assertion that there is found on the shores of the Vistula a grey amber which, on being rubbed with iron, will cause leaves lying on the ground to fly up to it, even if held a distance of two feet above them.

Cardan transcribes, almost literally, Agricola's list of things which the amber will attract, and then, for the first time, offers an interpretation purely physical. He speculates neither upon similitudes, sympathies or analogous principles, but boldly assigns a wholly material cause; namely, "that it has a fatty and glutinous humor which, being emitted, the dry object desiring to absorb it is moved toward the source, that is the amber. For every dry thing, as soon as it begins to absorb moisture, is moved toward the moist source, like fire to its pasture; and since the amber is strongly rubbed, it draws the more because of its heat." It is not necessary to criticise this theory, which was certainly as reasonable as any advanced either before or for the next hundred years. Its importance lies in the fact that, good or bad, it was the first hypothesis ever advanced to account for the phenomenon of the amber in contradistinction to and as different from that of the lodestone. There is no doubt as to its author's meaning, for immediately succeeding the theoretical statement, comes the making of the actual contrast in a passage of extreme historical importance, beginning with the unqualified assertion that "the magnet stone and the amber do not attract in the same way"—and thus squarely denying the assertions of all the philosophers of the past, and his medical brother of Verona in particular. Observe the reasons:

"The amber draws everything that is light; the magnet, iron only." He then had not been misled by the amber's attraction for finely-pulverized iron.

"The amber does not move chaff when something is interposed: the magnet nevertheless will attract iron." An age had gone by since St. Augustine had recorded the last. It was to a rejuvenated world that Cardan thus brought the first suggestion of electrical insulation.

"The amber is not mutually attracted by the chaff: the magnet is drawn by the iron." This was intended as a blow at Fracastorio, and his notion of analogous principles. Here one wishes that the details of his experiment had been given, even as Porta would have recorded them.

"The amber does not attract at the end: the magnet attracts the iron sometimes at the North and sometimes at the South." It is with the permanent polarity of the magnet that the distinction is here drawn.

"The attraction of the amber is greatly aided by heat and friction: that of the magnet, by cleaning the attracting part." The important point here lies in the implication that, while the amber effect can be augmented by heat and friction, that of the magnet can not. The cleaning of the magnet to which he alludes is probably the removal of foreign matters or a scale, of which he elsewhere speaks, attached to the natural lodestone, which he supposed impaired its force. He knew that the power could not thus be cut off.

This having been said in the sixteenth century and not in the nineteenth, the necessity obtruded itself of reconciling his plainly experimental results with his previously-announced theory. He sees that the link must be a physical and not a metaphysical one. Instinctively the leech finds his analogy in an instrument of torture still lingering in the chirurgical armament. The hotter the amber, he thinks, the more it draws—just like the action of "the cupping glass due to fire and hot things;" and there he rests content, oblivious of the hopeless inconsistency of this notion with that part of his theory which accounts for the phenomenon by the attraction between dry and moist bodies. The cupping-glass idea was his own—the rest of the analogy he borrowed from the ancients.

That Sarpi knew of this remarkable differentiation of magnet and amber, is hardly to be doubted; and it may be surmised that his master mind perceived the consequences, which others pointed out. But there is nothing in Porta's reflection of Sarpi's light to support this; and the loss of Sarpi's writings leaves the matter probably forever in obscurity. Porta himself seems to have attached no importance to the subject, although he shows abundant familiarity with Cardan's work. Indeed, at times he fairly revels in disputing the assertions of the Milanese physician, in terms so much more vigorous than refined, that it is not difficult to imagine that the Neapolitan philosopher had imbibed his notions of Cardan from those life-long rivals who had furnished the older scholar abundant basis for his epigrammatic definition of envy as "mild hate."

Cardan, for example, avers that iron is the magnet's food; so not only accounts for magnetic attraction, but insists that a magnet is best preserved in iron filings: which may be perfectly true if the filings are packed in a dense mass to form an armature or keeper. Porta, however, as has been recounted in the last chapter, retorted by burying a magnet with iron filings and occasionally digging it up to see how much of the latter the magnet had devoured: a literal interpretation of Cardan's directions, which, it is needless to add, was not attended with results. On the other hand, Porta cordially agrees with Cardan, that the virtue of the magnet cannot be destroyed by garlic nor by the presence of the diamond.

I have now reached the end of that epoch which immediately precedes the earliest attempt to systemize electrical and magnetic knowledge and thus to reduce it to a science. In the rise of that knowledge through the centuries we have seen the conception of the soul animating the amber and the magnet give place to more material hypotheses—indeed to many of them in turn—and ultimately become degraded to a mere physical emanation or to an appetite. We have found the phenomenon of magnetic attraction, familiar for centuries to the western world, and that of magnetic polarity known for as long a period to the nations of the east, and yet that there was practically no interchange of this knowledge. In time, however, we have seen this interchange take place, and in tracing the separate items to their coalescence, we have at the same time followed the evolution of the first great electrical invention—the mariner's compass.

We have seen the enormous advance in human progress directly owing to this instrument. We have perceived that, although the amber phenomenon found no practical application to the uses of man, still the inherent mystery, the unexplained nature of it, was sufficient to impart to it all the vitality inherent to a problem which constantly and automatically forces itself upon generation after generation for solution. After the compass had begun its great work, after it had revealed the New World to the Old, the alliance, in the general mind, of the amber to the magnet ended forever the possibility of the questions concerning the nature of either sinking into oblivion. But the modes of dealing with these questions we have found to change with the changing times.

The tendency to speculate and to account for facts by theories, which seems implanted in the race, slowly, very slowly, even in the individual under the discipline of education, loses its energy. So, in those old days, it was only as men began to question Nature, and not their own brains, that they began to perceive the conditions which Nature had actually imposed on them, and to recognize the difference between these and the imaginary conditions which their speculations had sought to impose upon Nature. Gradually the new logic of experimental demonstration gathered momentum, the phenomena of magnetic polarity and of induction became recognized, and so came about the first crude conception of the magnetic field of force. A more exact knowledge of the magnet led inevitably to the perception of the differences between the effects produced by the lodestone and by the rubbed amber; and at last to the drawing of a clear line of demarcation between them.

And then the Sphinx of the centuries follows the flies and the reptiles into the golden recesses of the amber, and there enthroned poses once more the nature of the amber soul as a new riddle. There is no kinship between this evanescent energy drawn from these yellow depths and the stolid pull of the dull stone—no similarity between the wayward and mastering spirit which seizes upon anything within its strength and the unrelenting tyranny with which the magnet enforces servitude only upon the stubborn iron. What then is this genius which is called forth by the friction of the amber, even as the Afrite was summoned by the rubbing of Aladdin's lamp? Thus the question first asked twenty-two hundred years before was renewed: and now impressed with greater urgency than ever upon the newly-awakened human intellect.

During this great period the attraction of magnet and amber had been dealt with, first by the philosophers, then by the priests, now by the physicians. To the theologians of the last three centuries of this era, the subject is a favorite mine of metaphor from which saints and popes have not disdained to draw. The metallurgists, headed by Agricola, make both the stone and the resin the subject of their didactic description; and, contrariwise, the mystics and magicians heap upon the already-existing mystery of it, new and endless mysteries of their own devising. Only occasionally does a master mind, dominating all known sciences, like that of Sarpi, or some keen student of nature ahead of his times, as Robert Norman, achieve genuine progress.

A review of all that has been handed down to us makes it clear that to the members of the medical profession more than to those of any other, is due the impetus which, at the end of the sixteenth century, brought the world to the point where the next step beyond meant the incoming of electricity as a new science. Yet it may well be doubted whether the work of searching out and establishing it could have fallen into hands less adapted thereto by past training. Medicine is an inexact science. In no field of human endeavor has the imagination been more severely taxed to frame hypotheses to accord with or account for seemingly endless adventitious phenomena. "Medicine," says Bacon,[1] speaking of it as it existed in his time, "is a science which hath been more professed than labored, and yet more labored than advanced; the labor having been, in my judgment, rather in circle than in progressing. For I find much iteration, but small addition." And as to its practitioners, he says, "in the inquiry of diseases they do abandon the cures of many, some as in their nature incurable and others as past the period of cure; so that Sylla and the Triumvirs never proscribed so many men to die, as they do by their

[1] Bacon: *De Augmentis,* ii., x, 3.

ignorant edicts: whereof numbers do escape with less difficulty than they did in the Roman proscriptions."[1] Yet the reproach brought against the Asclepiades that they "resigned themselves to visionary speculations, and obeyed the instincts of their understandings rather in crude meditations on the essence of things, the origin of the world, the nature of God and the soul of man, than in developing a practical and useful system of medicine"[2] often repeated against the mediæval physicians, has little justice in it. Like all knowledge depending upon physical investigation, that of the human body lay under the ban of ecclesiastical control. Ceremonies and relics and consecrated specifics, amulets, miracle-working images, and a celestial faculty recruited from the ranks of the saints—such were the means too often relied upon to meet the fearful diseases which flourished under conditions which favored every form of contagion and infection. "Afflictions sent by Providence" and "demoniac possessions" were terms which readily veiled the density of the existing ignorance. Man, it was insisted, must not investigate the structure of his own frame with the scalpel, since this argued contempt for the doctrine of final resurrection. Medical practice must be first of all orthodox. Supernaturalism must prevail, and the struggling lunatic dealt with through book and holy water, rather than through remedies ministering to the mind diseased. Progress in any department of the healing art could hardly be expected in such circumstances.

Hence, while extended allusion to the therapeutic employment of both the magnet and the amber in the Middle Ages has been made in the preceding pages, it would be incorrect to infer that the advancement of magnetic or electrical knowledge was materially accelerated by such use. In fact, so long as the principal value of the lodestone lay in its utility as "a means of expelling gross

[1] Bacon: De Augmentis, ii., x, 5.
[2] Meryon: The History of Medicine, London, 1861.

humors," as Dioscorides and Galen averred, the world was none the better for the attention bestowed upon it by these fathers in medicine. It was when the physicians ceased to deal with it, however, as physicians, and began to deal with it as physicists, that real advances began. It was the leaven of the inductive method of Hippocrates which worked for good in them—Hippocrates, who had asserted demoniac possession to be "no wise more divine, no wise more infernal, than any other disease," and the sturdy common sense of whose precepts had refused to be destroyed by the magic of the Persians, or the dreams of the Asclepiades, or the numbers of Pythagoras, or the atoms of Democritus, and which even asserted itself free of the entangling meshes of the Aristotelian Matter and Form.

The priests of Samothrace sold magnet rings to cure rheumatism and gout. A thousand years later the fact was so far forgotten that when Aetius, in the fifth century, compiled all the medical knowledge of his predecessors, and announced that "those who are afflicted with gout in their hands or feet or with convulsions are relieved by holding a magnet in their hands," the discovery was regarded as wholly new, despite the writer's cautious prefix of "they say" to his asseverations. How the magnet in the hands of the arch impostor Paracelsus became the foundation of speculations as wild and as fantastic as ever man conceived, has already been told, and some reference has been made to the vagaries of Raymond Lully concerning it. The knowledge of the embryo science did not advance because of the visionary theories of these people, but despite of them—just as it grew in the works of Cardan and Porta, where the statements of great discoveries in it are jostled by the descriptions of alleged phenomena as false and as absurd as anything which the veriest charlatan could devise.

Nevertheless it is to be remembered, that there was hardly a medical writer of any eminence, from the time of Oribasius onward, who did not refer to the magnet in some way—often writing utter nonsense about it, sometimes interspersing his rumors and vagaries with truths frequently the more forceful for the re-telling in a new manner. If in a multitude of counselors there is wisdom, if the truth resides in numbers of witnesses, surely we may ascribe some of the progress effected to the mutual cancellation of the mistakes and misstatements repeated and reiterated in the works of the old medical writers. The subject was sifted through the books of the Arabs and by their greatest leeches, Hali Abbas, Avicenna and Serapion the Moor; while in Europe, in the fifteenth and sixteenth centuries, physicians of commanding eminence hasten to contribute their observations or speculations concerning it, to the general fund of knowledge. Fernel and Dupuis in France, Amatus in Portugal, Thomas Lieber (or Erastus) in Germany, Fallopius, Fracastorio, Costaeus and Cardan in Italy—such were the men who, with an abundant crop of tares, cultivated the harvest which, meagre as it was, increased a thousandfold within the next hundred years.

At the end of the sixteenth century, the Italians were far in advance of all other nations in their medical attainments, and the English well in the rear. I have encountered no writings by English physicians of that century which entitle them to any credit for either preserving or advancing electrical or magnetic knowledge. The practice of physic did not pass from the active control of the priesthood and become an independent profession in Britain until Henry VIII., in 1518, granted its charter to the Royal College of Physicians in London. The names of Dr. Linacre and Dr. Kaye (Shakespeare's Dr. Caius) then come into prominence, but chiefly as leaders in the struggle of the college to put down quackery, and to impose qualifications upon the medical practitioner, to maintain itself against the pretensions of the clergy, who still arrogated to themselves the right to license, and to assert its own privileges and dignity.

If it had been known that the reduction of the electrical and magnetic knowledge of the time to a science, coupled moreover with new discoveries of extreme importance and brilliancy, was predestined to come from a medical faculty, common consent, as well as the evidence to be derived from all written records, would infallibly have pointed to that existing in Italy; perhaps in Milan or Padua or Bologna. But no one could have foreseen that so startling an event could have originated in England, could have been the unaided work of an English doctor; and, perhaps least of all, of the particular physician who, at the time of its appearance, presided over the destinies and troubles of the much-vexed and hard-fighting college in London.

The rise in electricity had slowly taken place throughout all Europe, indeed, all the world, and therein many nationalities had taken part. It was now destined to move with a new and marvelous vigor, through the transcendent genius of an Englishman and on English soil.

CHAPTER X.

WILLIAM GILBERD (or Gilbert, as the name is more commonly written) was born in the year 1540, in Holy Trinity Parish in the town of Colchester, England.[1] He came of excellent family, and was the eldest of the five sons of Jerome Gilbert, at one time town recorder. Of his individual history there is but scant record. He was a physician, but the great work which has insured his immortality has no necessary relation to the healing art. No important discovery in medicine is known to be his, and he appears therein only as a teacher and an expounder. And this is the more remarkable, since, in dealing with a different branch of science, he displays not only a marvelous originality of thought, but intolerance of accepted opinion to a degree which ordinarily leads most men to revolutionary extremes in any field of action in which they may be placed.

Something of the difficulty which is encountered in reconciling the dual intellectual lives of Shakespeare the poet and Shakespeare the player, of Bacon the philosopher and Bacon the advocate, is again met when those of Gilbert the physician and Gilbert the discoverer are contrasted. We find, on the one hand, the hard-working London doctor, renouncing matrimony through simple devotion to his art, and year in and year out teaching a little band of students at his house hard by St. Paul's, until the

---

[1] Cooper: Athenæ Cantabrigiensis, Cambridge, 1858. This contains a very full list of works in which reference to Gilbert is made. Of the older biographies of him, that which is especially full appears in Biographica Britannia, London, 1757. Among later memoirs may be noted one by Prof. S. P. Thompson, London, 1891, and another by Mr. Conrad W. Cooke, London, 1890.

queen called him into her service; on the other, a philosopher of overshadowing genius pursuing, despite his arduous professional labor, and in the very teeth of the fixed beliefs of the world of his time, the first researches seeking to establish physical science on a philosophic basis, and which revealed and co-ordinated the amber-electricity as a new and distinct phenomenon of nature.

The archives of the University of Cambridge, of the Royal College of Physicians, and the meagre statements of his epitaph high up on the church wall in his native town, tell us the official honors which Gilbert won. But scores of other good and useful men whose fame never traveled beyond their birthplaces, who adopted liberal professions, rose in them, secured their rewards and departed, have left records equally respectable. There is nothing in the writings of his time which reveals to us any clear view of other manifestations of the living force which drove Gilbert to the accomplishment of the great task so controlling, so novel, and yet so foreign to his daily round of toil. True, it was not uncommon, in those days, for the physician to follow some other art or practice more to his fancy than his calling. "For you shall have of them," records the great Chancellor caustically,[1] "antiquaries, poets, humanists, statesmen, merchants, divines, and in every of these better seen than in their profession; and no doubt upon this ground, that they find that mediocrity and excellency in their art maketh no difference in profit or reputation towards their fortune; for the weakness of patients, and sweetness of life, and nature of hope, maketh men depend upon physicians with all their defects." But the official honors which Gilbert received included all which his profession could give; and, as none of the foregoing influences, however much they might have conduced to his material support, imply the Presidency of the Royal College of Physicians, and the *ex officio* status of professional primacy, it may safely be con-

[1] Bacon: Advt. of Learning, b. ii., c. x., 2.

cluded that, despite his great work in another art, he was none the less a good and skilful doctor, who rose to high places because he deserved them.

So the memorials of him fail. We can only read between the lines of his book and draw inferences, and perhaps measure his thought-power by noting its effect upon the thought inertia of his contemporaries: we can quote what this or that philosopher said about him, which is no safe criterion, for where is there less toleration of truths of to-day than in minds filled to saturation with the truths of yesterday? But no Boswell attended his steps, and no relics have been found of that voluminous correspondence which he is said to have opened with the learned men of Italy. One portrait[1] of him which hung in the house of the Royal College of Physicians was destroyed in the Great Fire of 1666, and another which he bequeathed to the Bodleian Library became decayed, was removed, and disappeared unaccountably during the last century. The sole vestiges of him are a few scraps of doubtful handwriting, and the old house in Colchester where he once resided. His fame rests upon the contents of two ancient and yellow-paged volumes,[2] one of which Peter Short printed for him nearly three hundred years ago; the other[3] his surviving brother lovingly collected from his scattered papers, and it lay in manuscript for half a century after his death.

In the dark days of Queen Mary, the town of Colchester, famous then and since for its oysters and Dutch weavers—being a "sweet and comfortable mother of the bodies and a tender nourse of the souls of God's children"[4]—the latter, so styling themselves, much affected the common inns as their meeting places. Consequently Protestantism flourished sturdily, until the Smithfield

[1] Evelyn's Diary, Oct. 3, 1662.
[2] De Magnete. London, 1600.
[3] De Mundo Novo Sublunari, Philosophia Nova. Amsterdam, 1561.
[4] P. Morant: The History and Antiquities of Colchester. London, 1748.

fires spread thither and burned it out. There was not much in the atmosphere of a place where half a dozen rank Gospellers went to the stake of a morning and as many more in the afternoon, to encourage free thought in a boy even of Gilbert's mental strength; nor was eight hours' work a day over the Sententiæ Pueriles, or the Accidence (which Mr. Robert Wrennald, in consideration of six pounds, thirteen shillings and four pence annually paid him, taught in the school which King Henry VIII. had founded)[1] especially calculated to expand the faculty of original ideation in any one.

At the age of eighteen, Gilbert matriculated at St. John's College, Cambridge. The condition of the University, then and for several years afterwards, was anything but one likely to promote the scholarship or foster the natural abilities of its students. It had fallen far below the high standards of Ascham and Cheke; it was destitute of leaders capable of stimulating others by their example to honorable exertion; its undergraduates were disorderly, insubordinate and even riotous, addicted to gaudy clothes, the taverns and the gambling houses, while religious dissensions ran high between the sympathizers with Rome and the adherents of the new Puritanism which had found lodgment chiefly in the colleges of Trinity and St. John's. Whatever Cambridge then achieved in advancing real knowledge was the outcome of individual genius rising superior to the prevalent influences of the culture which surrounded it. To science and its votaries, the great University then offered no permanent home.[2]

Gilbert's progress was unremittingly upward. He attained his bachelor's degree in 1560, became a Fellow on Symson's Foundation in 1561, "commenced" M. A., in 1564, and during the two years following, was mathematical

[1] P. Morant: The History and Antiquities of Colchester. London, 1748.
[2] The University of Cambridge: Mullinger. Cambridge, 1884. v. ii., pp. 100, 573, 574.

examiner of his college. Then he studied medicine and reached his Doctorate and a senior Fellowship in 1569, when he terminated his eleven years' connection with the University.

The facilities for studying anatomy and clinical medicine in England, at that time, were not comparable with those to be obtained on the Continent. Vesalius, of the University of Padua, had written, not long before, his famous work based upon actual dissections of the human body, and pointing out the errors into which Galen had fallen through studies said to have been made upon the organs of apes. Eustachius, then living, was continuing the work of his greater contemporary upon the foundations of the science of anatomy. The discoveries of Fallopius were still new and arousing the keenest interest. Cardan was teaching in Bologna. The chemical medicine of Paracelsus was creating widespread controversy. For a student such as Gilbert, whose turn of mind was of the most practical nature and who possessed a keen taste for experimental research, the opportunities for such study available outside of England furnish abundant reason for his sojourn of four years abroad, and make it needless to picture him as simply making "the grand tour" which, in those days, formed a part of the educational course of well-to-do people.

Although the habits and mode of thought acquired during his period of study in the foreign universities had much to do with the development of his later achievements, Gilbert was no one's disciple. No one even played for him the part of a Southampton or an Essex, unless *sub silentio* the Queen herself. Even the dedication to the young Prince of Wales, who never wore the crown, which prefaces his posthumous volume, was penned by his brother, and not by himself. Nor is any especial influence recognizable which can be said to have aroused in him a spirit of emulation and so to have directed him into his chosen path of discovery. Galen and Dioscorides, in fact all of the ancient writers, treated of the lodestone as a part of the *materia medica;* the more modern authors dwelt much also upon its occult powers, and Paracelsus had rejuvenated but recently the superstitions of the old Greeks and had opened the Pandora's box of delusions and deceptions concerning it, which have plagued the world ever since. Gilbert, while showing abundant familiarity with these and other authorities on the medical uses of the magnet, disposes of their labored speculations with scant respect and few words. Therapeutically he thinks the stone has some uses, not however dependent upon its magnetic quality. As for Paracelsus, he observes that headaches can no more be cured by a lodestone applied than by a steel hat, and he singles out the apostle of laudanum and mercury for especial scorn. That he owed nothing to the accumulated magnetic wisdom of his professional ancestors—saving perhaps the knowledge of a host of errors to be avoided—is clear. His greatest debt, as I shall show hereafter, lay to Peregrinus, to Cardan and Fracastorio as philosophers rather than as physicians, and to Sarpi through Baptista Porta's transcriptions.

Gilbert's medical reputation must have preceded him, for upon his return to England, he was at once made a Fellow of the Royal College of Physicians. He began practice in London, and established himself in a house on "St. Peter's Hill between upper Thames Street and Little Knight Rider Street." As Dr. Linacre is known[1] to have given a house on Knight Rider Street to the college as its first abode, it may be that Gilbert took up his residence in the college building. At all events, it seems that he led an all but cloistered life and taught medicine at his dwelling to a number of students. More probably, however, this gathering was modeled on the Italian ridotto, or was something after the fashion of Porta's suppressed society, the Otiosi, having for its object not only didactic instruction, but free discussion and interchange of opinion. It

[1] T. Allen: A New History of London, 1883, iii., 573.

was the first association of its kind in England, and the precursor of the Royal Society.[1] That it was popular among the students is attested by John Chamberlain,[2] who lived with Gilbert and who speaks of "the town as empty as if it were dead vacation, nobody at the Doctor's." Later, when Gilbert was called to Whitehall, Chamberlain predicts the disbanding of the society, saying "I doubt our college will be dissolved and some of us sent to seeke our fortune;" and still later, after Gilbert's departure, "the covie is now dispersed," he chronicles somewhat ruefully, "and we are driven to seeke our feeding further off, our Doctor being aredy setled in Court." Meanwhile Gilbert was elected to the office of Censor of the College three times, twice to that of Treasurer, then he became, in 1597, Consilarius in place of Dr. Giffard, and finally, in 1600, the same year in which his famous work appeared, he reached the Presidential Chair.

In the last-named year also, as Chamberlain records, Queen Elizabeth appointed him one of her body physicians, a merely perfunctory office, for she detested doctors and would have none of their drugs. Perhaps her unlucky experience with the Jew, Rodrigo Lopez, whom she covertly favored and allowed to prescribe for her, until he was detected trying to give her poison (being thereunto incited, so it was said, by Spain) and duly convicted, shattered her faith in the medical profession: perhaps, in her last years, she believed in her own sarcastic remark that the people would say that the physicians killed her if she died of old age after following their counsels: perhaps she

[1] It is generally stated that the organization of the College of Philosophy instituted in London in 1645, which immediately preceded the Royal Society, was due to the scheme of Solomon's House described by Bacon in the New Atlantis—and first suggested in his Praise of Knowledge, published in 1593. Gilbert's society, however, appears to be of still earlier establishment. It may have been the first medical "quiz" class in England.

[2] Letters written by John Chamberlain during the reign of Queen Elizabeth. London, Camden Society, 1861, pp. 88, 102, 103.

drew no distinction between Giffarde, Caius and Caldwell, and the barber surgeons or the leeches turned loose upon her people, through the Heaven-sent discernment and selection of his Grace of Canterbury. At all events, as is well known, she refused to take physic to the last, and grimly flouted her doctors from her pile of cushions as long as fierce will and frail body remained together; and then, with characteristic inconsistency, left to one of them—and that one, it is said, Gilbert—the only substantial bequest whereby she remembered any of her personal attendants. The chorus of execration which arose from the ignored royal household is historical; but the great work of Gilbert had then been written and laid at her feet. The book itself was not without a spice of ingenious flattery for herself, and so it is not difficult to imagine that the Queen was willing to give to him, in order to carry on labors whereof she saw the value, the pension which she was equally ready to deny even to the most sycophantic of her court satellites.

The laudatory address of Edward Wright, the mathematician, which is prefixed to Gilbert's first and chief volume,[1] wherein all his magnetic and electrical experiments and discoveries are recorded, says that it was held back from the press for nearly twice the Horatian period. This places the time of its inception shortly after Gilbert became Censor of the Royal College of Physicians—the acquirement of which dignity, and the fact that he was enabled to undertake a task requiring so great an expenditure of time and labor in addition to the duties imposed on him by his profession, fairly indicate that in the decade which had elapsed since his settlement in England, he had achieved no small measure of success. The statement has

---

[1] The title in full is as follows :
Guilielmi Gil / berti Colcestren- / sis, medici londi- / nensis, / De Magnete, Magneti- / cisque Corporibus, et de Mag- / no magnete tellure ; Physiologia Nova, / Plurimis et argumentis, et expe / rimentis demonstrata. / Londini / Excudebat Petrus Short Anno / MDC. /

been made that his magnetic experiments involved an actual outlay of over £5,000 sterling,[1] and Gilbert himself avers in his preface that, in his endeavor to discover the true substance of the earth, he examined matters obtained from lofty mountains, sea depths and hidden mines—from which it may be inferred that he had made a large collection of rare substances, which, in those days, must have involved great outlay. So also, in describing one of his experiments, he speaks of testing the supposed magnetizing effect of seventy-five diamonds. It is evident, moreover, that all of his practical research was made with the utmost attention to detail, that his tests were repeated over and over again, sometimes with very slight variations. They form a great multitude of experiments and discoveries dug up, he says, with much pains and sleepless nights and at great cost; and "all of them done again and again under my own eyes."

It was impossible for such an intellect as that of Gilbert not to draw comparisons between knowledge based on the magnificent discoveries of the Italian anatomists, and that founded on the pedantic re-readings of Galen about which the English physicians ceaselessly wrangled; or between the intelligence which sought, at the bedside, the best modes of assisting the *vis medicatrix naturæ*, and the quacks, whom he denounces as prescribing gold and emerald and practicing wretched imposture for money. To him who has learned the art of questioning nature, there belongs a potent armament adaptable to all needs. The study of the obscure functions of the human organs and that of the equally obscure phenomena of the lodestone involved, in both instances, "sure experiments and demonstrated arguments"—the same care "to look for knowledge not in books but in things"—and the handling of bodies "carefully, skillfully and deftly." The skill trained to one task was inevitably trained in all essentials

---

[1] Fuller: Worthies of England, 16. Morhof: Polyhist. Lit. Lubeck, 1732, Vol. II., 3d ed., 409.

to the other, and thus properly directed, the genius of Gilbert moved forward in the path of new discovery, perhaps as nearly in a right line as any fallible human effort can so proceed.

In 1543, the year of his death, Nicolas Copernicus ventured "merely as an hypothesis for their better explanation" to publish the cosmical discoveries which he had made thirty-five years earlier. It did not become a Polish Catholic canon and prebendary to do more than cautiously suggest, even at the eleventh hour, a theory which would have been perilous to advance at an earlier time. Yet, the hypothesis of the earth's revolution about the sun was no new one. Among the ancient philosophers Heraclides of Ponticus, Ecphantus, Nicetas of Syracuse, and chiefly Philolaus, had all affirmed it, and it had found its first modern support during the fifteenth century at the hands of Cardinal de Cusa, who asserted, without qualification, "jam nobis manifestum terram in veritate moveri," although he offered no more proof of the fact than did his predecessors. Copernicus, however, took "the liberty of trying whether on the supposition of the earth's motion it was possible to find better explanations than the ancient ones of the revolutions of the celestial orbs," and concluded that " if the motions of the other planets be compared with the revolution of the earth, not only their phenomena follow from their suppositions, but also that the several orbs and the whole system are so connected in order and magnitude that no one part can be transposed without disturbing the rest and introducing confusion into the universe."

This doctrine was brought into England by Giordano Bruno of Nola, one of the last martyrs of philosophy, whose statue, erected within late years, marks the spot in the Eternal city where his too aggressive wit was expiated under the all-embracing name of heresy. In 1583, he held public disputations with Oxford doctors, and subsequently formulated his metaphysics in his treatises. From Bruno it may be presumed that Gilbert imbibed the ideas which made him not only the first of English Copernicans, but from his very nature an active defender of the new theory, with the original tenets of which he coupled his belief that the diurnal rotation, as well as the polarity, of the earth is due to the magnetic nature or Form of a so-called terrene matter of which he regarded the globe as composed.

The ultimate aim and object of Gilbert's work was therefore to substantiate the doctrine of Copernicus by entirely new arguments and experiments: and this at a time when the opinion of the world—or what was then nearly the same, the opinion of the Church—was inflexibly arrayed against it. To have published such a theory in the England of Mary would have inevitably resulted in the consignment of the book, if not its author, to the flames; but the Roman arm did not extend to Elizabeth's England, and the Queen's physician might safely brave the power which, in his boyish days, for the utterance of heresies far less pestilent and subversive, he had seen hale his townsmen and neighbors to the stake. So he printed what he had excogitated, not in barbarous monk-Latin, bristling with contractions and packed into a dumpy octavo, after the fashion of most scientific works of the time, but in language which, if not entirely Augustan and betraying its English origin in its sturdy assertiveness and bluff invective, is far from destitute of rhetorical grace; and replaced the incubus imprimatur of the Holy Inquisition with pictures of the Queen's Arms, and her monogram and her falcon badge of maidenhood, inherited from the ill-fated Anna Boleyn, and her rising phoenix—*semper eadem*—which, a dozen years before, had soared to glory over the wreck of the Invincible Armada. It may have been chance which transferred to Gilbert's pages the same emblazonments which appear in those of Darcie's History of England of earlier date, which ends with the story of the magnificent victory in the Channel; it may have been that the Cupids and flowers which entwine the Royal monogram were put there because of their assumed pleasant significance to the "fair vestal throned by the West;" but no one will deny the singular appropriateness of the emblems of England's grandeur impressed upon the first great scientific treatise of modern times, and flaunting anew the challenge of the free Anglo-Saxon in the field of thought as in that of arms. Rome denounced the book; but there is no record that along with the treatises of Galileo, to which they had lent inspiration and in comparison with which they were the greater offender, the Italian hangman burned the pages which bore the English rose.

I have stated that Gilbert's physical researches were intended to support the Copernican theory. This he sought to do, not directly, but by founding upon his experiments a so-called "magnetic" hypothesis, whereby he believed that the earth's motion could be explained. A brief review of this speculation is, at the outset, desirable. Afterwards I shall note the unfavorable reception which it encountered, and the possible temporary disrepute of Gilbert's entire work because of his errors concerning dip and variation. As resulting in the first great physical investigation, depending upon the inductive method, some consideration of Gilbert's mode of philosophic thought is also necessary: all of the foregoing being a prelude to the review of the discoveries which underlie the modern science of electricity.

The fundamental arguments which Gilbert advances in support of the heliocentric theory do not differ essentially from those which had already become known among the Continental philosophers. He regards the geocentric doctrine as best refuted by the suggestion of the immense rapidity with which the spherical heavens must revolve— the extravagant whirling of the *primum mobile*—if the earth be regarded as the motionless centre of the universe (a notion which had occurred long before to the Venetian Benedetti, the first clear refuter of Aristotle's mechanics) and which Burton quaintly describes as so great that "an arrow out of a bow must go seventeen times about the earth whilst a man can say an Ave Maria."[1] Such earlier objections as this to the Ptolemaic system he epitomizes with characteristic perspicuity; but when he undertakes to present affirmatively his own supposed magnetic proof of the earth's motion, he becomes both doubtful and obscure; doubtful, inasmuch as he leaves it questionable whether he intends to accept the idea of an annual motion of the earth about the sun, and obscure, in his explanation of the manner in which the magnetic quality of the earth in his belief causes the diurnal movement on its axis.

In developing his cosmical theory, Gilbert, following the precedent of earlier co-believers, makes his main point of attack the theory of Aristotle, that the earth is spherical and has its center coincident with the center of the universe about which the heavens revolve: and more particularly the Peripatetic argument that the earth does not move, first, because it is at the center of the universe, to which all heavy bodies gravitate to find a position of rest, and second, because a rotary motion would not belong naturally to the earth itself, but would pertain equally to each portion of the earth, whereas such is obviously not true, all of these portions being carried in a straight line to the center.[2] Against this Gilbert maintains that the earth is not a chaotic spherical mass, but one having definite poles which are not merely mathematical expressions, but which, on the contrary, are set at fixed points, whereat the greatest verticity of the earth is manifested, and whereon, he holds it is magnetically demonstrable, the earth revolves. This rotation is diurnal, for none else will account for the attending phenomena. The existence

[1] Anatomy of Melancholy, part 2, § 2, Mem. 3.
[2] Aristotle: De Cœlo, ii., chap. xiv.

of these poles is due to a creative act whereby forces primarily animate were implanted in the globe in order that it might steadfastly take direction (in space), and in order that the poles might be opposite, so as to serve as the extremities of an axis on which the earth turns. The direction in space is such that the North pole of the earth constantly regards the Pole star ; so that, if that pole were turned aside from this steadfast position it would go back thereto.

It will be apparent that this doctrine rests upon the conclusion that the earth itself is a freely movable magnet, having poles and amenable to the same laws as the compass needle. How this was reached will soon be shown.

Thus far Gilbert's theory is not difficult to follow; but when he comes to explain, not conditions under which an asserted rotation of the earth on its axis might take place, but how such rotation, through magnetic means, actually does take place, difficulties arise. Obviously any tendency of the earth's axis to return to normal position when diverted therefrom cannot account for the revolution of the globe itself. But, says Gilbert, the *whole* earth regards the Pole star, and similarly, each true part of the earth seeks a like place in the world (universe) and turns with a circular motion to that position. The natural movements of the whole and of the parts are alike; hence, since the parts move in a circle, the whole has circular motion, and hence the whole earth is adapted to such movement.

This is not only inconclusive, but, on *prima facie* showing, appears to be nothing more than the theory of Peregrinus (that the magnet is directed, not solely by the poles of the heavens acting upon the poles of the stone, but by all parts of the heavens acting upon all parts of the stone) which Gilbert has applied to his huge magnet, the earth. But so to assume would be to involve Gilbert in the fatal inconsistency of both denying and affirming the existence of a rotary heaven ; for, according to Peregrinus' doctrine, unless such be present, there is no force acting to rotate the poised stone, while of course, in the Copernican system, a rotary heaven has no place.

There is, however, to be added Gilbert's further hypothesis that every magnet, the earth included, is surrounded both by an "orb of virtue" which includes the whole space through which the magnetic action extends; and by an "orb of coition" which includes all that space through which the smallest magnetic body is moved by the magnet, and beyond which, in other words, the magnet can produce no motion in solid matter. In modern terms the orb of virtue may be regarded as the whole magnetic field capable of recognition as such, and the orb of coition that part of the field in which a selected extremely small magnetic body is attracted. These orbs or spheres which Gilbert speaks of as "effused," and as produced directly from the earth's exhalations, are magnetic because so generated; but such phenomena are by no means limited to our globe alone. Thus, he considers that all heavenly bodies, and especially the sun and moon, have such effused spheres, which are capable of acting upon other bodies and other effused spheres. Hence, not only does the earth, as has been said, remain in its place by its own magnetic virtues, but "by a confederation of the adjacent globes through the connected effluent strengths, it is directed harmoniously to its neighbors;" it is moved by "the conspiracy of motions of other bodies and by their effused forms moving together, especially by the sun and moon, by which it is bounded and limited."

It seems therefore that Gilbert, besides apprehending the existence of a field of force around the earth, also pictured to himself the action of that field upon other fields, and of other fields upon it; a conception so far in advance of his age that nothing but his unequivocally direct statements make one willing for a moment to entertain the belief that he ever harbored it; a conception which finds an every-day illustration in the electric motor—in fact, in every electrical apparatus in which mechanical motion is caused by the reaction of fixed and moving fields of force. It cannot, of course, be affirmed that Gilbert conceived of the rotation of the earth in the fields of the sun and moon in any such way as we regard the rotation of an armature in a magnetic field; but that he certainly did regard the earth's diurnal rotation as somehow due to the confederacy and conspiracy of the earth's effused Form acting on, and being acted upon by, the effused Forms of other celestial bodies is plain.

Of course all this probably intensified the obscurity of Gilbert's theory at the time of its production. And an obscure hypothesis, intended to substantiate another which, according to prevalent opinion, was not philosophical argument, but pestilent, soul-destroying heresy, had not much way-making power even among those who disputed theological conclusions and were inclined to tolerate truth regardless of the finger-posts at Rome. Hence it may readily be imagined that even to the Copernicans themselves Gilbert may have seemed a doubtful auxiliary, while there was manifestly not much heart of grace to be taken from his long category of experiments and arguments, however individually true and interesting they might be, so long as they seemed in respect to their aim and object merely a ladder leading nowhere.

There was a more serious trouble in Gilbert's work, however, than even the advocacy of proscribed astronomical doctrines, and that lay in his erroneous notions concerning the dip and variation of the compass. At the time he gave these to the world, the English seaman was rapidly merging the pirate in the merchant adventurer, the naval supremacy of England was established, there were no more Invincible Armadas to be feared, a great trade was to be wrested from Spain and Portugal and Italy, and the exploits and discoveries of Drake and Raleigh and Frobisher were setting the heart beating and the fancy aflame of every youth in whose veins ran the blood of the men who had sunk the Danish harriers, and who longed to voyage to Virginia and the strange lands of the New World, and perchance to find fortune in some gold-logged galleon on the Spanish Main.

Errors likely to wreck ships, made when the eyes of all Englishmen were turned to the sea, were likely to prove doubly harmful, especially as there was nowhere the correct knowledge necessary to recognize their fallacy, or even to prevent people presumably expert in such matters from endorsing them as truths. Edward Wright, in his prefatory address to Gilbert's treatise, although looking askance at the magnetic theory of the earth's rotation, (nevertheless he does "not see why it should not meet with indulgence") grows eloquent over Gilbert's mistaken idea that the dip of the magnetized needle "differs in the ratio of the latitude of each region," and that hence, the dip being once determined and the latitude observed, "the same place and the same latitude may thereafter be readily found by means of the dipping needle even in the darkness and fog." Equally erroneous was his opinion that compass variation is nothing but the deviation of the needle to more massive or elevated parts of the globe, and that it is constant at the same place. When these beliefs were proved to be wrong, it must have seemed to many that not only was Gilbert's general theory vague and indefinite, but that even his especial practical applications of it to the purposes of the navigator were misleading, and more likely to invite the perils of the sea than to prevent them.

Before the extent of his errors and uncertainties was generally perceived, his work was generally praised; but later it seems to have fallen into oblivion. Small wonder, when to the many who believed it heretical were added more who thought it open to the charge of teaching delusions. But in time the world separated the delusions from the truths; it rehabilitated Gilbert, not for his speculations, not even because he rescued the study of the magnet from the atmosphere of mysticism which surrounded it: but because, in celebrating the man, it likewise celebrated the beginning of the removal of all natural science from the quicksands of empiricism and speculation, and the placing of it upon the solid basis of actual experiment, the evidence of the senses and philosophical thought.

Although Gilbert constantly revolts against the physical theories of the Peripatetics, it is none the less clear that his mind was deeply tinctured with the logic and metaphysics of Aristotle. But he stood at the dividing line between the old philosophy and the new. To the rules of the Stagirite he could conform his speculations; but he drew his conclusions under the rules imposed by Nature. The control of Aristotle over mental processes did not imply with him a corresponding control over the interpretation of physical facts; and this being so, he definitely established, for the first time in the world's history, the truth that metaphysical arguments alone are incompetent to explain Nature's workings or to detect her immutable laws.

This appreciated makes fairly clear the method of investigation which he endeavored to follow, and sheds light on many of his statements otherwise obscure or self-contradictory. His treatise contains much of what Aristotle calls exoteric discourse—a process of noticing and tracing out all the doubts and difficulties which beset the enquiry in hand, along with the different opinions entertained about it, either by the vulgar or by individual philosophers, and the various reasons why such opinions may be sustained or impugned.[1] After doing this, still following the procedure of the Stagirite, he begins to lay down and follow out affirmative principles of his own, thus passing from the dialectic to the didactic stage. But

[1] Aristotle: Topica, i. (Grote, Aristotle, i., 68).

when he comes to recording his experiments, to testing the results by negative arguments and contradiction, to rising from particulars to the general, and thus deducing new conclusions, we shall find in practical application the principles of inductive reasoning which his great contemporary and critic, Francis Bacon, a few years later, formulated for all time and all men. The discussion of Gilbert's relations to Bacon, however, must be deferred to another chapter, in order that our present review of his theories may progress in an orderly way.

If the magnetic earth-rotation theory be, for the moment, laid aside, Gilbert's primary thesis is that the globe consists of a certain solid homogeneous substance, firmly coherent and endowed with a primordial actualizing Form. The various substances which appear on the surface of the globe through contact with the atmosphere, waters, and by influence of the heavenly bodies, have become more or less deprived of the prime qualities and true nature of this terrene Matter. But the lodestone and all magnetic bodies contain the potency of the earth's core and of its inmost viscera, in virtue of which the earth itself remains in position and is directed in its movements. Thus the earth is in fact a huge magnet, or the lodestone is a fragment of the magnetic earth possessing the primal Form of things terrestrial. Between Matter and Form he drew substantially the Aristotelian distinctions.

The investigations made by Gilbert in support of this theory, consisted first in determining what is a magnet, second, the cause and character of magnetic attraction, or as he preferred to call it, coition, and third, the nature of its polarity or directive quality, or to use his own word, "verticity." Having found certain phenomena of the lodestone true of the earth, and conversely certain terrestrial phenomena true in a miniature earth made of lodestone, he concludes the globe to be itself a magnet, and thence proceeds to the researches wherein he not only passed in review all preconceived notions of magnetism, and probably tested every experiment thereto relating of which he could find record, but made a remarkable number of new discoveries. More than this, he took up, for the first time for systematic study, the phenomenon of the amber—not primarily for the purpose of inquiring into its nature, but really as a digression, and with the object of showing that it was totally different from that of the magnet.

The research begins with a comparison of the poles of the heavens, the poles of the earth and the poles of the lodestone; and the proposition is at once laid down that the poles of a magnet on the earth look toward the poles of the earth, move toward them and are subject to them. This was the first statement of the truth that the compass needle is governed—not by the heavens nor by the Pole star, nor by the poles of the heavens—but by the magnetic quality of the globe itself.

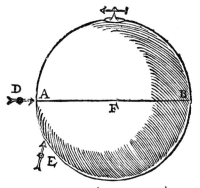

GILBERT'S TERRELLA.[1]

In order to prove the like nature of the earth and the lodestone, Gilbert carved a piece of the stone into spherical form; because, as he says, that shape is the most perfect, agrees best with the earth, which is a globe, and is

---

[1] From the first edition of his treatise De Magnete. A and B represent the earth's poles, F the earth's centre and D and E pivoted compass needles applied to the lodestone ball.

better adapted for experimental purposes. This miniature earth he calls an earthkin or terrella, and upon this he makes his experiments, mainly by placing near to it pivoted iron needles or iron plates, and noting the directive or attractive force exerted by the globe.

The close similarity of this course to that followed by Peregrinus will at once be apparent. But Peregrinus regarded his spherical magnet as a miniature representative of the celestial sphere: Gilbert regarded it not merely as a representative of the spherical earth, but actually *as the* earth; in the sense that it was physically a fragment thereof, possessing, though in less degree, the same potencies and energies. Peregrinus considered the magnetized needle as influenced by the poles of the spherical heavens represented in the lodestone globe : Gilbert, by the actual poles of the small spherical lodestone, in precisely the same way as by the actual poles of a greater spherical lodestone—namely, the earth.

Yet undeniably both Peregrinus and Gilbert performed exactly the same experiment and with the same thing. Natural phenomena are not changed by the names men give them, and whether the lodestone globe be regarded as a miniature earth or a miniature heavenly sphere cannot alter in the slightest either the nature of the object or the effects produced by it. I may even go further and, as I have already suggested in discussing the experiments of Peregrinus, point out that the analogy of the lodestone globe to the earth may, from Peregrinus' language, be fairly inferred as not unperceived by him. But, there is all the difference in the world between approximating a result, however nearly, and actually attaining it; while there is no argument more frequently specious and hence more perilous than that which seeks to establish conclusions as foregone after the event. Granting that Peregrinus perceived an analogy between his globular lodestone and the earth—he did not see them as one and the same thing differing only in magnitude. Gilbert did: he made his terrella, as he thought, of the earth, as the earth and in the shape of the earth; judged of the whole from the part, and thus attained the conclusion which Peregrinus did not reach—namely, that the globe on which we live is a huge magnet.

But the details of the initial experiments on the terrella, its manufacture on the lathe as lapidaries turn and polish crystals, the modes of finding its poles and magnetic meridians by short bits of iron, the greater attraction of the poles for these pieces, their erection at the polar points, their varying inclinations when supported in different parts of the field, the practical demonstration of the laws of magnetic attraction and repulsion, and the distinguishing of the magnetic poles, Gilbert takes directly from the famous Letter of Peregrinus, at times almost verbatim. He even copies figurative expressions which Peregrinus uses, such as the comparison of the magnet in its floating bowl to a sailor in a boat. The fact that Gilbert makes no acknowledgment of Peregrinus' achievements in all these vital matters some may find explicable by the disregard for the amenities which characterizes his entire work. Others again will find it difficult to reconcile his appropriation of Peregrinus' discoveries with his immediately following statement that the whole philosophy of the magnet is ill-cultivated even in its elementary principles.

It is true that this systematic habit of not acknowledging the effective work of his predecessors makes it no easy task to distinguish with certainty the true extent of Gilbert's accomplishments, even in the light of the review of past progress which has already been presented. He rarely mentions an earlier writer except to dispute conclusions, which may perhaps be due to the influence exerted upon him by Aristotle, who, as Bacon repeatedly remarks, "as though he had been of the race of the Ottomans, thought he could not reign except the first thing he did he killed all his brethren."[1]

[1] Advt. of Learning, Book 2., c. viii., 5. "And herein I cannot a little

He cites Porta more than any one else, Cardan next and then Fracastorio, mainly to exhibit their errors; but he draws freely upon Porta, for example, in stating the relations of divided lodestones and the behavior of magnetic bodies in the field of force, and upon Cardan for differentiation between lodestone and amber. If, by chance, he happens to express a favorable opinion of these philosophers, he always manages afterwards to reverse it. He begins by calling Porta a philosopher of no ordinary note, and ends by denouncing his statements as the maunderings of a babbling crone. Fracastorio is an ingenious philosopher and also a reckless speculator. As regards Cardan, the importance of whose brilliant differentiation of lodestone and amber we have already seen, Gilbert is at least consistent, for he never permits himself any praise at all of the famous Milanese, who he asserts reasoned solely on the basis of vague and indecisive experiments. And as for the other philosophers, whether writers on medicine, or on navigation, or on astronomy, ancient or modern, Platonist or Peripatetic, charlatans, such as Paracelsus, or scholars, however of learned repute, all are included in censure and often abuse, which last perhaps reaches its lowest level in a bitter anathema against Taisnier, the plagiarist of Peregrinus; although, from one point of view, it may be urged that the only difference between Taisnier and Gilbert himself is that Gilbert's plagiarisms from the same source are much the more complete and accurate.

The sole exception to be found in this wholesale condemnation is an accordance of honor to Aristotle, Theophrastus, Ptolemy, Hippocrates and Galen, whence he says came the stream of wisdom, and who, he is per-

marvel at the philosopher Aristotle, that did proceed in such a spirit of difference and contradiction toward all antiquity; undertaking not only to frame new words of science at pleasure, but to confound and extinguish all ancient wisdom; insomuch as he never nameth or mentioneth an ancient author or opinion but to confute and reprove: wherein for glory and drawing followers and disciples he took the right course." Advt. of Learning, B. 2, c. viii., 2. Bacon himself did the same.

suaded, would gladly have embraced many of the new things brought to light since their departure, had they known of them; and his mention of St. Thomas Aquinas, (who seems to have anticipated his notions of magnetic coition), as a man of god-like and lucid mind; a tribute which, by reason of its solitude, engenders the suspicion that his failure to contradict it subsequently was due rather to oversight than design.

Following the general enunciation of his conception of the earth's magnetism, and his repetition of the experiments of Peregrinus, Gilbert enters upon the researches which are plainly original. Then he rises to an eminence so lofty, that his contemptuous criticisms of his predecessors soon resemble the scorn of the eagle for the flights of the sparrows. If then he is intolerant, it is that intolerance which every man who sees the truth, however obscurely, feels for others who preach error or half truth. If he seems to belittle the achievements of his predecessors, it is due to that instinctive tendency of the mind to conclude that that which is false in part is false in all, rather than to impute to truth the greater leavening power. Consequently, when, at the very outset of his studies, he finds Cardan gravely asserting that a wound by a magnetized needle is painless—when he had only to prick his finger to learn the opposite—or Fracastorio that a lodestone will attract silver, or Scaliger that the diamond will draw iron, or Matthiolus that garlic cuts off magnetic attraction—all susceptible of easy disproof, which disproof he actually makes and sees, he says, in the uneuphemistic terms characteristic of his day, not that these people are mistaken, but that they wilfully falsify. After that the mental process is easy. Anything proved true, he undoubtedly argued to himself, if drawn from that sink of mendacity, redounds, not to the credit of the sink, but of him who rakes it out. Therefore he did everything anew—not, as he says, for the purpose of refuting prior falsehoods or overturning old delusions, but to build up the new physiology of the magnet from the very foundations.

If then he plagiarizes Peregrinus, it is not the *ipse dixit* of Peregrinus which he copies blindly, regardless of whether right or wrong, but he tests the experiments of Peregrinus crucially, finds them true, and, for that reason, adopts them as a part of his structure. It is immaterial to him who originated the experiment which he tests—whether himself or some one else. The great question which he seeks to solve is, is the result true—"Not in books, but in things themselves?" he says, "look for knowledge." Every experiment—not merely those which he first conceived, but those of any and every origin—"has been investigated and again and again done and repeated under our eyes."

Gilbert's philosophy, resting, as I have said, upon that of Aristotle, is imbued with the distinction between Form and Matter, which the Greek makes of fundamental import in his Philosophia Prima, and diversifies with an infinitude of subtleties. Borrowed by the Stagirite from the familiar facts of the sensible world—that matter has always some shape and shape has always some matter, and that we can name and reason about matter without distinguishing its shape, and equally name and reason about shape without attending to the material shaped or its various peculiarities—the doctrine assumed the abstract signification of two correlates inseparably implicated in fact and reality in every concrete individual that has received a substantive name, yet logically separable and capable of being named and considered apart from each other. Matter is the lower, inchoate conception—the unactual or potential. Form actualizes this into the perfect or complete, and furnishes the energizing principle. Matter is a cause co-operative. Form is a cause operative. Matter is to Form as brass is to the statue, wood to the couch, and the body of man to the soul.[1] Form in the

[1] I have followed in the foregoing the close analysis of the Aristotelian

writings of the schoolmen was synonymous with attribute: "it is that by which a thing is." An angel to a schoolman was a Form not immersed in Matter. "Angeli sunt formae immateriales," says the Angelic Doctor.[1]

The Matter of the earth, according to Gilbert, is endowed with Form or efficient potencies which give to it firmness, direction and movement. Of these the principal feature is "verticity"—a word which he coins to signify the self-directing capacity or directive polarity of the globe. Just to the extent that it loses Form—as by the terrene Matter becoming combined with base or excrementitious substances—so it loses verticity. Ultimately he draws a somewhat subtle distinction between this unique and peculiar Form which he ascribes to the earth and the *prima forma* of Aristotle, by limiting the first to a particular variety or kind of Form which keeps and orders its own globe—giving a specific Form to the sun, another to the moon, and so through all the heavenly bodies. Thus he reaches his general conclusion as to the magnetic nature of the earth, and at the same time differentiates his theory from the older hypotheses.

This nature is not derived from the heavens as a whole,

conception, given by De Grote in his discussion of the De Anima. Aristotle, Vol. II, p. 181 et seq.

"The implication of the two (Matter and Form) constitutes the living subject with all its functions, active and passive. If the eye were an animated or living subject, seeing would be its soul; if the carpenter's axe were living, cutting would be its soul; the Matter would be the lens or the iron in which this soul is embodied. It is not indispensable, however, that all the functions of the living subject should be at all times in complete exercise; the subject is still living, even while asleep; the eye is still a good eye, though at the moment closed. It is enough if the functional aptitude exists as a dormant property, ready to rise into activity when the proper occasions present themselves. This minimum of Form suffices to give living efficacy to the potentialities of the body; it is enough that a man, though now in a dark night and seeing nothing, will see as soon as the sun rises; or that he knows geometry, though he is not now thinking of a geometrical problem." Aris., *De Anima*, II, i., p. 412, a. 27.

[1] St. Thomas Aq.: Sum Theol., 1, q. 61.

he says, neither is it generated thereby through sympathy, or influence, or other occult qualities: neither is it drawn from any special star; but the earth has its own proper magnetic vigor or Form, just as sun and moon have theirs. Consequently, as a fragment of the sun would arrange itself under solar laws to conform to the shape and verticity of the sun, or a fragment of the moon under lunar laws to conform to the shape and verticity of the moon, so a fragment of the earth under terrestrial laws, being endowed with the same magnetic vigor or Form, will dispose itself correspondingly to the earth. Now as a lodestone is not merely a fragment of the earth, but is of the inmost earth and possesses the primal Form of things terrestrial and the whole impetus of magnetic Matter, therefore it has the fixed verticity, and the innate whirling motion of revolution, inherent to the earth.

The notion that the lodestone is both a fragment of the earth and is polarized by induction therefrom, is not inconsistent with modern ideas; but that of the earth rotating because of its magnetic quality reduces itself, as I have already pointed out, to mere guess-work and to proof of the strength with which the speculative tendency asserts itself, even in a mind which repudiated "probable conjectures" as a basis of reasoning, and despite the belief that it recognized no control save that of "sure experiment and demonstrated argument."

Not only does Gilbert explain the existence of magnetism through the Peripatetic conceptions of Matter and Form—the last, as we have seen, somewhat modified in particulars—but he recurs, ultimately, to the same source for ground-work for interpretations of special magnetic phenomena. Aristotle applies the term "nature" to a constant which perpetually tends to renovate Forms as perfect as may be, and invariably acts in a uniform way, producing phenomena which are regular and predictable. In opposition to nature stands variability or chance, which interferes with and impedes the work; so that, although results which have taken place in the past can be definitely stated and recorded, those still in the future defy all power of prediction.[1] One example may be cited which will serve to show how Gilbert applied this hypothesis, while incidentally it may indicate how, being as I have said on the middle ground between the old and new philosophies, he wandered, even in the face of the simplest experimental proof, from the path of logical inductive reasoning.

He repeats, in exactly the same way, the experiment of Peregrinus, showing the mutual repulsion of like poles of two parts of a divided lodestone floating in water. "By such a position of the parts," he says, "nature is crossed and the form of the stone is perverted. But nature observes strictly the laws which it imposes on bodies, hence the flight of one part from the undue position of the other, and hence the discord unless everything is arranged exactly in accordance with nature."

This obviously is the Aristotelian idea of necessity—the constant sequence or conjunction—the fixed means through which the fixed ends of nature only can be obtained. To place like poles in juxtaposition is to place them wrongly, and then Gilbert avers nature is perverted, and the Form of the stone disturbed, and hence there is discord: nor can there be any compromise but only war until the stones acquiesce as nature decrees. He does not assert that under given circumstances, shown by a multitude of experiments, like magnetic poles mutually repel, and that thence a general law may be inferred from which their similar behavior under similar circumstances may be predicted; but that, when everything is arranged exactly according to nature—that is, unlike poles juxtaposed—then these parts attract one another. It has all been "settled by nature."

Gilbert speculated, as I have said, with the logic of Aristotle, but he made experiments and interpreted the

[1] De Interpretatione: Grote, cit. sup., vol. i., 166, book i, chap. vi.

results of them by their own logic. Unable, struggle as he might, wholly to divest himself of that reverence for antiquity peculiar to all the philosophical thought of his time, a half-defined belief persisted in him that the spirit of the Greek, notwithstanding the lapse of two thousand years, was still competent in some way to define, to elucidate and to account for all that the human mind might find obscure; but, despite such belief, he saw and knew, even though he might not admit the knowledge to himself, that it was not the intellect of Aristotle, but his own, which was making for straight thought. And where he erred, it was because he courted the ancient influence and ignored, even if he did not fail to perceive, the plain deductions from the facts before him. None the less he, first of all men, systematically replaced the great doctrine of words by the greater doctrine of works. It was only when he thought it incumbent on him to reconcile the teachings of man's books and nature's books, and just in proportion as he allowed the first to obscure the second, that he landed in inevitable contradictions and fallacies.

From this general and necessarily brief showing of Gilbert's mode of thought, I now pass to his actual experimental work and the physical discoveries resulting therefrom. To record all of the facts relative to the magnet which Gilbert first brought to light, and to show their relation to the modern science, would involve explanations far too extended, if not too didactic, to find place here. Nor is it necessary to do so: for I am now approaching the period when we may begin to trace the independent development of amber-electricity as distinguished from that of magnetism, and need therefore in future allude to the latter only in so far as the discoveries made in it may have directly conduced to such progress. And here it may be recalled that there is no necessary relation between the advance or rise in a specific branch of knowledge during a given period of time and all of the discoveries pertaining thereto made within the limits of that period. On the contrary such a rise is apt to be determined by a comparatively few salient achievements, which being more readily appreciated and understood than others, are more promptly turned to useful account. In every stage of the world's progress the making of discoveries "ahead of the times" has been going on; and of these perhaps the greater proportion remain mere items of abstract knowledge for years, perhaps for centuries, until thought advancing to new points of view so discerns their practical utility: or some keener intellect sees in them possible applications to which other minds have been blind. It will be apparent therefore that in tracing historically such an intellectual rise as is here chronicled, a more or less arbitrary selection must be exercised, and matters often in themselves important, but which appear to exert no active influence thereupon, must be omitted. Otherwise the work reduces itself to the gathering of chronological annals.

After having declared the origin and nature of the lodestone on the strength of initial experiments, which, however he interpreted them, were in fact drawn mainly from Peregrinus, Gilbert takes up the problem of the iron magnet; for here was plainly a substance having the properties of the lodestone and yet differing from that primary terrene Matter, although of like Form or vigor. He evolves the theory that the earth gives forth humors or exhalations, which coalesce with solid materials to form metals, and, if these materials be the more homogeneous or internal Matter of the globe, the result is iron or lodestone, which is nothing but a noble iron ore; if they be the globe Matter, in an altered or baser state, or efflorescences, then other metals are produced. Iron ore is, therefore, the homogenic telluric body to which the earth humor has been added; but the latter does not destroy the potency of the earth-Form existing therein, and hence it remains, or may be rendered, magnetic. Lodestone is the same body concreted with a stony Matter; and both magnetized iron and lodestone conform themselves to the globe of the earth.

So much for the hypothesis, fanciful enough in itself, and yet, if not directly leading to Gilbert's practical discoveries, at least not serving to conduct the investigator directly away from them, as many an older assumption had done.

GILBERT'S ARMED LODESTONES.[1]

There is no perspective in Gilbert's record of these researches save that leading to his supposed magnetic proof of the Copernican doctrine. Hence the difficulty in disentangling from his often prolix restatements the really novel and important achievements. His own attempt to do this by marking large asterisks beside the descriptions in his book of those experiments which he regards as of more importance is of little aid, since this does not imply that the matters noted are of his own inception, and, in some instances, they are plainly taken from Porta. Nevertheless, it is possible to distinguish, as probably original with Gilbert, the following remarkable discoveries in magnetism:

That the strength of a magnet can be augmented and preserved by placing upon its pole an iron helmet or cap—the effect, as now regarded, being to collect and converge the lines of force. This was the first suggestion of the armature or keeper.

That the magnetic attraction will not be cut off by any substance except, as he says, by an iron plate.

That the earth is a huge magnet, and has magnetic poles.

That the compass needle is directed by the earth's mag-

[1] From the first edition of his treatise De Magnete.

netism, and disposes itself in the line of a great circle passing through the poles; or, in other words, in a magnetic meridian, a term also first used by Gilbert.

That iron or steel acquires magnetism from the lodestone and is thus itself a magnetic body, capable of attracting the stone as the stone attracts it, so that the two come together by forces mutually exerted, and not by the one-sided attraction of magnet on object. This was, to some extent, pre-suggested by Cardan, and, long before him, by St. Thomas Aquinas and Cardinal de Cusa.

That the magnetic force moves from one end of an iron rod to the other. It travels through all bodies, he says, and is continued on by them. Here was the first notion of magnetic conduction; the first suggestion of the possible movement of the force from point to point.

That magnetization of iron occurs with great rapidity. "It is there in an instant," he asserts, "and is not introduced in any interval of time, nor successively, as when heat enters iron, for the moment the iron is touched by the lodestone it is excited throughout."

That the lodestone most strongly attracts the best and purest iron—that the best iron is derived from the lodestone or magnetic ore—that the strongest magnets are made from the best iron—and that the best iron, even if not magnetized, acts like the lodestone in directing itself to the earth's poles, through induction from the earth.

That iron can be magnetized by simple placing in the plane of the magnetic meridian—or, better, by being hammered or wire-drawn, or heated and cooled while so disposed; and that maintenance of a magnet in the same plane conserves its properties. Thus Gilbert says that iron bars which have been fixed in buildings for twenty years or more in north and south position acquire verticity, and thus he explains the magnetization of the iron rod taken down from the church of St. Augustine in Rimini; a phenomenon which Giulio Cesare had accidentally remarked several years before. There is a world of difference between such a physical interpretation as this, and the assumption that the iron by sympathy or similitude had become converted into a lodestone.

That two lodestones fitted with armatures, so as to have a common pole-piece, exert a much greater lifting force than either separately. This is the compound magnet.

That bodies, to use the modern term, can be saturated

MAGNETIZING HOT IRON BY HAMMERING IT WHILE HELD IN THE MAGNETIC MERIDIAN.[1]

with magnetism. "Magnetic bodies," he says, "can restore soundness (when not totally lost) to magnetic bodies, and can give to some greater powers than they originally possessed; but to those which, by their nature, are in the highest degree perfect, additional strength cannot be given."

I have already made some reference to the orbs of virtue

[1] From the first edition of Gilbert's treatise De Magnete. The workman is directed to place himself facing the north, and to hammer the hot iron so that it will expand or elongate in a northerly direction.

and coition—the effused strengths or Forms which surround the lodestone, and which illustrate Gilbert's conception of the magnetic force. A more detailed examination of his theory shows that he regards the force of the terrella as sent out in all directions, attracting whatever iron or magnetic body may come within the sphere of influence; and the nearer the iron to the lodestone, the greater the force by which it is drawn. The shape of the field, he thinks, conforms to that of the emitting body, and he compares its physical characteristics, as Porta had already done, to those of light; but he goes a step further and regards it as merely soliciting bodies that are in amicable relations with

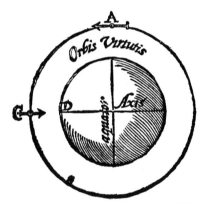

GILBERT'S NOTION OF THE ORB OF VIRTUE AROUND THE MAGNET.[1]

itself, without actually exerting any motive energy upon them. In fact, he is inclined to regard the magnetic field not merely as revealed by the presence of bodies of magnetic material placed in it, but as in some way subjectively connected with such bodies, preventing either the force being imbibed, or given back to its original source. He finds, however, that the lines of magnetic force of his terrella are meridional and numberless, and concludes that the center of the terella is the center of force, although the

[1] From the first edition of his treatise De Magnete. A is a compass needle at the equator, and C another needle at the pole D.

energy is concentrated at the poles. All his deductions, however, lead to the conviction, on his part, that the magnet emits no true effluvium—nothing corporeal—and that its whole action, whether attractive or directive, depends upon its capacity to impart its Form to the iron. As soon as the metal comes within the lodestone's sphere of influence, even if at some distance from the stone, the Form—the soul—of the iron is renewed: that which before was dormant and inactive becomes lively and active, and the Form, being now arranged and ordered, again joins forces with the lodestone, and the two bodies enter into alliance, "whether joined by bodily contact or standing within their sphere of influence."

The most curious conclusions to which Gilbert's ideas of the magnetic field of force led him, are those which are recounted in his posthumous volume.[1] Here he asserts that the earth's orb of magnetic virtue extends to the moon, and ascribes the moon's irregularities to the effects which it produces; that the moon is magnetically bound to the earth because its face is always turned earthwards, and that there is a magnetic coition between our globe and its satellite, the seas being drawn toward the moon and the moon reciprocally to the earth. For the effused lunar forces he says reach to the earth and act on fluids, while the magnetic virtues of the earth surround the moon, both bodies agreeing and consenting in motion, although the earth, by reason of the greater mass, predominates.

Perhaps more interesting than all else is his assertion of a relation which the greatest modern minds have suspected, have sought to prove, but so far with only negative results. All that is of the earth and is homogeneous with it, says Gilbert, belongs to it—so of the sun, and the moon, and other bodies. Such belongings adhere to and do not spontaneously leave their globes; and if they are removed by external force they seek to return, because each

[1] De Novo Mundo, Amsterdam, 1651.

globe, by its own virtues, attracts them. Otherwise, the dissipation of the universe would necessarily follow.

Now this, he avers, is not an appetite or inclination to position, or to space, or to a boundary, but is to the body, the source, the mother, the beginning where all are united and safely kept. Thus the earth attracts all magnetic bodies, besides all others in which by reason of material, the primary magnetic force is absent; and this inclination to the earth in terrene substances is commonly called gravity. The gravity of a body then is inclination to its source, and all things which come of the earth return to it.

Thus, repeating himself in many ways, not uninfluenced, perhaps, by the recollection of the return of all flesh to the dust, he suggests the correlation of gravity and magnetism —a thought still burning, a question still unsolved. More than two centuries afterwards another great student of nature, facing like problems, conceived of the same relationship; and it was while endeavoring to penetrate into its mysteries, the one by speculation, the other by experiment, that both William Gilbert and Michael Faraday each reached the *ultima thule* of his life-work.

The amber phenomenon had begun to detach itself in men's minds from that of the lodestone, as Cardan's differentiation plainly shows. Gilbert now made the separation complete, and not only brought electricity—so termed as distinguished from magnetism—into the sphere of human thought of his times, but gave to its intellectual progress an impetus which has ever since continued, and with growing force.

"Those unobvious, delicate and often cumbrous and tedious processes of experiments which have thrown most light upon the general constitution of nature," says Mill,[1] "would hardly ever have been undertaken by the persons, or at the time they were, unless it had seemed to depend upon them whether some general doctrine or theory which

[1] Mill: System of Logic, ii, 18.

had been suggested, but not yet proved, should be admitted or not." It was to sustain his cosmical theory that Gilbert accomplished the achievements which will render him forever known as the father of electrical science. What he did, and how he did it, I have now to relate.

Gilbert's cosmical system is based, as I have endeavored to show, upon his own application of the results of his experiments to the Copernican doctrine. It was open, therefore, to his opponents to attack him either by disputing the sufficiency of his experiments, or by showing that there were other phenomena similar to those of the magnet, and presumably of like nature, which could not be accounted for by his explanations, and hence that the latter fell short of universal application and so failed to satisfy the conditions of the problem to be solved. Naturally the assault would be directed upon the new and specific support provided by him for the Copernican heresy rather than upon the theory itself, against which the forces of the prevailing theology and philosophy were already turned; and this, it may be fairly presumed, no one appreciated and perceived the need for anticipating better than did Gilbert himself.

Of the two before-noted objections, that which went to the sufficiency of the experiments was the least to be feared, for he could point to such a multiplicity of tests—and practically did so, marking the records of some two hundred of the principal ones by asterisks on the margins of his pages for the express purpose of attracting attention to them—that in those days, when from the slenderest physical occurrence unbounded speculation often flowed, it would require a more than ordinarily bold disputant to challenge the thoroughness and exhaustive quality of his work. As a matter of fact, as we shall see later on, such an antagonist did arise; but this was years after Gilbert's voice had become forever silent.

The other possible criticism was, however, more serious and immediately pressing. Gilbert knew that ostensibly at least it was well founded: he knew that the difficulties involved must be met and overcome, or avoided simultaneously with the presentation of his main argument, and he knew that anything less than complete destruction or avoidance of them would inevitably result in his own confusion.

The peril which thus menaced him came from an unsolved problem of the ages: the same which had vexed Thales twenty-two centuries before; the same which had persisted to mystify men's souls ever since; the riddle of the Amber Sphinx, which now, Oedipus-like, he must solve, or fall.

Let us recall two facts: first, that the world in general classed the amber and the magnet together, and saw no difference in their respective attractions upon other bodies; and second, that Cardan, nevertheless, had drawn a clear distinction between them and had contrasted their behavior. With the popular opinion and with that of Cardan, Gilbert was fully familiar. He saw that the effect of the first would be at once to lead people to attempt to apply his magnetic theories to the amber attraction, while that of the second was an authoritative impress upon his own mind of reasons why the discovery of discrepancies would follow. Granting, for example (he perhaps argued to himself), that the magnet and the amber are alike in attractive power, they are not so in verticity; and, if the attractive capacity shows that both contain the same assumed primordial terrene Matter, how is it that the Form which determines self-direction in the one is absent in the other? What is this primordial Matter which can exhibit such totally different physical characteristics as are seen in the light and brilliant resin and the heavy and dark stone? Even if the Form be the same, is the Matter identical in both? If verticity is absent in the amber, is this because the latter is an "efflorescence" and hence "impaired" primordial Matter? But what impairment is it which excludes one essential quality (verticity) and not the other (attraction)? Why are not both qualities equally impaired, or both absent, as in inferior magnets or in most substances? Why does amber or jet (then supposed to be black amber) alone, out of all the vast number of terrene bodies, exhibit this strange attraction? How is the drawing of all light bodies by amber to be reconciled with the selective property of the magnet, which enables it to draw only iron and steel? How is it that the magnet, being wholly or mainly primordial terrene Matter, can effuse and excite a new Form in iron alone; when amber, which, if primordial terrene Matter at all, lacks a chief capacity thereof, is able, on the same reasoning, to excite a new Form in any substance which is light and minute in size? If amber does not excite such a Form in the thing attracted, where is there coition in the attractive action? if it does, why is there not some residual attractive power left in the straw or chaff, such as the lodestone leaves in the iron? If there be two effused Forms respectively different, one proceeding from the amber and the other from the magnet, which is the true Form effused by primordial terrene Matter? Which the true effused Form of the earth? How is it that this attractive capacity is always present and inherent in the lodestone, and not so in the amber unless the resin be excited? If magnetic attraction is a primordial terrene characteristic, implanted by creative act, why is human aid necessary to develop it in a certain substance? What is the effused Form of a heavenly body if it undergoes attrition in space? that of amber, or that of lodestone, or a combination of both?

It is needless to multiply such questions, for, the instant the doubt fell into the placid pool of theory, it roughened the surface in circle after circle, ever widening until the smooth quiescence was gone. The issue no longer was one limited by the mere observation that a bit of amber attracts a particle of chaff only when rubbed, and a bit of lodestone a particle of iron always; but to Gilbert it was a universal problem, dealing with the relations of worlds and the structure of the universe. He saw that it touched the very heart of his whole cosmical hypothesis.

Thus Gilbert came to study the amber, not for the first time in the world's history, but for the first time by the methods which he had brought into use in finding out the laws of the lodestone: methods which ultimately led, not to the futile utterance of "corn" or "millet" before the closed door, but of the magic "Open Sesame." But what had he before him indicating where to begin his quest?

The new facts which had been added to the knowledge of the ancients concerning the amber had been noted by Cardan, who had not only drawn the suggestive distinction between the amber and the magnet, but had agreed with Fracastorio in the averment that the amber quality also resides in another and totally different substance, namely, the diamond. Nor did this capacity of the diamond lack apparent corroboration from other philosophers. Scaliger had alluded to it in his commentary. Porta had specifically asserted that an iron needle rubbed with a diamond would turn northward, as when rubbed with the lodestone. And Fracastorio had not merely recorded the drawing of "hairs and twigs" by both amber and diamond, but in the very passage from his work which Gilbert quotes, he ascribes the effect to a principle inherent in and common to both resin and gem. Nevertheless, it is not likely that Gilbert seriously considered these assertions, much less tested them by experiment, without something of a mental struggle. His antagonism to Cardan and all his works is profound. For Scaliger he has only contemptuous indifference. Porta's assertion he put to specific trial with great elaboration and at no small expense, for he says he tested, before many witnesses, the frictional effect on iron of seventy-five diamonds, with the result of completely refuting the Neapolitan philosopher's averment. Still the doubt remains. Cardan had said to rub the diamond itself—not to rub iron with the diamond—and the attraction thus produced could not be explained away by inconsequent suggestions that Porta had been misled by a similarity in names, and had confused *adamas*, the diamond, with *adamas*, the lodestone. Besides, however much Gilbert might flout Cardan, or refute Porta, there remained the clear statement of Fracastorio, whom he knew to be neither a charlatan nor a mere transcriber, but, on the contrary, a philosopher of commanding eminence and fame.

In the end Gilbert probably rubbed some of his seventy-five diamonds and found Cardan and Fracastorio to be right. But, as he was not seeking to establish their reputations, he did not trouble himself to record the fact, but left the famous Italians pilloried with the other philosophers as "word-mongers" and "chattering barbers"—a species of comparative vituperation which came not unreadily from the student of Vesalius and Fallopius, already overflowing with fine scorn for the blood-letting and tooth-drawing knights of the lather and basin, who in England were contesting the right to practice surgery with the regular professors of the healing art.

The great point gained was not perception of the fact that something else beside amber would attract in the same way, but the proof of it. The immediately following questions were: are there any other substances having this same capability? If so, how many? Are they so few that the behavior of all can be lightly explained away as a *lusus naturæ*, and the general hypothesis so saved? Are they so numerous and of such importance that another theory, not inconsistent with the first, may be predicated, which will subsist concurrently by satisfying the peculiar physical conditions of the amber and its cognates, while not extending to the great cosmical application of the magnetic hypothesis? Are they so overwhelmingly many as to destroy the cosmic theory *in toto* by reducing its magnetic foundation to insignificance.

These, or like questions, I believe, led to the first deliberate, orderly effort to study electricity as a separate and distinct entity in the economy of nature. The second chapter of the second book of De Magnete opens with a characteristic onslaught upon the whole tribe of commentators, theologians and metaphysicians; or, perhaps more correctly, upon that variety of them who spent their lives in glossing one another's errors, or in spinning cobweb learning from their own brains and entangling their wits in self-contrived labyrinths. For especially keen reproach, however, are singled out the modern authors who had written about amber and jet because they had contented themselves with stating the attractive qualities in an occult way and never presented any experimental proof of them. This is sweeping enough to include Fracastorio, but whether it properly applies to Cardan, who, however occult he might have been in describing other things, was undeniably explicit and straightforward in his description of the amber (and who, moreover, in his De Subtilitate, makes a strong plea for more experimental proof than was customary among his congeners), may be fairly questioned. But, as Gilbert had evidently determined not to recognize Cardan in the matter of the diamond discovery, the casting of him into outer darkness, in respect to more debatable achievements, was not difficult. Hence, he makes no reservations in favor of the Italian or of any one else. All are embraced in one inclusive "they."

The famous announcement which begins the modern science is as follows:

"For not only amber and jet, as *they* think," he says, "attract corpuscles, but so also do (and now he sets first foot upon the great new field which still stretches so far before us) the diamond, the sapphire, the carbuncle, the iris stone, the opal, the amethyst, the vincentina, the English gem or Bristol stone, the beryl, rock crystal, glass, false gems made of crystal or paste glass, fluor spars, antimony, glass, belemnites, sulphur, antimony glass, mastic, lac sealing wax, hard resin, orpiment, rock salt, mica and rock alum."

It was an astounding discovery—this prevalence of the amber-soul. It meant that the spirit which men, through all ages, had supposed locked in the amber along with the dead flies and bees there imprisoned, had never been so confined. This was an Ariel which had not been bound in the cleft pine, now at last set loose by the magician's hand, but a sprite which had always been free to play a part among the things of heaven and earth undreamt of in man's philosophy. But Gilbert was no poet, nor ever "waxed desperate in imagination." Even when his inner vision pictured the eternal motion of the rolling spheres, their silent music never reached his thought. Besides, in the present instance, he was vitally concerned with the bedevilments of his theory, which seemed likely to follow; and a clear, practical and definite understanding of the physical cause was what he needed, and least of all any befogging of it by poetic imagery or idealization.

What could be more different than the substances which this force seemed to animate—what more contrasting than sulphur and the sapphire—or the true gems and the false? There were no such dissimilarities between the various kinds of lodestone, or even between the lodestone and the iron; so that the attracting capacity possessed by these involved no great diversity of substance. But here was attraction existing in bodies so totally unlike that to assert that all of them contained a primordial terrene magnetic Matter, would be to ascribe to that assumed substance a Protean capacity for change which would virtually argue it out of existence.

It was plain, therefore, that the amber quality was not something exceptional pertaining to the resin, but depended upon some cause hitherto unrecognized yet widely prevalent. Equally plain was it also to Gilbert, that so far from the difficulties of bringing this phenomenon into harmony with his magnetic hypothesis being diminished by the discovery of such prevalence, they were so greatly magnified as to render the effort obviously futile. A few years earlier it would have been easy to attribute everything troublesome to the influence of the stars or any other "occult" control, and, in fact even then, books on "the miracles of nature" jostled the commentaries on Aristotle on the shelves of every philosopher. But nothing could have been more repugnant to Gilbert than such a course. The amber effect, he saw, must be accounted for, and now, by an hypothesis which would be consistent with, though different from, the broad theory which, at all hazards, was to be maintained. Such was the path which now opened before Gilbert.

Far back in mediæval times there arose that curious divagation of the human mind, based, perhaps, in some degree, on the ascendency of the Aristotelian philosophy of words, of seeking to explain things not understood by giving to them new names. Later, this was carried to extremes by Paracelsus, and the same course has since been followed by charlatans generally. It was also in Gilbert's day the custom of the alchemists, and, to some extent, that of all scientific students, to hide discoveries and modes of operation in arbitrary words and phrases, often the merest gibberish, of which only the users knew the meaning. Thus there came into existence a pedantic terminology.

"A Babylonish dialect which learned pedants most affect,"

which invaded every department of knowledge and which, in some branches of science, though much modified and more logically conceived, still flourishes.

Gilbert, from his own professional experience, was well aware of the dangers which word-manufacture involved on the one hand, and the temptations which it offered on the other; for, no matter how sure his experiments and well-demonstrated his arguments, the necessary learning of a new vocabulary would be almost an insurmountable barrier to the very minds to which his appeal lay from the schoolmen and philosophants. But when he unearths matters that are genuinely hidden, and which must be identified somehow in speech, by marks which enable them to become the subjects of discourse, Gilbert has no hesitation in naming them, and the orbs "of virtue" and "of coition" already alluded to, are instances of such designations. These, with something of the same care which is found in the definitions of terms used in a modern British Act of Parliament, he groups together and elucidates in a separate and commendably brief glossary prefixed to his De Magnete.

The discovery of many substances partaking of the amber quality raised at once the need of a generic term including and fairly describing all, by which they might be spoken of and thought about without repetition and circumlocution. The property which all had in common was that of attracting corpuscles. And this attraction was not similar to that of the lodestone, but similar to that of the amber: similar, because, whatever its true cause might be, it was certainly ostensibly exerted in like manner to the amber attraction. Gilbert's treatise being in Latin, he frequently translates the English word "amber" by the Latin "electrum"—a derivative from the Greek ἤλεκτρον—and, on this basis, originates the term for the new genus. The word which he so coins is "Electrica"—translatable as "electrics"—which he defines as signifying "quae attrahunt eadem ratione ut electrum" (those substances which attract in the same manner as the amber). Thus the father of the science—by right of paternity—gave to it its name; for the subsequently-invented word "electricity" simply refers to the condition or state prevailing in an electric.[1]

I have now to outline the course of Gilbert's experimenting and the principal results which he achieved. Trying his electrics on many different substances, he soon reaches the conclusion that they will all attract, not only

[1] Further on I have noted the origin of other similarly derived words such as "electrical," etc.

straws and chaff, but metals, woods, leaves, stones, earths, even water and oil—"everything which appeals to the senses"—provided it be not aflame or in a too rarefied state. He is working from the vantage-ground of the isolated facts observed by others, and thus he moves beyond the implication of Fracastorio that the amber attracts only "hairs and twigs," and incidentally seizes a congenial opportunity to anathematize Alexander of Aprodiseus for drawing an absurd conclusion to the effect that the resin exercises an occult selection in attracting only the stalks and not the leaves of the garden-basil. In like manner he passes beyond the bounds of Cardan's discovery that the amber attraction may be cut off, and shows that a screening effect happens on the interposition of moist breath, a current of humid air, a sheet of paper, water, linen cloth, and the silk gauze known as "sarsnet."

He is not satisfied with merely stating that he has

GILBERT'S ELECTROSCOPE.[1]

proved all this by actual experiment. So anxious is he to avoid even the appearance of the prevailing mysticism, so careful to forestall any possible charge of concealing his mode of operating, so Faraday-like in his desire to leave behind him his ladder for the use of others to come, that he invites a repetition of his tests and a reverification of conclusions, and describes the simple apparatus which he has employed. He calls it a versorium—in modern terms it is an electroscope—made of a light metal rod centrally poised on an apex like the needle of a compass. It turns to the rubbed electric when the latter is brought near its

[1] From the first edition of Gilbert's treatise De Magnete.

end, and so shows the attractive effect. Before he devised this he seems to have made the electric draw to itself the attracted object bodily; but he found that the attractive force, in some substances, was too weak to overcome both inertia and frictional resistance, and the pivoted needle, the position of which a very small drawing force could easily disturb, was therefore contrived.

This is the first of all instruments depending upon amber-electricity. And again what is it essentially but the compass needle? Not the freely-movable magnet needle turning itself under the influence of the earth's magnetic field and yielding itself to the earth's attraction, but simply a freely-movable needle of any substance turning itself under the influence of the electric field of the rubbed amber and yielding itself to the amber's attraction. Here was the first electrical invention beyond the mariner's compass, the adaptation of the same physical means (the balanced needle rotating on its pivot) to the recognition of the fact of a field of force. Gilbert had shown how that colossal magnet—the earth—governed the compass needle, and how the same control was exerted upon the needle by the miniature earth—the terrella. Also he had shown that the excited amber would attract any substance provided the latter were light in weight and so within its exertable strength. A step further, and the excited glass or sulphur and the compass needle—the two things that lay respectively at the beginning of the new advance and at the culmination of the old—came together: and the needle, (immaterial whether magnetic or not, so long as it were light and easily controllable,) moved in response to the call of the electric.

By means of this instrument, Gilbert says, he detected his electrics, and thus suggests the amount of patient labor which he brought to the task. How many substances he procured, rubbed and carried to his needle, only to see it remain motionless, we can but surmise. In the list which he gives of things which are non-electrics, because they failed to move the versorium, are emerald, agate, carnelian, pearls, jasper, chalcedony, alabaster, porphyry, coral, marbles, coal, flint, bloodstone, emery, bone, ivory, hard woods (such as ebony, cedar, juniper, cypress), the lodestone, silver, gold, copper and iron.

It is no reproach to say that such experimenting was merely empirical. In the nature of things at the time, it could not have been otherwise.[1] He, doubtless, tried every available substance over and over again, making many an inconclusive test, until he discovered that a body might appear as an electric at one time and not at another, and that changes even in atmospheric conditions might easily lead to its entry into or exclusion from the electric category. Nor could he have found a much worse place for such researches than foggy London, where the prevailing dampness probably many a time frustrated his most careful efforts. At last, however, he learns that the best electrical effects are obtained when the weather is cold, the sky clear and the wind in the east, and that on overcast days when the breeze is southerly the indications of the quivering versorium are not to be trusted.

The unexpected revelation of so many substances partaking of the amber property made it plain that the field upon which Gilbert was now entering was wholly new and untrodden. That he had reached its border through the devious ways of his magnetic hypotheses, that his further advancement upon it would be but a digression from his chosen main path, that he had come to it in pursuit of a special object—all these considerations are immaterial. To all intents and purposes his advent as the first explorer might have been owing to any other influences, or to none, save the merest arbitrary selection of the amber attraction as an inviting subject for inquiry. Thus we reach a perception of the simple fact, clear of its surroundings, that

---

[1] The history of the development of some modern electric appliances is not altogether free from instances of a similar course commending itself to the nineteenth-century intellect.

here is a man at the end of the sixteenth century undertaking the study of a natural occurrence which had never before been systematically studied at all, and which no one understood.

How did he set to work? The ordinary course of procedure of the contemporary philosopher would be the gathering of a few isolated examples, not necessarily correlated, although ostensibly applicable to the same subject, and the making of a speculation or several speculations of more or less ingenuity about them. Nothing could differ more widely from this than the strikingly original course now followed by Gilbert. Despite the overwhelming authority of Galen and Avicenna, he brushes aside their guesses at the causes of attraction, as wholly inadequate to explain, and then, for the first time in the history of modern philosophical thought, he systematically gathers negative instances and undertakes affirmatively to discover and separate out the truth by proper rejections and exclusions—something which "had not been done or even attempted," says Bacon, " except perhaps by Plato."

The attraction of electrics he finds is *not* caused:

By heat, because heating alone, even up to the flaming point, will not produce it.

By a mode of operation analogous to that of the cupping glass, as Cardan suggests, because of the contradictory character of Cardan's own explanations, which we have already noted.

By the seeking of other bodies by the electric as food, because the attracted body would then diminish while the electric would grow.

By the attractive force of fire, because the non-electrics, when heated by fire or the sun, show no attraction.

By draught of displaced air (the cause assigned by Lucretius to magnetic movements), because that effect could not produce attraction in the open atmosphere.

By hot objects or by a draught of hot air, for neither an iron rod at white heat nor a candle-flame brought near the

THE NATURE OF THE ELECTRIC. 307

versorium, although the flame certainly produces a heated current, will cause the needle to turn.

By any peculiar property of amber or special relation between it and other bodies, because very many other substances partake of the same electric nature.

By similitude or likeness, because all terrestrial things, whether like or unlike, are attracted by the electric.

Nor has the electric attraction any resemblance to the drawing of moisture by plants, the purging of a morbid humor by a drug, the removal of water from a stoppered bottle when covered with a heap of wheat, or the mythical sucking up of water by elephants' tusks.

Then follows the list of solid non-electrics already given, and to this are added many substances which either fall to pieces or grow sticky by rubbing, such as pitch, soft resin, camphor, galbanum, ammoniacum, storax, asa, gum benjamin and asphaltum.

Having thus cleared the ground negatively, Gilbert proceeds to draw his affirmative conclusion as to the physical nature of the electric. The earth, he says, is made up of two kinds of Matter; moist and fluid, or watery, and dry and firm, or terrene. Any given substance consists either of both kinds of Matter or of a concretion of either kind. Amber and jet are concretions of water—so are all shining gems—and electrics generally have their origin in humor or watery Matter. This humor can even be driven out by heat and discharged as vapor. But electrics have certain necessary physical characteristics; namely, that they are firmly concreted so that they shine on being rubbed, and retain the "appearance and property of fluid" in a firm, solid mass. These conditions present, they attract all bodies, whether humid or dry, by a force which likewise has its origin in the humor.

The next step is to account for this attractive force. The attraction of the magnet, it will be remembered, he supposes to be due to its effused Form awakening an inert Form in the drawn iron, so that the thing attracting and the thing attracted mutually come together by a movement of coition. This effused Form (field of force) is wholly incorporeal. It is the animating energy, or as Thales looked upon it, the "soul" of the magnet.[1]

The attraction of the electric, however, he concludes to be due to a diametrically-opposite cause. The force is not awakened until the substance is rubbed, and then the substance is altered—that is to say, it attains a moderate heat, becomes shining or polished, and finally gives out an effluvium. This effluvium is corporeal—it is the original Matter in another condition—like a vapor that is given off from a fluid, or as if the body were dissolved into an exhalation.

Now as to the qualities of this effluvium, he says that the effect of moist breath, or a current of humid, atmospheric air, or a sheet of paper, or a linen cloth, interposed between the electric and the object attracted, is to choke its powers. Thus the electric differs entirely from the magnet, which attracts through any obstacle. Barriers such as the foregoing therefore act physically to stop the progress of the material electric effluvium, while they are perfectly transparent to the immaterial, effused, magnetic Form. In order to produce this effluvium, the heat generated in the body itself, not heat contributed by other bodies, must act; and a gentle and rapid friction must be used, not force applied violently and recklessly, to cause the finest effluvium to arise from a subtle solution of moisture—an exceedingly attenuated humor, much more rarefied than the ambient air. To explain how such a humor could be obtained from so dense a body as the diamond, he instances odoriferous substances which exhale fragrance for centuries; having in mind, perhaps, the still-persistent odor

---

[1] See Spectator, No. 56, May 4, 1711, for this same comparison. Addison describes Albertus Magnus as placing the lodestone on glowing coals and perceiving "a certain blue vapor to arise from it which he believed might be the substantial Form: that is, in our West Indian phrase, the soul of the lodestone."

of the musk which was mingled with the mortar in the building of the Mosque of St. Sophia in Constantinople.

How then does such an effluvium attract? Does it set the air in motion, and is the air-current followed by the attracted bodies; or are the latter directly drawn? If an air-current moves the objects, how can a minute diamond of the size of a chick-pea pull to itself so much air as to sweep in a corpuscle of relatively large dimensions, seeing that the air is drawn by only a small portion of one end of the stone? Clearly it is not the air which is moved, for then clearly the attracted body must stand still or move more slowly before coming in actual contact with the amber, on account of the heaping-up of the air on the surface and its rebounding after collision. And, furthermore, if there be a variation in the character of the effluvia, if they go forth rare and return dense (as with vapors), then clearly the body would begin to move a little after the beginning of the application of the electric. But—and here is the first statement of that marvelous speed of transmission which, in the telegraph and telephone, annihilates distance "when rubbed electrics are suddenly applied to the pivoted pointer—*instantly* the pointer turns."

New ideas now crowd fast one upon the other. The increased attractive power of the electric, as the attracted body approaches it, is recognized; the motion of the body is seen to be quickened, "the forces pulling it being stronger." At once Gilbert perceives the similarity in this respect between electric and magnetic attraction, and it seems that almost of necessity he must be led to interpret this as a most untoward result, tending to show the identity of the very phenomena which he was hoping to differentiate. But note how he dealt with it. Not only, he says, is this quickened motion, this augmenting force, true of the magnetic and electric attractions, "but of *all natural motions.*" The great generalization of the correlation, not only of magnetic and electric attractions with one another, but with the other forces of the universe, is here suggested

—a conception which, emanating from a mind of the sixteenth century, is an inspiration and a marvel.

Then he says that if the attracted body were moved by an air-current, it would remain in contact with the electric but for a moment. On the contrary, that this attractive power persists "sometimes for as long as five minutes, especially if the weather is fair." Such is the first statement of the electric charge.

That the amber does not attract the air, but the body, is shown by its drawing the particles on the surface of a drop of water into a cone, and not moving the whole drop. But this landed him in another paradox; for how could the electric thus attract water if, as he had already found, water directly applied to the electric destroyed its attractive power? So he concludes that it is one thing to suppress the effluvium at its rise, and another to destroy it after it is emitted. Hence, he discovers that to cut off the attraction completely it is necessary not merely to interpose a silk texture midway between the electric and the object, but quickly to lay it over the electric directly after friction. This is the first suggestion of insulation applied directly to the charged conductor—the prototype of the coating which covers the wires which convey the currents through our streets and dwellings, and prevents leakage of them on the one hand while guarding us from their dangers on the other.

While Gilbert's experiments often end in genuine discoveries, and involve conceptions far in advance of his time, it not infrequently happens that his deductions and conclusions are vague, speculative and obscure. This not only occurs when (as he says himself in his preface), after having described his magnetic experiments and accounted for the homogenic parts of the globe, he turns to the general nature of the whole earth, and then proceeds "to philosophize freely," but even in his statements as to what his experiments specifically prove. His notion of electric effluvia finds its true limit when he describes the emanation—very much as the Chinese Kouopho had done centuries before—as a breath proceeding from the electric and reaching to the attracted object. But when he essays to account for the actual movement of the latter, his explanation is based, not on the observed behavior of the electric, but on the gravitation of bodies or bubbles floating in water, which he believes come together through some effect of the liquid between them. Water, he considers, is a moist or humid link between the bodies, and so is the electric effluvium, although the last is much rarer, and all things come together because of humor. He fails to perceive that, even if the effluvia be regarded as material arms which permeate the air without moving it and grasp straws, etc., no explanation is thus afforded why or how these arms draw the attracted object.

Nevertheless, in his own mind, this theory was sufficient for the differentiation which he sought. And he sums this up finally by asserting that electric motion is one of matter toward concretion, while magnetic motion is that of arrangement and order; and thus he assigns to electric action the bringing and holding together of the materials of the earth, while to magnetism he believes the verticity or direction of the globe in space and also its rotation to be due. Ultimately he attempts to distinguish the characteristic natures of gravity, magnetism and electricity, while suggesting their generic resemblance. By gravity, parts of the earth are borne to it by natural inclination. By magnetism, bodies are borne to one another mutually. By electricity, corpuscles are carried to the electric.

I have dwelt upon Gilbert's theories because they serve to make clear the originality of the man in philosophic thought, and the onward momentum which he gave to it. Nor, if we are to accept the dicta of the apostle of the inductive method, is Gilbert's merit any the less because later and wiser generations may regard his speculations as to the magnetic relation of the planets as mistaken. "Truth," says Bacon, "emerges more readily from error than confusion." Better a wrong hypothesis than none at all.

But mistakes mislead, and erroneous theories obscure the vision for new discovery. Perhaps for this reason, perhaps because he did not regard them as of sufficient importance, in view of the object sought, Gilbert failed to observe many electrical facts which were well within his horizon. He knew nothing of electrical conduction. Magnetic conduction he realized easily under the assumption of the change in Form occurring throughout an elongated, magnetized body from one end to the other. But he never carried his electrical effluvia, even in imagination, through solids, nor, in fact, could he logically do so under his assumption that they were corporeal emanations capable of being dammed by a sheet of paper. He speculated somewhat concerning terrestrial electricity, but only as a means of uniting and holding the earth Matter. Electric repulsion he not only did not observe, but he denies its existence, asserting specifically that "electrics neither repel nor propel." Nor is this notion, for him, inconsistent, in view of his belief that the placing of like magnetic poles together was an unnatural disposition of them, which nature would proceed to set right. Magnetic repulsion was therefore merely a preliminary rotation of one magnet, so that both might come together "perfectly according to nature.".

The practical character of Gilbert's work is well indicated by the inventions which he makes. Thus he describes the first filar suspension of the needle of an indicating instrument, and even advises that silk filaments be used, twisted differently and not all in one direction, so as to eliminate the torsional effect: the first instrumental magnetometer (an iron versorium), and incidentally points out that the stone which from the greatest distance causes the needle to turn, is the best and strongest. He first determined the directive strength of a pivoted magnet, by noting the frequency and extent of its vibrations before coming to rest. His method of magnetizing iron is still in common use, and his counsels as to keeping compass-needles away from other magnets, and of placing all magnets, during storage, in definite position with respect to the earth's magnetic meridian are universally followed.

Finally the magnetic rocks—those mythical wanderers from distant Cathay, by way of the Red Sea to the Arctic Ocean—which Fracastorio had relegated to the hyperborean regions, and made them the sole cause of the northing of the needle—which Maurolycus had deprived of that high office, and imprisoned on a small northern island with no function save to disturb the compass—these, under Gilbert's magic touch, grew to fill the entire globe and lost their identity in the great earth-magnet.

Briefly recapitulated and freed from his astronomical theories, Gilbert's contribution to physical science, and to the philosophical advancement of mankind, was as follows:

He was the first: to investigate natural phenomena philosophically and systematically, and by a true inductive method, for he interrogated nature by actual experiment and from the particulars thus ascertained rose to correct generalizations; to recognize electricity (as distinguished from magnetism) as a new natural condition or force, and to study and name it; to extract the facts and laws of magnetism from the existing mass of speculation, mysteries and delusions, and to reduce them to a science; to suggest the correlation of gravity and magnetism with other natural forces, and a relationship between gravity, magnetism and electricity; to formulate a definite conception of the magnetic field of force, and to attempt to show its extent; to suggest the reaction between two fields of force, and mechanical motion of the inducing bodies resulting therefrom; to recognize that the earth is a great magnet, capable of magnetizing iron and iron ore by induction; to determine the magnetic polarity of the earth, and in the directive tendency thereof to reveal the true reason for the verticity of the compass-needle; to discover magnetic screening, conduction and saturation, the compound magnet, the mutual attraction or induction of lodestone and iron, the pole-piece or armature, the effect of induction on soft iron, and magnetization by molecular disturbance; and to discover electrical charge and its permanence for a considerable period of time, and that it can be retained by covering the excited body with certain substances. He invented the first electrical (as distinguished from magnetic) instrument, the first electrical indicating device, the first magnetometer, filar suspension, and the ordinary method of magnetization.

CHAPTER XI.

AFTER the death of the Queen, Gilbert was continued in his office of Court Physician by James I. He survived his royal mistress, however, by but seven months, his decease occurring in November, 1603. The year was a plague year, and London suffered with even more than usual severity; but whether Gilbert succumbed to that terrible disease or to some other malady is not known. His books, papers and collections, which he had bequeathed to the Royal College of Physicians, were all destroyed in the Great Fire. He was buried in Trinity Church, Colchester, where a tablet to his memory, bearing an epitaph far beneath his deserts and couched in doubtful Latin, still remains.

That Gilbert intended the De Magnete to be his final and greatest work, or that he designed submitting his discoveries and his assumptions to the judgment of the world only through its pages, is, I am persuaded, far from the truth. The concluding book of his treatise is at best but an outline of his cosmical theories; and, as the establishment of these was his chief aim, it is hardly supposable that he would have contented himself with so brief a statement of conclusions after so many years of experiment and study. The volume was edited and supervised while in press by Edward Wright, who at the time was a lecturer on Navigation for the East India Company,[1] and who takes occasion in the prefatory address to praise Gilbert's supposed discoveries concerning dip, compass variation and the finding of a ship's position at sea; so that it seems possible that Wright, because of his belief in the import-

[1] Ridley: Magneticall Animadversions, London, 1617.

ance of these practical achievements, induced Gilbert to give the work to the world before it had reached completion and with its parts disproportioned. That Gilbert designed making additions to it is proved by the single letter written by him now known to exist, which Dr. William Barlowe, Archdeacon of Salisbury, appends as a sort of testimonial to his own little essay on the magnet, which appeared in 1613. In this letter, which was probably written early in 1602, Gilbert speaks of adjoining "an appendix of six or eight sheets of paper to the book after a while," which was to be descriptive of some new inventions; and probably of two instruments for finding latitude at sea, which his friend, Thomas Blondeville, published and ascribes to him in a curious astronomical treatise entitled "the theoriques of the Seven Planets," which he produced in the last-named year. At all events, this addition to the De Magnete never was made; and Gilbert appears to have devoted himself to the preparation of a more elaborate exposition of his cosmical theories than that which terminates the earlier work. This he left, however, in a fragmentary state, only two books or divisions having been written; or, more properly speaking, sketched, for they show all the marks indicative of an intention to amplify at some future time. They probably bear a similar relation to the finished work as it would have been, as the two books of the Advancement of Learning to the final De Augmentis of Francis Bacon. This epitome, with the title-page belonging to it, forms the first part of the posthumous volume to which I have already alluded, and is called "A new philosophy of our sublunary world;" its contents being thus clearly distinguished in character from those of the De Magnete, which bears the general title of "A new physiology of the magnet, magnetic bodies and of the great magnet, the earth." The later work was manifestly intended to supplant existing cosmologies, and to inculcate the philosophy of the world's place in the universe which Gilbert believed that he had developed;

the De Magnete, on the other hand, contains the "certain experiments and demonstrated arguments," upon which the philosophy is based.

Appended to the new philosophy, is a treatise on meteorology "contra Aristotelem;" but this seems to be a distinct production, and not necessarily related to the first-named treatise.

I have referred somewhat at length to this posthumous work of Gilbert, which is now a literary rarity, because it has a remarkable history of its own, and because it forms the connecting link, so to speak, between Gilbert and Bacon.

It is but natural that the world should turn to the great English philosopher for the most authoritative of all contemporary estimates and opinions concerning the man whose fame waned amid his immediate posterity, and burst into brighter effulgence than ever three centuries after his death. With even keener expectancy does it seek to know how, at the hands of the apostle of the advancement of science, this new science of the magnet and of the amber found its impetus and promotion. I have yet to encounter any expressed opinion as to the manner in which Bacon dealt with Gilbert, which does not lay accusations at the door of the former, ranging all the way from a simple imputation of failure to understand Gilbert's magnetic and electric discoveries, up to direct charges of jealousy, malice and injustice; the characteristic common to all, however, being an absence of explanation of rational motive, so that one might well draw from them inferences not altogether consistent with the usual conception of Bacon's mental strength.

Throughout all of Bacon's philosophical writings there is no contemporary philosopher more frequently mentioned than is Gilbert; nor one for whose opinions Bacon shows any kindred respect. Even where he disputes and condemns Gilbert's conclusions, he leaves it in no doubt that they belong to Gilbert, and not to some Anonymous, for he writes Gilbert's name beside them. Nor does he satisfy himself with a mere expression of dissent, or even with a single bitter outburst of condemnation; but he comes back again and again, year after year, in his early works and in those written near the end of his life, always answering Gilbert, praising Gilbert, refuting Gilbert, condemning Gilbert—not Fracastorio, nor Cardan, nor Bruno, but Gilbert, "our countryman."

I shall now proceed to tell the history of the book which, as I have said, forms a connecting link between Gilbert and Bacon, and afterwards to examine the nature of the opinions which Bacon expresses regarding Gilbert's discoveries and hypotheses. In this way I shall endeavor to reach an understanding of Bacon's views and his reasons therefor, on which, perhaps, an impartial judgment of his course may be founded; and this, if throwing no new light on his character, may serve to heighten that with which some of its many sides are already illuminated. In this way also we shall see the working of one of the forces which for the time, so far from advancing the new science, tended rather to keep it in the slough of delusions and deceptions from which it was struggling to emerge.

The "New Philosophy" of Gilbert came to be published half a century after his death in the following curious circumstances. Within the period of apparently some two years after his demise, William Gilbert, of Melford, his elder brother, bearing, oddly enough, the same name ("nec sine causa ad rationes economicas spectante," says a later editor) found, among Gilbert's scattered papers, the fragmentary New Philosophy and the Meteorology. These (as he says, being governed by fraternal affection, as well as by an appreciation of the importance of the arguments advanced, whereof he felt unwilling to deprive the world), he arranged, caused to be translated into Latin, and prefixed to them a dedication to Henry, Prince of Wales, who died in 1612. That he intended to publish the book is clear; nevertheless, he departed, as its author had done, with his purpose unfulfilled.

In 1626 Bacon succumbed to the results of his ill-timed experiment in preserving chickens with snow, and bequeathed all his papers—saving his collection of speeches and letters—to his literary executors, Sir John Constable and William Bosvile—the latter better known as Sir William Boswell, sometime British Agent with the States of the United Provinces. The Bacon manuscripts were sent to Boswell's residence at the Hague, and there lay until Boswell, who died in 1647, confided them to the editorial care of Isaac Gruter, who culled from them nineteen essays and fragments, including the Cogitata et Visa, the Descriptio Globo Intellectualis, Thema Coeli and others, and published them all together in 1653. Among the papers which thus came into his hands, Gruter found the two manuscripts of William Gilbert, of Colchester, which William Gilbert, of Melford, had prepared, and these he edited and issued as before stated, in 1651.[1]

Gruter is unable to decide whether the treatises, thus brought to light, were written before or after the De Magnete. Mr. James Spedding, the learned biographer of Bacon, is of opinion that they were produced before 1604 "as the new star of 1572 is mentioned by itself, whereas later writers, as Bacon and Galileo, always couple it with the star in Ophiuchus first seen in 1604;"[2] and also conjectures that they are of later date than 1600, on the somewhat inconclusive authority of Bacon's remark[3] concerning Gilbert as one. who, "having employed himself most assiduously in the consideration of the magnet, immediately established a system of philosophy to coincide with his favorite pursuit." When the Meteorology was written is

---

[1] The Works of Francis Bacon, ed. by Spedding, Ellis and Heath, Vol. II., 196, Vol. V., 187, Boston, 1862.

[2] Ibid.

[3] Novum Organum, i., 54.

perhaps doubtful, but the internal evidence of the Philosophia Nova (presupposing, as its contents plainly do, a knowledge on the part of the reader not only of the magnetic but of the electric phenomena recorded in the De Magnete), leaves it, I think, beyond question that it was prepared after the writing, if not after the publication of the last-named work.

But the exact time of its production is of little moment. The significant fact lies in the possession of the manuscripts by Bacon during his lifetime. He studied them, he knew their contents. And in those great Monuments wherein he has invoked for his own fame the judgment of the next age, he attacks and condemns over and over again the opinions of a man who could neither speak for himself, being in his grave, nor be spoken for by the only written words wherein he had set them forth, and which, in the cabinet of my Lord Verulam, were as effectually silenced and entombed. The advocates of Bacon, who can reconcile his consignment of Peacham to the rack with the principles of natural law and the rights of the citizen which he so eloquently defended, may perhaps see in his dealing with the dead Gilbert's manuscripts no evidence of the meanness and baseness, of which others have professed to find in his character abundant proof. But possibly it may call for still further partisan ingenuity to discover the consistency of his suppression of this record of conclusions from an inductive research, and his severe strictures upon its author, with his simultaneous blazoning to the world of the value of the inductive method as the only means of discovering physical truth, and "hitherto untried."

The very persistence of his censure of Gilbert is of itself remarkable. Unlike the arraignment of Aristotle ("pessimus sophista"), or Galen ("canicula et pestis"), or Agrippa ("trivialis scurra"), or Paracelsus ("asinorum adoptiva") in the writings of his youth, which gave place to much more tempered expressions in those of his maturer years, the vigor and severity of the adverse judgments which he passes upon Gilbert's theories remained unabated from the beginning of his career to the end.

But to infer from the foregoing that Bacon's attitude toward Gilbert's achievements is always one of unqualified disapproval, is gravely to err. While the instances where he bestows praise are few, there are several in which he tacitly accepts the truth of Gilbert's discoveries; and if to this be added the further fact that toward one —and to us the most important—branch of these his real relation is substantially that of a passive disciple, it becomes evident that any correct conclusion as to the ultimate nature of his opinions must be based on careful discrimination between the matters to which he, at different times, refers. Between these, it is difficult to draw any precise dividing line which will enable us to say that with those on one side he wholly agrees, while he as completely disagrees with those on the other. No two categories can be framed in this respect which will not include serious exceptions. But, viewing all broadly, it will be found that when he acquiesces, it is in favor of Gilbert's direct conclusions from experiment; while on the other hand he seldom fails to condemn Gilbert's cosmical hypotheses and speculations. For Gilbert's chief effort, the attempt to base cosmical theories upon the outcome of magnetic experiment, his censure is without qualification; to that, every shaft of ridicule and disparagement is directed—it is vain, false, absurd, wrong in every particular —it is a generalization from wholly insufficient data—an attempt to build a ship from material not enough to provide the rowing-pins of a boat.

With this differentiation as a guide, we can now separate Bacon's opinions regarding Gilbert's magnetic and electric discoveries—which possess for us the more vital interest—from those which he formulates with reference to the broader, universal deductions.

He agrees with Gilbert in classing the lodestone as

among the things which work by the universal configuration and sympathy of the world—by the primitive nature of matter and the seeds of things—"by consent with the globe of the earth." Then, still following, he connects magnetism and gravity, the latter differing only in being by consent "of dense bodies" with the globe of the earth, and the magnetic motion "drawing both the iron to the magnet and heavy bodies to the globe."[1] He recognizes the production of the field of force; "immaterial virtues which pass through all mediums yet at determinate distances."[2] There is no doubt as to the signification which he attaches to the last phrase, for he asks himself the question "What may be distance?" and answers it almost in Gilbert's words "that which is not inaptly termed, orb of virtue, or activity."[3]

Gilbert's notion of the gradual diminution of the earth's attraction as bodies recede, he expressly affirms, adding that the downward motion "rises from no other appetite of bodies than that of uniting and collecting themselves to the earth (which is a mass of bodies of the same nature with them), and is confined within the orb of its own virtue."[4] His concurrence in Gilbert's idea of the earth's verticity takes the following, even cordial, form: "Now the diligence of Gilbert has discovered for us most truly that all earth and every nature (which we call terrestrial) that is not supple but rigid, and as he himself calls it robust, has a direction or verticity, latent indeed, yet revealing itself in many exquisite experiments north and south."[5] And again he agrees with Gilbert, whom he commends as having well observed it, that magnetic repulsion is not strictly an avoidance, but a conformity or attraction to a more convenient situation.[6]

[1] Nat. Hist., cent. x, 904, et seq.   [2] Nov. Org., B. ii, 37.
[3] De Augmentis, B. iii, iv.
[4] Des. Globi Intellect. Nov. Organum, B. ii, 35.
[5] De Fluxu et Refluxu Maris.   [6] Nov. Org., B. ii, 48.

But where an acceptance of any general theory advanced by Gilbert might lead, even indirectly, to a tolerance of the Copernican doctrine, which Bacon regarded as extravagant and claimed to be able to demonstrate as most false,[1] he is willing to go to great, if not illogical lengths, in his denials. He limits his sweeping endorsement of Gilbert's verticity doctrine by confining the assertion "to the exterior concretions about the surface of the earth and not extending it to the interior;" and then dissents from Gilbert's discovery that the earth is a magnet, which he ridicules as "hastily taken up from a very light fancy."[2] But observe the over-strained argument with which he supports this contrary opinion: "It is impossible that things in the interior of the earth can be like any substance exposed to the eye of man; for with us all things are relaxed, wrought upon and softened by the sun and heavenly bodies, so that they cannot correspond to things situated in a place where such a power does not penetrate." As Gilbert expressly says that lodestones vary in all degrees in purity, and hence in efficiency, through the primordial matter becoming more or less combined with other substances, it is evident that Bacon's answer to Gilbert is far from pertinent. Even more labored is his endeavor to avoid the conclusion of the earth's rotation, which he sees is liable to follow the admission of the verticity doctrine. "The upper incrustations or concretions of the earth," and not the whole sphere, he explains, "appear to correspond to the rotations of the heaven, air and water, as far as consistent and determinate bodies can correspond to liquids and fluids; that is, not that they revolve upon poles, but that they direct and turn themselves upon poles . . . so that the direction and verticity of the poles in rigid bodies is the same thing as revolving upon the poles in fluid," which may be left without further comment than that the most determined advocate of the Chancellor will probably find in it no higher evi-

---

[1] De Aug., B. iii, c. iv.   [2] De Fluxu et Refluxu Maris.

dence of genius than such as may attend an imaginable premonition of the relatively recent discovery of the flow of solids. Bacon's own opinion of it, as an argument, may perhaps be gathered from the following from the Novum Organum : "But if the motion of the earth from west to east be allowed, the same question (why bodies appear to desire peculiar situations) may be put; for it must also revolve around certain poles, and why should they be placed where they are rather than elsewhere? The polarity and variation of the needle come under our present head."[1]

Bacon's inefficiency in practical experimentation is so well known, that it need not be dwelt upon here. His little treatise of Inquiry on the Magnet is mainly composed of efforts to answer the questions which he suggests in the De Augmentis as subjects for experiment. They involve no special ingenuity, nor reveal any important discoveries. The principal conclusions are that the lodestone attracts steel-filings or its own dust as well as it does iron filings; a reverification of Gilbert's discovery of the effect of the iron pole-piece; that rubbing a magnet ("as we do amber") or heating it, does not increase its powers, and that the magnet attracts iron at equal distances through water, wine, air and oil. Perhaps the most interesting proceeding of all is the taking of a magnet to the top of St. Paul's Cathedral in London to see whether its power became diminished in consequence of its distance from the ground: another instance of the possibility of interconnection of gravity and magnetism making itself felt.

Despite Gilbert's electrical discoveries having been made in the course of a digression, it is clear that Bacon had by no means failed to perceive their novelty and importance. Among the "Physiological Remains" gathered by Tenison in 1679—the residue of the collection of Natural History notes and memoranda which Rawley had previously winnowed—there is a so-called catalogue of bodies attractive and non-attractive, written partly in English

[1] Nov. Org., B. ii, 48.

and partly in Latin, which it has been assumed, not infrequently, sets forth a series of electrical discoveries and experiments made by Bacon himself. The entire production, however, is merely an epitome of the famous second chapter in Gilbert's De Magnete, wherein the electrical matters are contained; no material fact being wanting, and the various facts being arranged in nearly the same order in which Gilbert presents them.

It is not an unreasonable inference that Bacon prepared this synopsis merely for convenience, intending at some future time to take up the subject of electrics for study; and this supposition gains support from his curt dismissal of the topic in his Natural History,[1] where he begins a paragraph as if he were about to discuss "emissions which cause attraction of certain bodies at a distance," but does nothing beyond excepting the lodestone from the category and noting his intention of considering "the drawing of amber and jet and other electric bodies" besides sundry other attractions under another title, which he appears never to have done. But if "imitation is the sincerest flattery," this is a shining example of it, which may justify the suggestion already made that in respect to this part of Gilbert's contribution to the world's knowledge Bacon's attitude is that of a disciple. He adds nothing to Gilbert's results—he does not dispute a single physical happening. But when he comes to the consideration of Gilbert's hypothesis of electrical action based on these experiments, then his inclination to dispute conclusions asserts itself. He will not accept Gilbert's assumption of effluvia—purely physical notion as it is. He prefers to go back to antiquity, and exhume one of those brain-spun abstractions, which it is his delight to condemn.

"The electrical operation, of which Gilbert and others after him have told so many fables, is none other," he avers, "than an appetite of the body excited by light friction which does not well tolerate the air, but prefers any-

[1] Nat. Hist., cent. x., 906.

thing tangible which it can find near by." No induction, and least of all, one based on precise rules, ever brought him to this conclusion. No conclusion ever contained, or was likely to lead to the acquirement of less of the "fruit" which the inductive method aimed to secure. It is easier to perceive in the assertion a *reductio ad absurdum* and a satire, than the inconsistency which otherwise obtrudes itself. He denied Gilbert's effluvia, and pointed his denial by suggesting, as a truer hypothesis, the notion of the magnetic appetite, which, none better than he knew, owed its existence to nothing but "the sterile exuberance" of ancient thought.

Macaulay likens the speculations of the old world in the realm of natural philosophy to ploughing, harrowing, reaping and threshing, with no better result than to fill the garners with smut and stubble. Bacon is fond of the parable of the farmer who directed his sons to dig in the vineyard for hidden treasure—the gold not being found, but the cultivation vastly increasing the yield of the vines.

His opinion of Gilbert's work accords with Macaulay's analogy, for he believed that it yielded no valuable harvest—on the other hand it falls within his favorite allegory, for Gilbert's digging—the experiments on the magnet and the amber—was in itself admittedly good and valuable. It was to him as if Gilbert had ploughed and harrowed to improve soil which had yielded, not grain, but weeds—not the vine loaded with bursting clusters, but the malignant creeper luxuriant with poisonous foliage. To Bacon the Copernican theory was a pestilent thing. Gilbert's tillage of the land could make it none the less noxious—rather the contrary : far less could he convert it into the fruitful vine. Nor, to change the figure, could the stones which Gilbert quarried suffice for a monument reaching to the skies. His pile ended in clouds—not in the heavens.

In distinguishing between Gilbert's physical discoveries and his cosmical speculations, Bacon regards the latter as of the higher import; and, in so doing, follows in the steps of Gilbert himself. It must be remembered that Gilbert's aim was not primarily the making of electrical and magnetic discoveries, but the establishment, through such means, of a great theory of the physical structure of the universe; that the actual facts proved by these experiments and Gilbert's application of these facts to support his hypotheses, were two entirely different matters. The last we have already seen to be in many particulars inconclusive and obscure. Nor was the general acceptance of the Copernican theory in any wise promoted by Gilbert's arguments; nor do the latter enter into any modern astrophysical doctrine; nor does any one maintain them now. What other position toward them could Bacon have taken, convinced, moreover, as he was of the error of the heliocentric theory, than that which he assumed? A false doctrine bolstered by wrong interpretations of experiments cannot be made true in the mind of any rational being so believing, by establishing the accuracy of the experiments *per se.*

That Bacon saw in Gilbert's hypotheses a flagrant example of the very errors resulting from incorrect generalizations, away from which he was seeking to lead the world, furnishes a probable reason both for the severity of his censure and the persistence with which he repeated it. To call Gilbert an empiric and a maker of fables, was merely to indulge in a style of vituperation in which he was far excelled in point of picturesqueness, vigor and fecundity by Gilbert himself, and besides to follow a fashion of the times, whereof the irascible daughter of King Henry was no weak exemplar. But Bacon's strictures were rarely in the form of hasty invective. They were painstaking—and years often elapsed before he found expressions for them which seemed to him entirely satisfactory and adequate. Of this peculiarity, two prominent instances are worth noting as typical.

In the Advancement of Learning, published in 1605, he

says that "the Alchemists have made a philosophy out of a few experiments of the furnace, and Gilbert, our countryman, hath made a philosophy out of observations of the lodestone"—this being in illustration of the proneness of humanity to generalize upon insufficient or incomplete data. The same statement is repeated in the Novum Organum, published fifteen years later, and in the De Augmentis (1623). And, finally, in the History of Heavy and Light Bodies, which did not appear until after Bacon's death, it takes a more severe form, declaring that Gilbert "has himself become a magnet; that is, he has ascribed too many things to that force and built a ship out of a shell." Apparently, it took Bacon as long to reach a final formulation of this judgment as it did Gilbert to make all the experiments in the De Magnete.

Another even more curious example forms a part of his attack on Gilbert, especially as a Copernican. In the Advancement of Learning, he speaks of the establishment of a Calendar of Sects of Philosophy, in which he proposes the setting down of the philosophy of "Gilbert, our countryman, who revived, with some alterations and demonstrations, the opinions of Xenophanes." The opinions of Xenophanes, who was the founder of the Eleatic school of Greek philosophy, concerning astronomy, were extravagant in extreme; but, as they included a wild speculation involving terrestrial rotation, he is commonly mentioned among the ancient prototypes of Copernicus. Bacon's statement, of course, had no foundation in fact, and was derisively intended. This is repeated with odd variations. In the Cogitata et Visa, we are told that "our countryman Gilbert," in order that he might examine the nature of the magnet, constantly sought, with great firmness, constancy of judgment and many experiments, "to start new sects in natural philosophy; nor did he hesitate to turn into ridicule the name of Xenophanes, to whose opinions he himself inclined." In the Redargutio Philosophorum this is changed to read that he turned the name of Xenophanes into Xenomanes, an allusion which finds its explanation in the History of Life and Death, wherein Bacon describes the Greek as "a man who wandered no less in his mind than in his body, so that, in consequence of his opinions, his name was changed from Xenophanes to Xenomanes." But in the De Augmentis, in a paragraph similar to that originally in the Advancement of Learning, Xenophanes is dropped out of sight, and Gilbert is charged with reviving the doctrines of Philolaus. The strength of the judgment which, while persisting over twenty years, can exercise such a keen discrimination is sufficiently apparent.

The Philosophia Nova of Gilbert contains, as I have already pointed out, the most comprehensive statement of his cosmical and astronomical views. The Meteorologia deals more particularly with natural phenomena, such as comets, the winds and tides, and the rainbow. To both of these works Bacon often refers. Thus, in the Descriptio Globo Intellectualis, he mentions Gilbert's notions of the revolution of the stars, the vacuum in the interstellar space, the scattering of opaque globes through the heavens, and, with especial approval, his mapping of the moon and his conceptions concerning gravity. In his History of the Winds he draws so freely upon Gilbert's chapters on the same subject that Gruter notes upon the margins of the Meteorologia the places whence he has taken his extracts. In fact, even in the absence of knowledge of the discovery of the Gilbert manuscripts among Bacon's literary remains, there is abundant evidence to show that he was at least one of the distinguished men whom Gruter says had access to Gilbert's writing in its unpublished form.

It is not necessary for the purpose of this work to extend this review of the relations of Bacon and Gilbert beyond the present limit. That Bacon recognized Gilbert's eminence as a philosopher and as a discoverer is clear. He certainly regarded him in the light of "a foeman worthy of his steel." That he was governed in his censure by personal animosity it is needless to assume, in view of the existence of other and wholly impersonal considerations of ample strength. His suppression of Gilbert's manuscript is a part of that "checkered spectacle of so much glory and so much shame" which makes up his life.

As to the statement that is often made that Gilbert practiced the inductive method before Bacon presented it to the world, some discrimination is requisite. The greatest philosophical critics have never agreed as to the exact nature of the induction which Bacon sought to engraft upon human thought, and therefore it would be presumptuous here to seek its definition. If his philosophy advocates induction as a mode of reasoning only in a broad and general way, it but follows Roger Bacon, and, more closely, Leonardo da Vinci. Indeed, the words of the great Italian, "My design is first to examine facts, and afterwards to demonstrate how bodies are constrained to act. It is the method that one must adhere to in all research into nature" . . . find a better application in the experiments of Gilbert than in the aphorisms of Bacon. "Recent induction, that of Mill and Whewell, Herschel, Faraday and Darwin," says Professor Nichol, "is the means by which great sequences of nature, called laws, are investigated by the aid of apt conjecture and by careful verification established;"[1] and such was the induction which led Gilbert to the conclusion that the earth is a great magnet. But this, according to the same authority, is not Baconian induction, for Bacon aspired to penetrate into the inner nature of things, and so hold them in command by the aid of a method which, from its exhaustiveness, he held to be as certain in its results as a demonstration of Euclid; a "conclusion of necessity," so mechanical that when once understood all men might employ it, yet so startling that it was to be as a new sun to the borrowed beams of stars; a method compared by its author to a compass which equalizes all hands, and enables the most unpracticed person to draw a more correct

---

[1] Nichol: Francis Bacon, his life and his philosophy. Edinburgh, 1889, ii., 181.

circle than the best draftsman can without it, and which is to level all abilities, and eliminate intellectual acuteness and the play of genius in the solution of the problems of nature. If such be the Baconian inductive method, Gilbert never practiced it, and it may be questioned whether any one has ever done so. That Gilbert, however, pursued the inductive method as truly, in kind, as it is followed in the scientific thought of to-day, seems beyond dispute.

It has been suggested, in order to account for Bacon's attitude not only toward Gilbert but toward Copernicus and Harvey, that he did not, in reality, initiate modern philosophy, but closed the philosophy of the Middle Ages.[1] Nor is this altogether at variance with the view taken by Lord Brougham in his fine summing up of the Baconian achievements,[2] with direct reference to Roger Bacon, Da Vinci and Gilbert, as the generalization and extension of their modes of investigation "to all matters of contingent truth, exploding the errors, the absurd dogmas and fantastic subtleties of the schools." So, in estimating Bacon's part in the intellectual rise in electricity, we find him near the boundary between the old and the new philosophy, and apparently influenced by the old mode of thought as well as by the new. Toward the science as Gilbert begun it, his position appears to have been interpreted by his contemporaries and immediate successors as one of disparagement, and for a time this acted to retard progress; while his failure to do Gilbert justice, certainly savors of mediæval intolerance. But in so far as he led the world to the investigation of all physical phenomena by direct experiment and correct induction, he became ultimately a power mightily working for the advancement of knowledge in the new field.

[1] Erdmann, History of Philosophy, London, 1890.
[2] Brougham, Address on Unveiling of Newton's Statue, 1855.

CHAPTER XII.

THERE is no period in the annals of England which is more captivating to the student than that which includes the years which close the reign of Elizabeth, and those immediately following the accession of James I. It was at this time, we are told, that there came a wonderful awakening of the national life, an unexampled increase in opulence, refinement and leisure. It was then that the glory of the new literature burst forth; and imagination, winged by the genius of Shakespeare, soared to its supremest height. Then the sails of Britain swept over the furthest seas, and the romances of the old minstrels became dull and vapid beside the tales which the weather-beaten mariner brought back of the flowery lands and golden shores which, beckoning so seductively, set the staid trader of foggy London aflame with cupidity and with enterprise. Then, it is said, arose a new impulse to classical study and a passion for the master literature of Greece and Italy. English commerce increased and wealth poured into the land, bringing with it new luxuries and a new demand for wines and jewels and rich apparel and sumptuous equipage and costly dwellings. The huts of "sticks and mud" which the followers of Spanish Philip had declared the peasants' hovels to be, gave place to houses of stone and brick; the grim and battlemented walls of feudal times to mansions graceful and beautiful, embowered in smiling gardens and decorated with the exquisite refinement of Italian art.

Such, briefly, is the picture, so often shown, following the recital of the great sea victory and the story of the years of fear and suspense and stagnation which preceded it, until it seems as if the smoke of the guns of Hawkins and Drake and Frobisher, like the gauze of the theatre, had obscured the stage when the fortunes of the play were darkest, only to be swept aside to reveal the glory of England's transformation.

But national progress does not depend solely upon the growth, however remarkable, of polite literature, nor even of commerce. It finds another and potent aid in the labors of the investigator and the inventor in the diffusion of a knowledge of physical science among the people, and in an environment wherein discovery and invention are certain to be appreciated, stimulated and fostered. As will now be seen, the intellectual conditions which existed in England at the beginning of the seventeenth century were far from favorable either to the development of inventive genius or the encouragement of physical inquiry.

Whatever of scientific knowledge there was in the country was restricted to the physicians, and to perhaps a few individuals who, like Lord Arundel, built for themselves huge magnets and other apparatus merely as playthings. Certainly it was not to be found in the Universities. Oxford and Cambridge were under the rule of the Star Chamber. Bruno describes the Dons, despite their gorgeous robes and insignia, as "devoid of courtesy as cowherds." Student life combined the seclusion of the monastery with the riotous dissipation of the tavern. The Protestant sects wrangled ceaselessly among themselves, or combined their jarring forces against Rome. Faith in Aristotle, so greatly weakened abroad, here stood in unimpaired vigor, and those who had not drunk deep at his fountain, were denied, by statute, a degree either in philosophy or theology.[1] There was no suggestion of new advance which was not flouted, no tolerance save for end-

[1] Official statutes declared that Bachelors and Masters of Arts who did not faithfully follow Aristotle were liable to a fine of five shillings for every point of divergence or for every fault committed against the Organon. Bruno wittily called Oxford "the widow of sound learning." Lewes: Biog. Hist. of Philosophy. New York, 1857.

less quibbling over names and words. The learning of the realm, as Bacon said, was but "an infinite chaos of shadows and moths wherewith our books and minds are pestered."

The mob detested foreigners and all their ways. The aristocracy aped the Italians under what Ascham called the "enchantment of Circe brought out of Italy to mar men's manners in England," in everything except education. In 1605, Nicholas Peiresc, a Frenchman of great learning, visiting England, finds nothing more worthy of record than a discussion with Camden as to the meaning of the names of French towns, and a summons from the King to relate the story of a drinking match.[1] In 1615 Scioppius denied that James could collect twenty learned men in all the realm. Brilliant rhetoric, casting the glamour of romance and poetry about the Elizabethan Age, may obscure the fact that the existing state of learning was one of degradation; but it cannot destroy the truth of it. Neither can a recital of the varied attainments of the Queen, and notices of the erection of new grammar schools and of increased interest in the ancient classics among the word-spinners, serve to make that which was in the mire appear to have been in the clouds.

"The reign of Queen Elizabeth," says the ingenuous Thomas Sprat in 1667, summing up the true condition of learning in Elizabethan and Jacobean England, "was long, triumphant, peaceable at home and glorious abroad . . . but though knowledge began abundantly to spring forth, yet it was not then seasonable for experiments to receive the public encouragement, while the writings of antiquity and the controversies between us and the Church of Rome were not fully studied and despatched. The reign of King James was happy in all the benefits of peace, and plentifully furnished with men of profound learning, but, in imitation of the king, they chiefly regarded the matters of religion and disputation, so that even my Lord Bacon, with

[1] Gassendus: The Mirrour of True Nobility and Gentry. London, 1657.

all his authority in the state, could never raise any college of Solomon but in a romance."[1] Such were the times during which the announcement of the electrical discoveries of Gilbert appeared.

As a Copernican, Gilbert, in his own country, had few co-believers; and as he had not merely linked his physical researches to the heliocentric doctrine, but had sought to substantiate the latter by them, it followed for this reason that his entire work stood discredited in the eyes of English scholars generally. But even if he had not adhered to the new theory, it may well be doubted whether there was sufficient knowledge of physical science existing in England to secure for his magnetic and electric discoveries even a superficial understanding by the learned classes. So far as written records prove, there were but two men in the kingdom, both his personal friends, who, had any special attainments in matters magnetical. These were Edward Wright[2] and William Barlowe,[3] and even their interest in the subject was mainly utilitarian, and depended upon the belief that Gilbert had discovered some new navigating instruments and simpler methods than were in existence for finding a ship's position at sea.

Gilbert had no practical knowledge of navigation, and his sea voyaging had begun and ended with the crossing of the English Channel when he made his continental tour. Wright, on the other hand, was probably the most skillful sea mathematician in all England. He had made long voyages, even to South America. He had plotted new charts and corrected old ones, and had even become involved in a dispute with the famous Gerhard Mercator, wherein he claimed the maps, made on what is now known as Mercator's projection, to have been of his own first devising. He had invented new methods of solar observa-

[1] Sprat: Hist. Roy. Soc. London, 1667.
[2] Bibliographica Philosophica.
[3] Wood: Athenæ Oxonienses, 1813; Biograph. Britann.; Le Neve: Fasti. Eccl. Anglia. Ed. Hardy; Stephen: Dicty. Nat. Biog. N. Y., 1885.

tion, had written books on navigation, was lecturer on that subject to the East India Company, and ultimately became a tutor of the Prince of Wales. Wright actively assisted Gilbert, not only in the gathering of material and the editing of his treatise, but is said to have prepared the twelfth chapter of the fourth book, in which there appears a table of the fixed stars.[1] He also wrote the address to Gilbert which is prefixed to the De Magnete. It seems not unreasonable to assume that his belief in the practical importance of the instruments which Gilbert describes caused him to advocate speedy publication of the work, in order to bring them into the hands of the merchant adventurers and navigators as soon as possible, and this may account for the brevity of the final chapters, wherein Gilbert develops his cosmical theories. For these speculations Wright, although a Copernican himself, had little fancy, and merely mentions them perfunctorily in his preface.

While Wright was a navigator who had learned his art at sea, Barlowe was one who believed himself to have acquired it in cathedrals. In 1597 he published a book entitled the Navigators' Supply, dedicated to the Earl of Essex, wherein he ingratiates himself with the sea-faring man by the following remarkable preface: "Touching experience in these matters (compasses, etc.) I have none. For, by natural construction of body, even when I was young and strongest, I altogether abhorred the sea. Howbeit, that antipathy of my body against so barbarous an element could never have hindered the sympathy of my mind and hearty affection towards so worthy an art as navigation is; tied to that element, if you respect the outward toil of the hand, but clearly freed therefrom, if you regard the apprehension of the mind." But the refreshing naiveté of this is even surpassed by his effort to neutralize its effect by claiming the especial consideration of the reader for his book because it "Was written by a bishop's sonne, and, by affinitie, to many bishops kinne"—

[1] Ridley: Magnetical Animadversions. London, 1617, p. 11.

thus betraying the innocent belief that the accumulated ecclesiastical influence which he wielded, because of his filial relation to one bishop and his fraternal relation to the four others whom his sisters had espoused, would secure for him, from the briny mariner afloat, the same sort of favor which, ashore, finally landed him in the comfortable Archdiaconate of Salisbury.

Barlowe, however, was far more deeply interested in Gilbert's magnetical experiments than Wright, because he was making similar researches himself. It has been claimed[1] for him that he had "knowledge in the magnet" twenty years before Gilbert's book appeared, and that he was accounted superior, or at least equal, to Gilbert as a "searcher and finder out of many rare and magnetical secrets;" but there is nothing to substantiate this in anything Barlowe ever published. He certainly was in no hurry to give to the world either his own magnetical researches, or to express his approval of those of his friend. He owed his earlier advancement to the friendship of Essex, whom, to his credit be it said, he did not desert in adversity, and to whom he ministered even on the scaffold; and then, in the next reign, he became chaplain to the Prince of Wales, on his way to his final preferment; so that, even if he had been a Copernican at heart, which he was not, it would have been to the last degree impolitic for him to have rushed into an endorsement of a work wherein the proscribed theory was so strenuously maintained. Besides, in 1605 came Bacon's earliest fling at Gilbert—the first English criticism of the De Magnete from an eminent source—and that this had a deterrent influence upon him may also be conceived. But there was a great deal of human nature in Barlowe, revealing itself with more than common transparency. He did not dare, in 1600, to challenge Gilbert's priority to himself, nor even then to make public his own alleged discoveries; but when, in 1613, another Richmond suddenly leaped into the field in the

[1] Wood: Athenæ Oxonienses, London, 1813, Vol. II, 375.

person of Mark Ridley, he rose in arms. It is of no moment to him that Ridley tells the world nothing more than Gilbert had already told it a dozen years earlier, and that he is merely attacking, at too late a date, Gilbert over Ridley's shoulders. He disputes Ridley *in toto;* and thus begins the first of the many controversies which have involved every material discovery and invention in electrical science.

Little is known about Ridley beyond his own description of himself on his title page as "Doctor in physicke and Phiilosophie, Latly Physition to the Emperour of Russia and one of ye eight principals or Elects of the Colledge of Physitions in London;" but his book is remarkable for its peculiarly practical advice and for the recommendation, in the preface, that the reader should provide himself with "such like forms of Magnets as I have described . . . as also of needles, wiers and waights of iron and steel," upon the procurement of which "*then* thou mayest read and practice the operations and demonstrations of this book."

There is a vast difference between this counsel, followed by pages of detailed instructions and pictures, and the apology which, not very many years before, prefaced Robert Norman's work. There it was feared that magnetic matters "may be said by the learned in the mathematicalles" to be "no question or matter for Mechanician or Mariner to meddle with," and it was begged that they "do not disdainefully condemne men that will search out the secrets of their Artes and Professions and publish the same to the use and behoofe of others."

Ridley's treatise, in the main, however, is in substance but an amplified review of most of the magnetic experiments which Gilbert records in the De Magnete. The Copernican doctrine is accepted somewhat hesitatingly, and with less reservation Gilbert's affirmation of the magnetic nature of the globe.

As allusion is also made to his cosmical notions, including the supposed magnetic attractions of earth, it appears that Ridley was also one of the illustrious men who Gruter says had access to the manuscript of Gilbert's posthumous Philosophia Nova, of course before Bacon suppressed it.

It is but just to Barlowe to state that he claims to have written the work which appeared in 1618 in reply to Ridley, some seven years earlier, and that the manuscript, having been delivered to his chosen patron, Sir Thomas Challoner, was, as he says, "either mislaied or embeseled." The book, as published, was dedicated to another politician, Sir Dudley Digges. Barlowe compares him to the magnet, because he thinks Digges maintains "so pleasing a carriage toward everie man, as causeth all good men which know you to love you by force of a natural sympathy," which was a new use of the old metaphor in its application to a politician, and confers upon the astute Ambassador of Elizabeth the honor of being the first of modern "magnetic statesmen."

I shall not pause to examine the magnetic experiments which Barlowe records, for they augment but little the facts already known. Nor does his brief reference to electric phenomena add anything, except the word "electrical," to the language. In fact, he translates Gilbert's "electrica," as "electricall bodies," and not "electrics;" and speaks of "electricall attraction," which he says is in "infinite other things both naturale and compound" besides those noted by Gilbert. But he gives no additional names of electrics, nor, despite his alleged extension of Gilbert's observations, has he the slightest notion of electrical repulsion or conduction.

Barlowe's assertion that his thunder had been stolen, provoked from Ridley a prompt and caustic reply under the title of Magneticall Animadversions,[1] in which Ridley avers that there is not a fact in Barlowe's treatise that was not well known long before his first manuscript was given

[1] Cit. sup.

to Challoner, and that what Barlowe had not purloined from Gilbert he had filched from him. But Barlowe is still eager for the fray.

"Except this Ridley had ploughed with my Heifer hee had not known my Riddle," he rejoins,[1] after asserting that Ridley had surreptitiously obtained a manuscript copy of his book, and identifying with careful precision much of his stolen property in Ridley's pages. And then, after sating himself with verbal scarification of Ridley, he essays to meet the acrimonious demand of the latter that he should state unequivocally the precise inventions he claims to have made, and specifies improvements in hanging the dipping needle, the magnetical difference between iron and steel, "the right way of touching magneticall needles," the piecing and cementing of lodestones, and that a "Loadstone being double capped must take up so great weight;" which we may pass by mainly because those which are of importance are not Barlowe's, and those which are probably his are not important.

The best thing Barlowe did was to draw a clear line between Gilbert's magnetic discoveries and Gilbert's cosmical theories, by distinctly affirming the first and as distinctly disaffirming the second—"Entreating of the motion of the earth," he says, "I think there is no man living further from beleeving itt than myself," thus setting himself right with the Anti-Copernicans; and then reconciling his apparent simultaneous belief and disbelief in Gilbert by quoting "Amicus Socrates, Amicus Plato, sed magis amica veritas" and "Nullius addictus jurare in verba Magistri." Ridley drew no such line because he was himself Copernican. It will be remembered that Bacon also separated Gilbert's discoveries and hypotheses, and that it is only after perceiving that fact that his diverse criticisms can be mutually reconciled; but unlike Barlowe he left the dividing boundary hazy and obscure.

[1] Barlowe: A Briefe Discovery of the Idle Animadversions of Marke Ridley. London, 1618.

Such was the reception which was accorded the Gilbertian discoveries at home; five years of neglect and probable ridicule, then Bacon's initial attack, then plagiarism of them, and finally a wrangle between the appropriators. Besides, and almost at the outset, there came from Holland the sneering comment of Scaliger the son—then professor of Belles Lettres at the University of Leyden—swift to repay the sharp criticisms of Gilbert upon the vagaries of Scaliger the father. "A certain Englishman produced a book on the magnet three years ago," he writes to Casaubon, "which has not justified the expectations formed of it." "It proved to be more his doctrine," he said at another time, "than the nature of the magnet.."[1]

> "Stare negas Terram: nobis miracula narras:
> Hæ cum scribebas: in rate forsan eras,"

was the sneering epigram written by John Owen.[2]

Between the state of learning in England during the period before noted and that existing in Italy, the contrast is impressive.[3] In the latter country, at the beginning of the seventeenth century, peace had reigned unbroken for forty years and as a consequence the advance in all the arts of civilization had been rapid. The universities of Bologna, Padua, Pisa and Pavia were attracting larger numbers of students than ever before, and of these no small proportion were devoting themselves to mathematics

---

[1] Scaliger: Epist. 200, and Epist. ad Casaubon. For these and other criticisms of Gilbert see Blount: Censura Celebriorum Authorum, Geneva, 1710.

[2] "This firm-set earth, you do deny.
   Perhaps when this you wrote
   'Twas not the sky that sailéd by,
   But only you, afloat.

[3] See Hallam's Literature of Europe, vols. II and III.; also Robertson's Fra Paolo Sarpi, London, 1893.

and medicine. While from such centres flowed whole rivers of learning, there sprang countless rivulets from the societies and academies which arose all over Italy. The purely literary gatherings which had met for many years and which had become a part of the social life of the country, were now promoting scientific culture and the interchange of philosophical thought. Such, for example, were the ridotti of Andrea Morosini the historian, of Paolo and Aldo Manuzio, "princes in the art of typography," and of the famous merchant Sechini, in Venice. And, of all, perhaps the most famous was that which Gian Vicenzo Penelli held in his magnificent house in Padua. Here the discussions took place between such intellectual giants as Fra Paolo, Galileo, Santorio, Fabricius of Acquapendente, Alpino, Mercuriale, Ghetaldo, Antonio de Medici and Fra Fulgenzio. Here was one of the finest libraries and scientific collections ever gathered by private munificence—a treasure-house of rarities, of globes, maps, mathematical instruments and fossils—presided over by a man who had made its establishment a labor of love, and had devoted to it a great fortune; in order, as Peiresc reports after visiting its marvels, to furnish "all the learned men of the age, both far and near, with such books and other things as they stood in need of."[1] Gian Francesco Sagredo maintained another museum in Venice, his house resembling a Noah's Ark, having in it, as he tells us, "all manner of beasts."[2] In Milan were the magnificent mineralogical and zoölogical collections of Aldrovandus. Finally there was the Academy of the Lyncei or Lynxes (so called with reference to its desire to pierce lynx-eyed into the depths of truth) devoted especially to physical science, and, although founded in 1603 by Frederic Cesi, then a boy of eighteen, soon numbering among its members such men as Porta, Galileo and Colonna.

In a country imbued with so great a taste for learning,

[1] Gassendus: Mirrour of Nobility, cit. sup.
[2] Celeste: Private Life of Galileo, Phila., 1879.

and possessing men whose attainments placed them far in advance of all other European scholars, it would have been indeed strange if the announcement of such discoveries as those of Gilbert had failed to arouse the liveliest interest. Nor did the acute minds of the Italians long delay the separation of the wheat from the chaff, for they quickly saw that Gilbert's recognition of the magnetic property of the earth, and the experiments underlying this discovery, were of far greater scientific importance than his notions as to the structure of the heavens.

Gilbert's treatise must have reached Italy with remarkable celerity for those days. In his letter, dated February 13th (presumably), 1602, which Barlow publishes, he speaks of being in direct epistolary communication with Sagredo, and says "that he hath conferred with divers learned men in Venice and with the readers of Padua, and reporteth a wonderful liking of my book." This was the verdict which Gilbert wanted—the praise of the men the extent of whose learning he knew and whose ability he honored. Beside this, the contemptuous silence of his own countrymen became a matter of indifference.

The name of Gian Francesco Sagredo has been rendered immortal by Galileo, who adopts it as that of one of the participators in his famous Dialogues. Nor was this distinction solely due to the fact that Sagredo, perhaps beyond all others, was the beloved disciple, as well as the ardent adherent and benefactor, of the great philosopher. He was a Venetian patrician, endowed with an ample fortune, which he spent profusely upon his collections of apparatus and curiosities. He had already studied the magnet and knew all that Fracastorio and Cardan had written concerning it. He was in immediate touch with Fra Paolo Sarpi, as well as with Galileo—the former the greatest representative of "the learned men of Venice," the latter a "reader of Padua."

In the fall of 1602, Sarpi is said to have written to Galileo referring directly to Gilbert's discoveries, and asking for explanations of them. But concerning this fact, as in regard to most other evidences of Sarpi's knowledge of magnetism, doubt has been thrown. Writers, probably influenced by the Church in its bitter hostility to Sarpi, insist upon the authenticity of the letter, and claim that it proves that the great Venetian had no such attainments in physical science as his advocates aver, and that hence Galileo's famous reference to him as "my father and master" should be interpreted as indicating that the Friar was merely the philosopher's spiritual guide.[1] The Florentine Society, in publishing the collected correspondence of Galileo, however, reject the communication as probably not written by Sarpi.[2] So, also, Sarpi's relations to Gilbert have been very differently regarded. Some biographers even assert that Gilbert learned from Sarpi all the magnetical discoveries which he subsequently presented as original,[3] and fix the time of communication as during Gilbert's foreign tour,[4] which is absurd, seeing that Sarpi was then very young and had only just attracted notice by his precocity in theological debate.[5] Sarpi himself, on the other hand, strongly praises Gilbert's work, adding: "I have not seen a man in this century who has written originally save Vieta in France and Gilbert in England"—an enconium which, as Hallam justly observes, he would hardly have passed without a hint to the effect that the discoveries were in fact his own.[6]

There is no doubt, however, that Galileo learned of Gilbert's discoveries from Sagredo, and repeated his ex-

[1] Nelli: Vita e Commercio Letterario di G. Galilei, Lausanne, 1793, i, 407.

[2] Opere de G. Galilei, Florence, 1851.

[3] Griselini: Vita de Fra Paolo Sarpi, Lausanne, 1760. Giovini: Ibid., Brussels, 1836. Fabronio: Vitæ Italorum, Pisa, 1798, xvii.

[4] Garbio: Annali di Serviti, Lucca, 1721, vii. Micanzio: Vita de F. P. Sarpi, Verona, 1750.

[5] Robertson: Fra Paolo Sarpi, London, 1893.

[6] Hallam: Lit. Europe, London, 1864, iii, 333.

periments very shortly after their communication to the world.[1] The great lawyer who wrote philosophy "like a Lord Chancellor" had already rendered his sarcastic judgment upon "Gilbert our countryman," who "hath made a philosophy out of the observations of a lodestone." The greater practitioner of the philosophy of works, writing to the Grand Duchess of Tuscany in 1606, had no compunction in overruling that judgment, and in announcing the advent of a philosophy confirmed by evident demonstrations, and "showing our earth to be in its primary and universal substance none other than a great globe of lodestone."[2] Nor did he ever waver from that opinion. A quarter of a century later it is re-asserted and amplified over pages in the famous Dialogue, which brought him into the clutches of the Inquisition.[3]

In 1607 began a remarkable correspondence[4] between Galileo and the reigning Duke of Tuscany, who had been, and to some extent still was, Galileo's pupil. Of all the magnetic phenomena which Gilbert had recorded, none, saving the theory of the earth's magnetism, appears to have impressed Galileo more strongly than the discoveries of Gilbert concerning the armed lodestone, and especially the notable increase in lifting power which seemed to follow the attachment of the iron helmet or cap to the pole. Gilbert had said that, by means of this cap or armature, a stone capable of raising but four ounces could be made to raise a weight of twelve ounces, and that when the poles of two such stones thus armed were caused mutually to attract, the joint action of both would lift a weight of

---

[1] "Galileo made many experiments upon the magnet, and both he and his favorite pupil, Sagredus, were moved to meditate thereon through having received Gilbert's book."—Nelli: Vita, etc., di G. Galileo. Lausanne, 1793, i., 103.

[2] Celeste: The Private Life of Galileo. Phila., 1879.

[3] Galileo: Systema Cosmicum, in quo Dialogis, iv., etc. Ed. Leyden, 1641, Dialog. iii., p. 296.

[4] Opere di G. Galileo. Florence, 1851.

twenty ounces; and he describes other experiments, all going to show the increased lifting power gained by the attachment of the armature. Galileo, it seems, for the purpose of repeating Gilbert's experiments, had prepared

GALILEO GALILEI.[1]

for himself a lodestone weighing about half a pound Tuscan, and this the duke wanted. Galileo thereupon wrote to the ducal secretary, stating that while everything that he owned was at the disposal of his sovereign, he ventured

[1] Reduced fac simile of the frontispiece of his Systema Cosmicum. Leyden, 1641.

to suggest that a friend of his (Sagredo) possessed a lodestone far more worthy of the notice of his Serene Highness, and which weighed fully five pounds, but for which the large sum of four hundred crowns was demanded.

The curious spectacle then followed of the Sovereign of Tuscany and the great philosopher keenly haggling for several months over the purchase price, until finally the duke's offer of a considerably reduced amount was accepted. Galileo then became uneasy lest the stone should not accomplish what he had stated that it would do, namely, lift its own weight, and thereupon he caused the magnet to be sent to him by Sagredo, in order that he might satisfy himself by experiment as to its efficiency. What he did with it he recounts in his letter of transmission. He fitted up the stone at his own expense with armatures, which he makes in the form of two little anchors (suggestive, as he says, of the fabulous notion that a magnet might lift a ship's anchor), for purposes of convenience, inasmuch as when the stock of the anchor is applied to the magnet pole other pieces of iron can be applied to the hook, up to the extreme limit of the strength of the magnet.

It is exceedingly interesting to note how carefully and ingeniously he proceeds to provide for the requirements of future experimentation. "I have not made the anchors of the great weight," he says, "which I have seen the stone to be able to sustain; first, in order to be sure that, without tedious trial, the irons suddenly presented to the poles of the stone would attach themselves, and, second, because I think that the same piece will not be sustained with the same force in all places of the earth." He thought that the magnet poles would be governed somewhat in attractive power by the proximity of the earth's poles, so that, in this way, "the stronger pole of the stone should sustain something more at Padua than at Florence or Pisa." Therefore, he is anxious to have this question tested, and to that end, while he makes the anchor armatures themselves of a weight not as great as that which the stone will sustain, he applies to them, in the form of separable pieces, numerous bits of iron which, with the anchors, aggregated a weight greater than the stone's sustaining power, as he carefully adds "in the condition in which I sent it": thus guarding against an apprehended possibility that the stone will not behave in Florence in the hands of others as well as it has acted in Padua in his own.

He provides the stone also with a strengthening piece, apparently arranged so that the armatures cannot be placed anywhere except at the proper places. Then he says that both armatures had better be applied at once, because he has found, to his great surprise, that "an iron so heavy that by itself it will not be governed by one pole, will become attached thereto if another iron is applied to the opposite pole." He also sends with the stone two extra pieces of iron, one of which is to be in the form of a cylinder and to be placed upon a smooth table, and the other to be applied to the stone at a marked point; and this cylinder, in some way which he does not very clearly describe, is to be first repelled by the magnet brought near it, and then attracted—a result evidently depending upon variations in the distance intervening between the strong lodestone and the rolling cylinder which, by induction, it weakly magnetizes. And then he adds the first announcement of the true effect of the armature as a keeper in actually invigorating and retaining the strength of the magnet, by being allowed to remain in contact with the poles, and suggests the provision of a support, so arranged that the armatures may always remain attached and in place. Finally he says that not only will the stone sustain its own weight, but a load four times greater, which, in a magnet of such large proportions, he regards as marvelous, and he expresses the opinion that if it were cut up into small pieces the latter might be made to hold iron aggregating six or eight times their weight.

Such were the interesting results of the study which

Galileo made upon Sagredo's large magnet. His investigations at that time went no further, for the following year saw his invention of the astronomical telescope, immediately succeeded by the magnificent discoveries in the heavens upon which his fame chiefly rests. When he took up the magnet cursorily in after years it was still to ponder over its attractive power and how this might be augmented, or to devise theories to account for the apparent strengthening effect of the armature. He supposed, in the end, that the iron was drawn to the armature with greater force simply because the two surfaces, being smooth and polished, presented more points of contact than could exist between iron and the rough magnet,[1] which was perhaps as good for the time as Gilbert's notion that the touch of the stone awoke a slumbering virtue in the armature, and that then both pulled with their joint forces; and fully as reasonable for example as some hypotheses accounting for the microphonic transmission of speech.

The huge magnet which was sent to the Grand Duke served its purpose as a toy for that potentate and his successors for many a day. Ninety years later it became lost, and then Leibnitz, writing to Magliabecchi,[2] deplores the disappearance of a relic which he says the scientific world would have prized beyond the most precious gem, a lament which he might equally well have made over the earlier destruction of Gilbert's terrellas in the great London fire.

The Italian philosophers were not so swift to appreciate Gilbert's electrics as they were his conception of the earth's magnetism, and it was not until 1629 that the earliest of their researches upon the former were made known. In that year Nicolaus Cabæus,[3] a Jesuit, then of Ferrara, and a philosopher of remarkable ability, who had maintained a school of philosophy, mathematics and theology in

---

[1] Galileo: Systema Cosmicum, cit. sup.

[2] Clavorum Germanorum, etc. Florence, 1746, Epistle xxvii.

[3] Sotuello: Bib. Scripta Soc. Jesu. Rome, 1676; Brucker: Hist. Crit. Phil. Cabaeus: Philosophia Magnetica, Ferrara, 1629, p. 18.

Parma, produced, in the first complete Italian treatise on the magnet, a record of study which had extended over many years, and which, among other things, resulted in the first electrical discoveries following those of Gilbert.

This is the same Cabæus to whom I have already referred in a preceding chapter, apparently as advocating the hearsay discoveries of Leonardo Garzoni against those of Fra Paolo Sarpi. His name is often mentioned in the historical retrospects of electrical progress which have appeared during the last century or so, apparently solely because of his having added some more electrics to Gilbert's list; these being "white wax and anything made of wax which is hard and may be rubbed . . . several gums, such as gum elemi, gum carab, gum from mastic, pix, which is called Spanish, and gypsum, not burnt," a slender enough addition to the science, although of interest as being a real advance beyond Gilbert. His principal discovery, however, is of very much more importance than a few additional electrics, although the fact seems to have remained unrecognized.

Cabæus, while admitting the accuracy of Gilbert's experimental work and of the physical distinctions which Gilbert points out between the amber and the lodestone, refuses to accept either Gilbert's theory of electric effluvia or his general dictum of the attractive quality of bodies concreted from humor. "His words," says Cabæus, "are put together with ornate elegance, but I do not see that they explain any mode of attraction. Plenty of things which are hard and yet are concreted of humor have no attraction, and many things attract which do not appear to be concreted of humor." Floating bodies do not attract by humor, but through "gravity and levity." If wet bodies do adhere, that is due to agglutinating action of the interposed liquid. Cabæus is not here attacking Gilbert's theories merely from a spirit of opposition. He has found some strange facts, and Gilbert's effluvium notion refuses to be squared with them. He does not understand these facts, and he interprets them wrongly; but dissent from Gilbert's hypothesis and the production of a new one in supposed accordance with the new data were inevitable.

What had he seen? That, when the face of a well-prepared electric is applied to the drawing of light filings or sawdust or similar corpuscles, they run strongly to the electric, and when they reach it they fly back, not falling off merely, but being thrown off afar to a distance of two or three inches. And that sawdust groups itself upon the electric "like masses of hairs," the ends of which fluctuate and waver, and finally these extremities likewise do not fall off, but are projected afar.

In brief, he had found electrical repulsion—the phenomenon which Gilbert said had no existence. He had seen, as any one may now see, the oppositely-electrified body move to the electric, become similarly charged and fly away from it. This plainly could not be accounted for by supposing material arms or rods grasping the attracted body, and in some unknown way bringing it to the electric; and so Cabæus framed a new hypothesis, wherein repulsion was in fact the fundamental feature. The rubbed electric, he avers, produces a most thin effluvium, which attenuates the air and vigorously impels it; and this attenuated air, in returning to the electric in a gyration, brings with it the attracted body. In other words, he thinks that effluvium is first "expelled," and thereby the air is "propelled" in a wind. The wind comes back, entraining with it the chaff—sometimes even with such violence that it seems to rebound from the electric. Such was the first recognition of electrical repulsion and the first theory proposed to account for it.

It will be recalled that Gilbert says that the rays of magnetic force emanate in all directions from the lodestone's centre, and thus form an "orb" or "sphere of virtue" around "that great magnet, the earth." Herein he differed from Porta, who had insisted that the diffused virtue emanated from the two poles of the lodestone. But this was one of the instances in which Gilbert allowed theory rather than experiment to guide him; for, when he carried his iron needle around the terrella, he saw plainly enough, as Peregrinus had seen centuries before, that it never pointed to the centre, except when it was exactly at the poles. In fact, this was one of Peregrinus' methods of finding the poles. At the equator, the needle stood at right angles to this position, and between the equator and the poles it assumed various inclinations to the latter. Of course, a needle placed successively in different places along a meridian of the terrella would map out, so to speak, the direction of the lines of force from pole to pole. But Gilbert did not perceive this any more than he saw the inconsistency between his theory and his experiments; for clearly, if the magnetic virtue emanated radially from the centre of the terrella, his needle should always point to the centre, and so take the same position at the equator as at the poles.

Cabæus, however, was more keenly alive to the logic of the experiment, and to the fact that it was at odds with Gilbert's supposition. He, in turn, moving the needle in different positions along the meridian, sees it gradually incline from the equator to the pole, until at last it stands upright; or, starting from the pole, sees it gradually incline in the opposite way, until at the equator it has moved over a right angle. Then he goes a step, but a long one, further. Instead of a single needle moved into different positions, take a great many needles, he says— little ones, mere particles of iron, iron filings—and put them around the stone. Look at them! At the equator they adhere "prostrate" to the magnet, but at the poles they "erect themselves like hairs." Hairs, branching and curving away from the poles as starting points. What is controlling them? Certainly the emanations from the magnet; and, therefore, they must be showing the true paths of those emanations leading, not from the magnet centre, but from both poles.

Thus, for the first time, the lodestone was made to write its own story, and it was Cabæus who first recognized, not that filings erected themselves hair-like about a stone, for Porta had done that, but that they grouped themselves in a definite way, branching from the poles, making what we now term the magnetic spectrum.

Acute as he was, Cabæus failed to see all that was thus written for him. He did not perceive that the filings curved from pole to pole. For him, they swept outward in paths ending always like hairs in a brush, and thus he depicts them. But there were two brushes; and that was

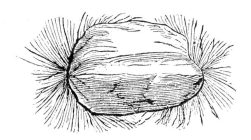

CABÆUS' PICTURE OF THE MAGNETIC SPECTRUM.[1]

enough to dispose of Gilbert's notion and so to serve his purpose.

Cabæus is the very Mercutio of philosophers. He is caustic and witty—his dialectic sword is ready and needle-pointed—his mental agility is swift. He flits around Gilbert like a wasp, stinging wherever he can. But I shall not follow him further, tempting as the task is. He may be dismissed for bias. He was of course savagely anti-Copernican. As a Jesuit he wrote to sustain Garzoni against both Sarpi and Gilbert; and also in the same capacity, and with characteristic casuistry, he denied that

[1] From his Philosophia Magnetica, 1629.

the earth is a great magnet, while stoutly averring that it is endowed with magnetic properties.

Thus far had the Italians advanced. They had undeniably made progress beyond Gilbert in the "new physiology," had stumbled upon electrical repulsion, and had attacked the Englishman, with more or less success, whenever they caught him wandering from the safe ground of sure experiments. But what had become of the great cosmical theory — the magnetic inter-relation of the heavenly bodies—the extension of the sphere of virtue into the heavens, and the government of the planets by the mutual reactions of their "effused" spheres—which Gilbert regarded as at once the flower and crown of all his labors? For that doctrine there was as little resting-place in the bosom of the Church, as in the inhospitable breast of Bacon. There was no lodgment for it, except among the Copernicans, and so it fell into the outspread arms of one of them—and there gently expired. John Kepler, casting about for clews, clutching at guess after guess in pursuit of proof of his great laws—trying to figure something like universal gravitation out of his inner consciousness—came upon this outcast theory, and administered upon its effects.

In his treatise on the movements of Mars,[1] we are told that the sun is a great rotary magnet carrying its—Gilbert's rather—sphere of virtue around with it. The planets are in that vast whirlpool and are carried with it. If it be asked—why should not this great solar magnet draw its satellites to destruction in its fiery mass? Because, is the reply, the field of force is made up of fibers— filaments—that are straight, that is, which surround the sun, so that the planets are dragged along in these magnet streams, like boats in a maelstrom.

"Sed proh Deum immortalem!" a few years later, shouts Athanasius Kircher, that irascible but omniscient philosopher of the Church Militant, losing his temper

---

[1] De Motibus Stellis Martis. Prague, 1609.

completely and banging the dust out of Kepler's immortal pages. "Quænam ista philosophandi ratio est?"[1]

But Kircher knew well enough that Kepler had been reading in the "Philosophia Magnetica of William Gilbert the Englishman," and that his argument thereon was that if the earth had magnetic properties, it was "neither incredible nor absurd" that the same might be equally true of other "primary bodies."[2] The Gilberto-Keplerian theory, however, had no more health in it than there was in the other tenet which the two philosophers held in common—namely, that the earth is alive and has a soul. No one cares to remember now the odd vagaries of the great student of the stars who overthrew the old astronomy. He willingly renounced many of them himself as his knowledge of phenomena grew wider; nor have they ever dimmed the glory which all the world accords to the finder of the laws whereby the planets move in eternal harmony with the Almighty Will.

"Having held and believed that the Sun is the Center of the Universe and immovable, and that the earth is not the center of the same and that it does move . . . I abjure with a sincere heart and unfeigned faith, I curse and detest the said errors and heresies and generally all and every error and sect contrary to the Holy Catholic Church," wrote Galileo Galilei, in mortal terror of the Inquisition; that was in 1633.[3] Twenty years before, under the protection of the Grand Duke of Tuscany, he had asserted the heliocentric doctrine, with no worse result than a friendly admonition from Cardinal Bellarmine, and he had agreed not to promulgate it further. But, as the world grew wiser, it smiled at the theological claims to infallibility in matters

[1] Kircher. Magnes sive De Arte Magnetica. Cologne, 1641.
[2] Kepler: Epitome. Ast. Copernic. Frankfort, 1635.
[3] Whewell: Hist. Induc. Sci. London, 1837, Vol. II., 133. Hallam: Lit. Europe. Part III., cviii.

of physics, and at last, in 1620, the Church itself yielded sufficiently to sanction the discussion of the Copernican theory as an hypothesis merely. This gave Galileo a safe opportunity, as he believed, once more publicly to reaffirm his belief therein. He went too far, and tried to prove it orthodox. However the ecclesiastical authorities may have intended to deal with others, the fact of his having violated, as they claimed, his earlier promise gave them a reason for coming upon him despite the permissory decree. He was the most shining of all shining marks. To crush him would do more to paralyze independent philosophical thought, at least within the pale of the Church, than any random anathema that Rome could hurl.

The effect upon all Europe was profound. The faithful, who found themselves in the van of philosophical progress, stopped and drew back. The blight of uncertainty fell upon them. If, after years of free discussion, Copernicanism had come to be heresy, inviting the dread visit of the Holy Office, what then might be safely taught and studied? The light laughter at the ecclesiastics, who sought to govern Nature's laws by theology, was heard no longer from the Protestant ranks; but instead the hatred to Rome leaped into new vigor, and sarcasm, invective, ridicule—fierce and bitter—came pouring forth.

But the blight persisted none the less. There was great force in it. "If the opinion of the earth's movement is false," said René Descartes grimly, locking the manuscript of his Principia in his cabinet, "all the foundations of my philosophy are also false, because it is demonstrated clearly by them . . . yet I would not for all the world sustain them against the Church." And so the book remained unpublished for ten years. But when it did appear there followed a revolution in the realm of thought.

So far, from Gilbert onwards, we have seen the students of the lodestone and the electrics dealing with phenomena, and seeking to derive laws from experiment. We have seen the inductive method, as it were, in the air and affecting the minds of all thinkers, crystallized and formulated in the language of Bacon, and then moving forward with renewed and concentrated force. But now, there appears a philosopher of the first rank, who tosses it aside as an instrument inadequate for the discovery of truth, and substitutes pure deduction; a man skilled in mathematics, wherein Bacon was most deficient, who regarded physics not as did Bacon, as the basis of all science, but as merely a reservoir of illustrations of his principles; who argued, not from effects to causes, known to unknown, but deduced effects from causes and explained things seen by reasons found by intuition. "It is not so necessary," said Descartes, "to have a fine understanding as to apply it rightly. Better progress can be made by walking slowly on the right road than by running swiftly on the wrong one." Bacon expresses the same idea, but the common ground is reached by paths leading from totally opposite directions. The ten years of delay in the publication of Descartes' great treatise perhaps gave him the opportunity to make it the almost perfect piece of scientific writing which it is. For unswerving directness of expression, for exquisite clearness, for pertinency of example, it has scarcely a rival in the whole literature of physics. We have now to see how the magnet and the electric were treated in the philosophy of Descartes—a philosophy essentially metaphysical, evolving first a clear hypothesis and then seeking to reveal thereby the causes of observed phenomena.[1]

Descartes, by his vortex theory, undertook to explain mechanically the solar system, the formation of planets, the relation of the tides to the moon, and to subject the laws of motion to scientific analysis.[2] He assumed[3] matter uniform in character throughout the universe, to be divided into polygonal masses. These having a circular

[1] Lewes: The Biog. Histy. of Phily., N. Y., 1857, Vol. II, 1445.
[2] Mahaffy: Des Cartes, Edin. and Lon., 1880.
[3] Des Cartes: Principia Philosophiæ (ultima editio), Amsterdam, 1692, Parts 3 and 4.

motion grind one upon the other, producing spheres and also filings or parings due to mutual abrasion of the polygons at their angles. By this means, there comes into existence the transparent substance of the skies (ether), the material of luminous bodies, such as the sun and fixed stars, and the material of opaque bodies, such as the planets. The motions of these parts are those of revolving

DESCARTES.[1]

currents or vortices, wherein the luminous body is at the center and the ether surrounds it. In the solar vortex the planets are immersed, and with it are whirled around. The similarity of Descartes' conception of a solar vortex to Kepler's notion of the magnetic whirlpool surrounding

[1] Reduced fac simile of the frontispiece of his *Principia Philosophiæ*. Amsterdam, 1692.

the sun is obvious. The difference is the substitution of Descartes' whirling matter, mechanically produced, for Kepler's whirling filaments magnetically produced.

Among the small filings which are ground from the revolving spheres by friction, are many which are compelled to escape through the interstices between the whirling particles, and these consequently are molded or shaped into the form of spirals. To the movements of these spiral particles through the pores or conduits of bodies adapted to receive them are due magnetic and electric phenomena. These conduits are shaped to receive the spiral particles, and extend through the bodies possessing them in a direction parallel to an axis. The spirals which can enter at one end of the conduits cannot enter at the other end— apparently on the principle that a right hand threaded screw cannot enter a left hand threaded nut; and also because, in the conduits, there are delicate protruding branches which allow the spirals to bend them freely, while moving in one direction, but become rigid and oppose their passage while moving in the other—something like the converging wires in an old-fashioned mouse-trap. The result is that the spirals, say from the North part of the heavens, can enter the conduits suitable to them at the South end of the stone, pass through these passages to the North end, and then returning enter the South end again—forming a whirlpool through and round the stone. Similarly, the spirals from the South part of the heavens can enter the conduits suitable to them at the North end of the stone and, in like manner, form a whirlpool.

In other words, Descartes plainly sees that there is a force, not merely radiating from the magnet poles, as Cabæus supposed, but traversing the stone from pole to pole in one direction and then traversing the external region around the stone from pole to pole, in the opposite direction. His spirals whirling about under that influence, were merely a device to render its effect thinkable—just as were Faraday's equally imaginary lines of force. What Descartes really had discovered was the endlessness of these apparent lines of magnetic force.

This idea of the spirals flowing in definite directions through conduits in the magnet, he applies to all magnetic phenomena, of which he finds therein an explanation—often with marvelous ingenuity. The stream of spirals flows more easily through the lodestone or iron than through the air or any other substance, because the conduits in the first-mentioned bodies are better suited to them. Wherever the streams enter and leave a body, there are its poles. If a magnet, free to move, presents its conduit entrances at an angle to the stream of spirals from the earth's poles, the force of the stream is sufficient to turn the magnet so as to bring the conduits in line with its path, and then so that the north entrances of the conduits are directed to the south pole of the earth, and *vice versa;* thus the directive tendency of the needle to the poles is explained. "There are always," he says, "more spirals around the magnet than elsewhere in the air, because, after they have left one end of the stone, they find in the air a resistance, which causes most of them to return to the other end of the magnet whereat they enter; and thus several remain around it, making a kind of whirlpool, the same as they make about the earth. So that the whole earth may be taken for a magnet not differing from others, unless it be bigger: and that on its surface where we live its virtue is not very strong." Thus the field outside of the magnet is accounted for—and a definite conception is suggested of a "resistance" to the force, compelling it to choose a certain path.

But there is still more in the foregoing quotation. If the earth is so vast and great a magnet, why is its virtue "not very strong?" The streams of spirals are generated in the earth in a certain region, which last is a spherical stratum. In passing to the earth's surface they encounter another and outer stratum of metals, etc., abounding in conduits suitable to them. Many of them pass through these conduits and back to the origin, hence but a small proportion of the total number of streams reaches the air. That is the first notion of "short circuiting" and "leakage."

Some of the explanations are curiously ingenious, such, for example, as that of magnetic attraction and repulsion. If two magnets are placed with unlike poles in proximity, the spirals from one may enter the conduit ends of the other. Then the air between the juxtaposed poles is driven out and forced around to the rear of the two magnets so that it pushes them together. If, on the other hand, like poles are opposed, the spirals from one magnet cannot enter the conduits of the other, and the spirals force the stones apart.

Iron is adapted to become magnetic because it has conduits suitable to receive the spirals; but it is not normally magnetic, because the little branches or projections in the pores are turned naturally in all sorts of directions. If, however, a magnet through which a strong stream of spirals is passing be approached to the iron, the force of that stream is enough to drive the spirals through the conduits in the iron, and in so doing to turn all the little branches in one way. After that the iron constantly receives streams, and is magnetic.

The mode of answering that standing puzzle, how is it that the magnet in communicating its virtue to large quantities of iron still retains its own unimpaired? is especially felicitous. "There happens no change in the magnet, because the spirals which leave its pores enter iron rather than some other body. In fact, they pass even more freely and in greater quantity through the magnet when there is iron around it than when there is none. Hence, instead of the magnet's virtue being in anywise thus impaired it is increased, besides being communicated to the iron." Yet he does not account for the strengthening effect of the armature in this way, but agrees with Galileo's hypothesis concerning it.

The foregoing will suffice to show the remarkable and novel character of the magnetic theory of Descartes. Apparently it seems to have had no other origin than the "scientific use of the imagination," but this is not entirely true. Induction from phenomena forced its way into his reasoning, despite his belief that he was dealing solely with his own intuitions. After he had explained, in his limpid style, the accordance of his hypotheses and the various phenomena of the magnet, which he sums up beautifully in thirty-four aphorisms, he betrays the material mechanism which really sets going all this speculation. I shall let him reveal it for himself.

"Now, if one should stop to consider how iron powder or iron filings thrown about a magnet arrange themselves, many things would be observed confirming the truth of what I have just said." (Observe the fallacy, post hoc, ergo propter hoc.) "For, in the first place, it will be seen that the little grains of this powder do not pack themselves together confusedly, but that joining themselves together lengthwise they form filaments, which are as many little tubes, through which the spirals pass more freely than through the air, and which, therefore, may serve to show the path of the spirals after they have left the magnet. But in order that the eye may recognize the curving of these paths the filings should be strewn upon a smooth surface, in which the globular magnet is half buried, so that its poles are in the same plane, as globes are supported in horizon circles; then on that surface the filings will arrange themselves in lines showing exactly the paths which the spirals take around the magnet and also around the earth."

Then he continues further and explains how the filings group themselves around the poles of the two magnets when attraction or repulsion takes place.

He had seen all that Cabæus did not see. He had recognized the whole magnetic spectrum, the complete magnetic curves, and that the lines of force or paths along which the imaginary spirals were urged were exactly mapped by the iron filings, a purely physical observation. The chief features of the field of force had been observed. The route of new discovery now lay toward its properties.

When Descartes reaches the electrics he shows some unwillingness to formulate theories about them, as it were, ex cathedra, as he had done in reference to the magnet. It is necessary, he says, to "say something" about these bodies—the electrics—it was not his original intention to

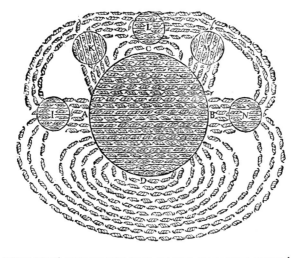

DESCARTES' REPRESENTATION OF THE MAGNETIC FIELD.[1]

do so—and then (lame and impotent conclusion for the man whose mind was the reverse of the Baconian medal) he is not fully certain why they act as they do until he shall have made "several *experiments* to discover their nature." Experiments! and by the apostle of deductive

[1] From his Principia Philosophiæ. This depicts a large spherical magnet (the earth) having its poles at A, B, with smaller magnets I, K, L, M, N, disposed in inductive proximity The lines of force in which the assumed spirals arrange themselves are clearly shown.

reasoning. Why not have deduced "their nature" from intuition?

Yet the speculation whereby he endeavors to account for electric attraction is one of the most remarkable of all. He begins by denying absolutely the notion of emanations of an apparently glutinous character which emerge, seize upon the chaff, and, on retracting, bring it back to the electric. There is no warrant for such an hypothesis, he thinks, and for it he substitutes the following:

The pores of the electric are slits of extreme narrowness. Nothing but the globules of the most subtle ether can enter them. But when they are filled these globules unify and form little ribbons (bandelettes), which move to and fro in the pores, and are molded to their shapes. They cannot of themselves leave the electric, because there are no passages in the air which they fit. But when the electric is rubbed it is heated. Its shape, and hence that of its pores, is deformed, and the ribbons are crowded out, and hence are moved toward other bodies. Not being able to find any suitable conduits in these bodies through which they can proceed, they engage merely in the pores of light chaff. As the electric, after rubbing, resumes its normal condition, they shrink back and bring the chaff with them.

It is a very far-fetched theory, and the ribbons are no better than Gilbert's effluvia. But there is something novel in Descartes' commentary upon it. After explaining how it is the nature of the element, whereof these ribbons are composed, to keep swiftly moving within the pores of the electric, he says, "and sometimes, on the other hand, they pass in a very short time to far distant places, never meeting a body in their path capable of stopping or diverting them. And then meeting afar like matter disposed to receive their action, they produce effects entirely rare and marvelous, such as causing the wounds of a corpse to bleed when the murderer approaches, exciting the imagination of those who sleep, and even of the waking, and creating in people thoughts which warn them of events happening far away, or presentiments of great afflictions and great joys or impending peril." And that was the first attempt, many and many a time since fruitlessly repeated, to explain the psychical things of heaven and earth, through the physical agency of electricity.

Descartes' Principia appeared in 1644, as I have said, after ten years' seclusion. Meanwhile it had become incumbent on somebody of greater ecclesiastical influence than Cabæus, and of more general eminence as a philosopher and theologian, to advance the arguments of the Church against the heresies of Gilbert, Kepler, Galileo, and other recusants of that stripe. The task naturally fell to Athanasius Kircher, to whom allusion has already been made ; a Jesuit, a Professor at Rome, and a man of encyclopædic knowledge, great gullibility, and the author, says Robert Southwell,[1] of twenty-two works in folio, eleven in quarto, and three in octavo. His treatise on the magnet was written about 1639 and issued in 1641. Many editions of it followed. It adds nothing to the existing knowledge on the subject, but it exhibits an astonishing collection of magnetic apparatus, from perpetual motion to the magnetic toys which are still sold everywhere. It was probably the vade mecum for the practical magnetician of the day, if any one pursued that calling.

Kircher has the honor of giving to the action of the lodestone its name—"Qualitatem Magneticam Magnetismus appellare placuit." (The magnetic quality may be properly termed magnetism.) And, what is perhaps a little surprising, he also invented the word "Electromagnetism"—heading one of his chapters on the electrics with "Ἠλεχτρο-μαγνητισμὸς—that is concerning the magnetism of electrics or the attraction of electrics and their causes."

---

[1] Boyle's works. London, 1744, v. 405. Kircher's genesis of the solan goose is classic. The eggs, he says, are laid in the Arctic regions; they mix with the sea and render it "eggified." Drops of sea water dash on the trees near the shore, and the specific egginess of the sea, the natural vegetation of the tree, and the influence of the sun, unite in hatching the goose.

Kircher was an admirable compiler, and, as a storehouse of doubtful facts, his work is interesting. Of course, he agreed with Cabæus that the earth is not a magnet, but magnetic. And his proof thereof is impressive. "How vast its mass—how prodigious would be its effects—what could resist its capacity?"—that is if it were a magnet, and not merely magnetic. He is a Professor of Mathematics, and to numbers he appeals. Comparing the size of the earth with that of a terrella a few inches in diameter, he staggers his reader with the assertion that if the terrella can attract one pound, the earth, if a magnet, can attract over three octillion pounds. That is sufficient for an exact idea of just what the earth can do: and so he returns to his Jeremiad.

"Woe to all iron implements," he thunders, "woe to all horses and mules (probably on account of their shoes), woe to cataphracts, woe to Gilbert's kitchen utensils." Why, the rocks and the precipices and the mountains would be bound in an indissoluble mass—everything would keep still! Instead of the motion Gilbert predicts, there would be utter quiet and the end of all movement. After that, denunciation follows naturally—and strong, sweeping denunciation too.

"Proprium est haereticorum res divinas et incomprehensas ingenio suo metiri, quas nisi comprehenderint nec credere velle videntur"—duly clinched with a quotation from Nazianzenus, Orat., 24.

But it did not do much good. The misbelieving Protestants wagged their heads in derision as usual, and the good sons of the Church took it all as a sermon, well enough in the abstract no doubt, but having no real immediate bearing upon the magnetic and electric problems which they were anxious to solve: certainly none comparable to that which would instantly be recognized in even a look askance from the Holy Inquisition.

Now back to England, where we left Barlowe and Ridley "animadverting" upon one another over their respective claims to Gilbert's experimental discoveries: and the navigators trying to turn Gilbert's nautical instruments to practical account. But the last were of no avail. Gilbert had made a fundamental error as to compass variation; "that the arc thereof continues to be the same in whatever place or region, be it sea or continent, and is forever unchanging." It was, however, soon detected, not by an Englishman, but in all probability by Gian Francesco Sagredo, who was Venetian Consul at Aleppo in about 1610, and who was then making observations himself there, and having others do the same at Goa in India. Vastly important as this subject was to the English sailors and merchants—for the safety of their ships and the success of their enterprises ultimately depended upon the truth of their steering-needles—little more than rumors of the changes in local variation seem to have reached the country for many years. Burton sums up, in a curious blending of the old legends of the magnetic rocks with the results of the new experimental observations, probably all that was then known. He asks whether there be a great rock of lodestone which may cause the needle in the compass still to bend that way, and what should be the true cause of the variation of the compass.

"Is it a magnetical rock, or the pole star as Cardan will; or some other star in the bear, as Marsilius Ficinus; or a magnetical meridian, as Maurolicus; *vel situs in vena terræ*, as Agricola: or the nearness of the next continent, as Cabæus will; or some other cause, as Scaliger, Cortesius, Conimbricenses, Peregrinus contend; why at the Azores it looks directly north, otherwise not? In the Mediterranean or Levant (as some observe) it varies 7 grad.; by and by, 12, and then, 22. In the Baltic Seas near Rasceburg in Finland, the needle runs round if any ships come that way, though Martin Ridley write otherwise that the needle near the Pole will hardly be forced from his direction. 'Tis fit to be inquired whether certain rules may be made of it as 11. grad. Lond. *variat. alibi*, 36, etc., and that which is more prodigious, the variation varies in the same place, now taken accurately, 'tis so much after a few years quite altered from what it was: till we have better intelligence, let our Dr. Gilbert and Nicholas Cabæus the Jesuit, that have both written great volumes on this subject, satisfy these inquisitors."[1]

Burton, however, has much to say about the Copernicans, and he knows Gilbert best as a defender of their theory, which he classes among the causes of melancholy. It is in "sober sadness," he says, that he finds Digges and Gilbert and Kepler defending the notion that the earth is a moon, a conception which makes one "giddy vertiginous and lunatic within this sublunary maze."

But Ben Jonson, in perhaps closer touch with London life than the Leicestershire clergyman, discovers that the making of terrellas into playthings for Tuscan Grand Dukes, like other fads Italian, was being widely copied among English aristocracy, and that magnetism and its wonders were therefore beginning to interest the people generally. Therefore he wrote his comedy, "The Magnetic Lady," wherein the heroine, "Lady Lodestone,"

> "Draws and draws unto you guests of all sorts,
> The courtiers, and the soldiers, and the scholars,
> The travelers, physicians and divines,
> As Doctor Ridley wrote, and Doctor Barlowe."

and which ends with the happy union of the magnet and armature.

> "More work then for the parson. I shall cap
> The Lodestone with an Ironside, I see"—[2]

[1] Burton: Anatomy of Melancholy, Part 2, § 2, Mem. 3. The first edition of this work appeared in 1621, and five editions of it appeared in Burton's lifetime, which ended in 1639. The reference to Cabæus in the last sentence of the quotation shows that this clause at least was written after the appearance of the Philosophia Magnetica in 1629.

[2] The date of this play is 1632.

But this last is the only new metaphor which Jonson bases on magnetism. He has not devoured Ridley and Barlowe with that insatiate appetite for knowledge which shows its results in his extraordinary mastery of occult subjects in the Alchemist. There was more poetry, perhaps, to be got out of alembics and retorts and receivers, from incineration and calcination and reverberation, than out of a stone. So, saving the two passages before quoted, the magnetic color, so to speak, of his play is factitious, and cheaply gained by giving his characters the technical names of Needle, Compass, Ironside, and so on.

But how deathless that figure of speech, here again recurring, which likens personal attractiveness to the drawing of the lodestone! It is almost as old as civilization. It never was more of a favorite than during the time of the literary awakening in England, and most of all with Robert Greene, who fairly strews it throughout his now almost forgotten novels and plays. Clarinda is "an adamant object to draw the wavering eyes of Pharicles." Love is "the adamant which hath virtue to draw," and "what adamants are fayre faces!"[1] Sometimes he deserts the lodestone for the amber attraction: "seeing you sit like Juno . . . I was by a strange attractive force drawne as the adamant draweth yron and the jeat the straw;"[2] to withstand the brunt of beauty is as impossible as "for the yron to resist the operation of the adamant or the silie straw the virtue of the sucking jeat."[3] So does Jonson once avail himself of the electrical simile:

"Your lustre too'll inflame at any distance;
Draw courtship to you as a jet doth straws."[4]

But Shakespeare never does. Only in Helena's reproach does he use the well-worn figure based upon the magnet—

[1] Greene: Mamillia. 1580-3.
[2] Ibid.: Menaphon. 1589.
[3] The Carde of Fancie.
[4] Every Man in his Humor. Act III., Sc. 2.

> "You draw me, you hard-hearted adamant,
> But yet you draw not iron, for my heart
> Is true as steel: Leave you your power to draw,
> And I shall have no power to follow you." [1]

Perhaps he thought that his rival, Greene, had already made the metaphor too common. Perhaps by putting the figure in its single use in the mouth of the love-sick girl he meant to satirize Greene's overworking of it.[2] Perhaps he was no believer in the "miracles of science," now so called. "They say miracles are past," he remarks, with something of the non-inquisitive superciliousness with which a peripatetic might stop in his lazy promenade beside the Lykeum to gaze on a gardener grafting a tree.

[1] Midsummer Night's Dream. Act I., Sc. 1.

[2] I append the following:
"Beautie is the Syren which will drawe the most adamant by force." —*Mammilia.*
"For the Adamant drawes by vertue though Iron strive by nature."—*Ibid.*
"Yet they have in their eyes adamants that will draw youth as the Iet the straw."—*Never too Late.* 1590.
"A woman's teares are Adamant, and men are no harder than Iron, and therefore may be drawn to pitie."—*Ibid.*
Here are two quotations in which the figure changes:
"Their hearts like Adamants that will turn no way but to one poynt of heaven."—*Never too Late.*
"For the fingers of Lifts (shoplifters) are fourmed of Adamant; though they touch not, yet they have vertue attractive to drawe any pelfe to them as the Adamant dooth the Iron."—*Notable Discy. of Coosnage.*
See Greene: Life and Works. Huth Library. 1881-83.
Jonson (Every Man out of his Humour. Act III., Sc. 2,) probably carries the attractive figure to the limits of hyperbole—

> Would to heaven
> In wreak of my misfortune I were turn'd
> To some fair nymph, that set upon
> The deepest whirlpit of the rav'nous seas,
> My adamantine eyes might headlong hale
> This iron world to me and drown it all.

Shakespeare (Troilus and Cressida. Act III., Sc. 2,) uses the simile "as true as iron to adamant," which may refer either to attraction or directive tendency.

"We have our philosophical persons to make modern and familiar things supernatural and causeless."[1]

If the study of the magnet in England had continued nothing more than a mere amusement, the conditions for the advancement and increase of both magnetic and electrical knowledge would have been vastly better than those which prevailed before half of the seventeenth century had ended. All physical science was under the domination of the theologians, and the rising Puritanism was scarcely more tolerant of it than was the Church itself. The divines, says Robert Burton, are "too severe and rigid, ignorant and peevish, in not admitting the true demonstrations and certain observations of the mathematicians," tyrannizing "over art, science and all true philosophy in suppressing their labors . . . forbidding them to write, to speak a truth, all to maintain their superstition and for profit's sake."[2] Nor does he bear any better testimony concerning the physicians, the only scientific body in the community; for the country, he says, is indeed overrun with mountebanks, quacksalvers, empirics in every street almost and in every village, calling themselves physicians, who serve to "make this noble and profitable art to be evil spoken of and contemned by reason of these base and illiterate artificers."[3] Obviously such an environment was the worst possible for the promotion of the truths of natural philosophy, or, what is the same thing, the very best possible for the cultivation of every item of popular superstition and ignorance.

Meanwhile there had arisen in Germany[4] a sect of fanatics calling themselves Rosicrucians—or brethren of the

[1] All's Well that Ends Well. Act II., Sc. 3.
[2] Burton: Anatomy of Melancholy, Part 2, § 2, Mem. 3.
[3] Ibid., Part 2, § 3, Mem. 1.
[4] Mackay: Memoirs of Extraordinary Popular Delusions. London, 1852, i. 262.

Rosy Cross. They followed Paracelsus in attributing occult and miraculous powers to the magnet, and established what is now known as the "faith cure;" or in other words, they worked upon the imagination of invalids, highly nervous persons and credulous people generally. They became known later as "magnetizers," and attracted to their ranks hundreds of the alchemists, whose calling was rapidly becoming disreputable, seeing that all their efforts to transmute the base metals into gold had invariably failed, and that they were now manifesting an inclination to swindle.

The magnetizers pretended to transplant diseases by means of the magnet and to cure them by applications of the magnet to the body—the latter, a delusion dating from the period of the Samothracian rings. That was "mineral" magnetism. "Animal magnetism," so called, developed from this, and did not necessarily involve the interference of any actual lodestone or iron magnet at all. As I have stated, the magnetizers derived many of their peculiar doctrines from Paracelsus; but their principal deception, the magnetic cure of wounds, rested upon the imaginary properties of an unguent originally invented by one Corrichterus, who was physician to Maximilian II. The peculiarity of this compound, which contained, among other gruesome ingredients, "the mossy periwig of the skull of a man destroyed by violent death in the increase of the Moon," was that no magnet was ever put into it; not even the powdered lodestone which the ancients and the mediæval leeches mixed in plasters to draw out iron from the body. Subsequently, it was made of less horrible materials, and eventually became nothing but iron sulphate in powder.

Among the leading Rosicrucians was John Baptist Van Helmont, a Flemish physician and chemist, still honestly renowned as the first to recognize the existence of different kinds of air and to use the term "gas," and as the reputed discoverer of carbonic acid. He had been a close student of Gilbert and was well familiar with Gilbert's magnetic experiments, references to which he mingles in his writings with his own falsehoods relative to the curative properties of the magnet, so as to make it appear that his absurdities somehow rest upon Gilbert's researches.[1] From Gilbert's theory of the amber effluvium, he evidently concocted the explanation of the effect of the magnetic unguent or powder (which, by the way, was never to be applied to the wound, but to either the weapon which inflicted it or to an ensanguined bandage), wherein he maintains that "the blood effused doth send out subtle streams to its fount," namely, the body; and these streams or "Magnetic Nuntii" carry with them "the Balsamick Emanations of the Sympathetick Unguent or Powder." So far as the actual electrical effect was concerned, it did not appear to enter *per se* into Van Helmont's curative agencies except as a direct means of drawing contagion out of the body, and "venome and bullets out of wounds;" but the passage in his work which prefaces this announcement has another and more noteworthy claim to fame. In the English translation of Charleton, it is:

"The phansy of Amber delights to allect strawes, chaffe and other festucous bodies, by an attraction, we confess, obscure and weake enough, yet sufficiently manifest and strong to attest an *Electricity* or attractive sign-nature."[2]

That was the first appearance of the actual word which is now the name of the science.

As the Rosicrucians increased in numbers, they became bolder in their assertions, insisting that magnetic agents not only transmit their spiritual energy into determinate patients, but do so "at vast and intermediate distances." The common people accused them of witchcraft, and believed them especially inspired by the powers of darkness. Helmont retorts with "experiments," and the following

---

[1] Charleton: A Ternary of Paradoxes of the magnetic cure of wounds, etc. 2d ed. London, 1650. (Trans. of Van Helmont)

[2] Charleton: Supra, p. 77

one in particular, which he declares "cannot but be free from all suspect of imposture and illusion of the Devil."[1]

"A certain inhabitant of Bruxels in a combat had his nose mowed off, and addressed himself to Tagliacozzus, a famous Chirurgeon, living at Bononia, that he might procure a new one; and when he feared the incision of his own arm, he hired a Porter to admit it, out of whose arm, having first given the reward agreed upon, at length he dig'd a new nose. About thirteen moneths after his return to his own Countrey, on a sudden the ingrafted nose grew cold, putrified, and within a few days drop't off. To those of his friends that were curious in the exploration of the cause of this unexpected misfortune, it was discovered that the Porter expired neer about the same punctilio of time wherein the nose grew frigid and cadaverous."[2]

"There are at Bruxels yet surviving some of good repute that were eye witnesses of these occurrences," he adds, gravely, oblivious of the difficulties of eye-witnessing events simultaneously happening in Bruxels and Bononia. "But," he demands, triumphantly, "is not this Magnetism of manifest affinity with mummy,[3] whereby the nose, enjoying, by title and right of inoculation, a community of life, sense and vegetation for so many months, on a sudden mortified on the other side of the Alpes? I pray, what is there in this of superstition? what of attent and exalted Imagination?"

Of course, there were people—the anatomists especially—who were not quite satisfied with such evidence, and demanded more definite and physical explanations. But Van Helmont was ready with the retort irrelevant, which in one form or another is still the most serviceable reply in the dialectic armament of the "magneto-therapist."

"Go to, I beseech thee!" he says haughtily. "Does

---

[1] Charleton: Supra, p. 13.

[2] This story is evidently the basis of M. Edmond About's novel, *The Nose of a Notary*. See, also, Tatler, Dec. 7, 1710, No. 260.

[3] The bodily humor of Paracelsus, see p. 222.

the Anatomist, our Censor, happily know the reason why a Dog swings his Tayl when he rejoyces, but a Lyon when he is angry; and a Cat when pleased advances hers in an erect posture. . . . The imbecility of our Understandings in not comprehending the more abstruse and retired causes of things is not to be ascribed to any defect in their nature, but in our own hoodwinkt Intellectuals."

This, of course, is delightfully subtle; indeed, to Hoodwink our Intellectuals, and then to say that we cannot understand the hoodwinking deception because our Intellectuals are "hoodwinkt" leaves Van Helmont perched on a pinnacle of effrontery which the modern promoter of the electric and magnetic nostrum has yet to climb. "His experiments need to be confirmed by more witnesses than one," says Robert Boyle,[1] in his solemn fashion, delivering the judgment of the next generation, "especially since the extravagances and untruths to be met with in his treatise of the magnetic cure of wounds have made his testimonies suspected in his other writings." Yet perhaps he deceived no one more than he deceived himself, for he invented an "Alkahest" as a remedy for all diseases, and claimed to have discovered the means of prolonging life far beyond its natural term; but none the less left the world in his sixty-seventh year.

The Rosicrucian delusions regarding the magnet were taught in England by Dr. Robert Fludd ("a Torrent of Sympathetick Knowledge," says Charleton) who began to practice medicine in London by virtue of a degree from Oxford in 1605. They made headway—why not, since after all they were in full accordance with so deep-rooted a national superstition as that the King's touch would cure scrofula? Why not, in a country rapidly nearing the vortex of Civil War, under conditions when differences in theology and politics made a man's neighbors his foes, and every man's sword his best friend? What were all the

[1] Boyle: Works, Ed. by Birch, London, 1744 (The Skeptical Chemist), Vol. i, 313.

magnetic discoveries of the philosophers since the world began, in comparison with the marvelous magnetic powder which stood ready to heal the wounds of Edgehill and Marston Moor? In fact does not Walter Charleton, King's Physician, positively tell of the cures "neer allied to miracles" wrought "by Sir Gilbert Talbot upon many wounded in the King's Army; chiefly in the Western Expedition?"[1] And thus, during the period when Charles and Cromwell were fighting, superstition and ignorance and war all united to bring the condition of scientific learning in England to perhaps about the lowest depths which it has sounded in modern times. Then, perforce, it had to rise.

Charleton, in his preface to his translation of Van Helmont, mentions Fludd, but regards as "the choicest flower in our garden" Sir Kenelm Digby. This was because Digby, being of fairly high station, was the promoter of the new cult at Court, and also because Digby had told so many and such variegated fables about the results produced by his vitriol powder as a cure for wounds, as to leave the less fertile Charleton lost in wonder and admiration.[2] He alleged that he had cured a person named Howel, who was pinked in the arm, by the simple expedient of rubbing Howel's garter with the magnetic powder, and that he could set Howel writhing in pain at will by dipping the garter in vinegar. But the new rise of science in England began in the person of Digby. It was very like that of a man clambering out of a mud-hole. The adhering filth was most in evidence.

[1] For by his side, a pouch he wore
Replete with strange hermetic powder,
That wounds nine miles point blank would solder.
—*Hudibras*, ii, 225.

[2] Poudre de Sympathie, Discours fait . . . par le Chevalier Digby. Paris, 1660.

Digby was adventurer, conspirator, naval commander, and diplomatist. He rejoiced in probably one of the most extensive collections of personal enemies ever gathered. They included the Pope, the King, Parliament, afterwards the Lord Protector, and so on through all sorts of people, down to and including his wife's relations. The last accused him of murder. Nevertheless his manners were charming. When Parliament locked him up, his co-prisoners said that he turned the jail into "an abode of delight." His natural winsomeness accounts for his success in gaining the greatest beauty in Europe as his wife, and in inducing the Queen Dowager of France to wheedle Parliament into permitting him to retain his forfeited head.

But his estates, such as they were, were confiscated, and he went into exile in France, and there produced, in 1644, a treatise on the nature of the soul,[1] intended, he says, for the instruction of his son, in which he appropriated as much of Descartes' theory of the magnet and the electric as served his purposes, and presented it as his own.

Digby was by no means without ability, as his career amply proves. And in point of scientific attainments he ranked high for his time. He was the first to observe the importance of oxygen to plant life, and he was the first Englishman to write of the magnet and the electrics in the light of the knowledge gained from the continental philosophers. If he had made his work completely a compendium in English of the discoveries and theories of the latter, as it was in part, he would have rendered a service of great value.

In place of Descartes' spirals coming from the heavens and moving through the pores of the magnets, Digby substitutes atoms, caused to rise from the torrid zone of the earth by the sun's heat, to be replaced by others borne on the heavier air which flows to the equator from the poles.

[1] Digby: Two Treatises, in the one of which The Nature of Bodies; in the other the Nature of Man's Soule is looked into in way of discovery of the Immortality of Reasonable Soules. Paris, 1644.

When cold polar atoms and dry equatorial atoms meet they conglomerate, sink to earth and form stone, which retains the original north and south flowing tendency of the atoms, and hence is magnetic. Then follows an attempt to account for the magnetic phenomena on this theory, palpably modeled on the similar effort of Descartes.

His electric hypothesis is that the electric, being heated by rubbing, breathes out steams which, as they come into the cold air, are condensed and spring back "in such manner as you may observe the little tender horns of snails use to shrink back if anything touched them, till they settled in little lumps upon their heads." These steams, meeting a light body, pierce into it and settle in it, and if it be of "competent bignesse for them to wield," they bring it back with them. It will be observed that Digby's steams behave exactly like Descartes' ribbons. Both make the emanations fly out when the electric is warmed. Descartes brings them back by the cooling of the electric; Digby, by the cooling of the air.

A revival of scientific learning was taking place in France, and Digby had the advantage of being there. The ridotti of the Italians were being copied. Societies for the discussion of scientific subjects were gathering at the houses of Mersenne, Thevenot and De Monmor. There Digby met Descartes, and besides, such men as Gassendus, Paschal, father and son, Hobbes, Roberval and others of less eminence.[1] From that membership came the historic gathering of mathematicians in the Library of Colbert, in June, 1666, and the founding of the Académie Royale des Sciences.

Thus, Digby had no lack of sources of information, and, if the generality of his countrymen could have been induced to believe him, he would have come down to us, perhaps, as a great rejuvenator of English science. But, unfortunately for him, this was not to be. Evelyn, who knew him, speaks of him in his diary as an "arrant mounte-

---

[1] Fontenelle: Eloges Historiques des Acad., Vol. ii.

bank;" and Lady Fanshawe delicately says that, while he made scientific experiments, he had an "infirmity of lying about them." Still he is entitled to the credit of producing a philosophical work in the English language, which unquestionably had no small effect in helping the onward progress of his country in knowledge of natural science, and of probably creating a renewed interest in the physical phenomena of electricity and magnetism—an accomplishment which, although retarded by the errors which he taught, was at least not neutralized by them.

He figures prominently in the scientific literature of his time, and occasionally his opinions are quoted with much deference; but this was probably less due to their originality and merit than to their author's rank and position. It was a new thing for a "man of quality" to be interested in such matters, and still more of an innovation for him to pose as an authority thereon. But the fashion spread, and to Digby is due the honor of leading in a path into which not many years afterwards the king and all the court rushed pell-mell.

The example of the French societies, modeled, as I have said, on the Italian ridotti, was soon followed in England. Shortly after the breaking out of the civil war, an assembly of learned and curious gentlemen, "in order to divert themselves from those melancholy sciences, applied themselves to experimental inquiries and the study of nature." This was the so-called "Invisible College," which began its meetings in 1645 in Dr. Goddard's lodgings in Wood Street, chiefly because there was an artisan in the house able to grind glasses for telescopes and microscopes.[1]

It was the second scientific society instituted in England —the first being the little gathering of students and friends which met, as already noted, in the house of Gilbert, hard by St. Paul's. The new assemblage met to discuss pretty

[1] Boyle: Works, cit. sup. Thomson: Hist. Roy. Soc. London, 1812.

much everything except theology and state affairs. It was unwise to deal with the first—still more unwise to meddle with the second, especially during the Protectorate, seeing that most of the members were devoted royalists.

In the following year Dr. Thomas Browne, another London physician, published his famous Enquiries into Vulgar and Common Errors[1]—a book which represented immense labor in experiment and the collection of curious facts, and which had many editions during its author's lifetime. The popular reception which it encountered is a significant commentary on the changed conditions of the times. Not only was England flooded with copies of it, but it was speedily translated into foreign languages. It was the first systematic and deliberate onslaught upon the popular superstitions and beliefs which had been accepted as true for centuries, and was itself an expression of the skepticism not alone of the author, but of the age.[2] Browne had already written the "Religio Medici," a work which is now classic, and in so doing, had become involved in controversy with Digby, who had explosively replied in a book written in twenty-four hours, part of which time was spent in procuring Browne's work and part in reading it; a proceeding which brought down upon his multitudinous inconsistencies and infirmities the later censure of Browne.

A collection of all the ancient blunders and traditions concerning the lodestone fills Browne's quaint pages.[3] Some of them he refutes in a sensible way; others, in a manner which leaves confusion worse confounded. He records no discoveries of his own, and his theories are borrowed. He followed the Jesuits in the belief that "the earth is a magnetical body," and he adds a bare suggestion to the earlier ideas concerning the tendency of the earth's magnetism to fix its position in space, by saying that the globe "is seated in a convenient medium," thus implying

---

[1] Browne: Pseudodoxia Epidemica. London, 1646.
[2] Buckle: History of Civilization. N. Y.. 1877, i., 263.
[3] Ibid., Chaps. iii. and iv.

that there is some coaction between the medium and the magnetic virtue of the earth, which results in the directive tendency of the globe, and that the latter does not depend upon the effused magnetic force acting in some unknown way. He has no definite theory as to the magnetic virtue, but regards the hypothesis of Descartes, and the notions of Digby founded thereon, as equally worthy of belief. He ridicules Digby's magnetic powder, and all magnetic unguents for the cure of wounds generally; and then, with characteristic shrewdness, puts his finger at once upon the real reason which underlay the healing effects which seemed to follow the use of these nostrums. It is not necessary, he thinks, to conceive of spirits to "convey the action of the remedy unto the part and to conjoin the virtues of bodies far disjoined," because only simple wounds are ever healed, and these, when "kept clean, do need no other hand than that of nature and the balsam of the proper part." In other words, he had noted the rigid requirement of all magnetic healers—that, while the weapon was to be anointed and dressed, the wound was to be simply brought together, bound up in clean rags, and, above all, to be let alone for seven days; and he had seen that simple flesh injuries under this treatment healed themselves by "first intention."[1]

Browne's experiments on the electrics are repetitions of those already well known. He thinks that electric effluvia behave like threads of syrup, which elongate and contract, and, in contracting, bring back the attracted objects. He arrives at the conclusion that no metal attracts, "nor animal concretion we know, although polite and smooth." But the "animal concretions" which he has tried are extraordinary. They are elks' hoofs, hawks' talons, the sword of a sword-fish, tortoise shells, sea-horse and elephants' teeth, and unicorns' horns"—indicating that he had made up his mind that all ordinary substances had already withstood the test of experiment, and further

[1] Paris' Pharmacologia, 23, 24. Mill: System of Logic, Vol. ii., 402.

investigation ought only be undertaken among the contents of museums.

Browne, however, made one experiment which is of especial interest, and which requires from us a glance backwards, and hence a brief digression.

It has already been noted that John Baptista Porta refers to possible communication "to a friend that is at a far distance from us and fast shut up in prison" by means of "two Mariner's Compasses having the Alphabet writ about them."[1] This is the first known suggestion of a possibility which fretted men's minds for many years; namely, that by reason of a supposed sympathy between magnets, the movements of one would be copied by those of another, no matter how great the distance between them; and that, hence, it was necessary only to dispose alphabets around two widely-separated pivoted needles, which had both been magnetized by the same lodestone, to cause the letter to which the needle at one station is moved to be indicated simultaneously by the needle at the other and distant station. Of course, this would now be termed "telegraphy;" and it would not be difficult to find modern dial telegraph instruments operating in accordance with a very similar process.

Porta's idea appears to have been improved upon by Daniel Schwenter, who, in 1600, devised an apparatus of some complexity. He divided the compass card, in each of the widely-separated compasses, into compartments each containing four letters of the alphabet. The needle in signalling was intended to move first to the compartment containing the letter, and then to indicate the especial character desired by one, two, three or four vibrations. Just how the needles were to be worked by the bar magnets or "chadids" which were employed is not clear; but

[1] Porta: Magia Nat., 1589, Book vii. Natural Magic (Eng. Tran.), 1658, Book vii, p. 190.

attention was to be called by the needle striking against a small bell placed in its path.[1] Schwenter's plan did not, of course, bring telegraphy into the world, centuries ahead of its time. In fact, he tacitly repudiates it himself in a later publication,[2] wherein, after learnedly explaining how Claudius in Paris and Johannes in Rome can thus communicate with one another, he denies that any magnet in the world has sufficient strength for the purpose; although he says "Thomas de Fluctibus" (probably meaning Fludd) describes a secret stone in his works which is possibly powerful enough, but neglects to mention "where it was found and who found it."

At the same time, no earlier instance having been encountered, it appears that we may accord to Schwenter the credit of the invention of the first apparatus for (presumably) causing a bell to be sounded by the moving armature of a magnet.

Nine years later, the feasibility of magnetic communication was elaborately disputed by the celebrated lapidary and mineralogist de Boodt,[3] who says that the notion that the magnetic needle can communicate the secrets of thought between friends fifty-five leagues distant is an error, "because it is very certain that the magnet which has touched an iron needle can cause it to move only through a certain and very small interval, perhaps three or four feet." After that, no one seems to have much faith in the idea; and probably because of his own perception of its absurdity, Famianus Strada selects it as the subject of his parody upon the poem of Lucretius, which, it will be remembered, abounded in references to the magnet. This he presents with burlesques of Claudian, Lucian and other ancient poets in the Prolusiones Academicæ[4]

[1] Schwenter (De Sunde): Steganologia et Steganographia. Nurnberg, 1600.
[2] Schwenter: Deliciæ Physico Mathematicæ. Nurnberg, 1636.
[3] De Boodt: Gemmarum et Lapidum Hist. etc. Hanoviæ, 1609.
[4] Strada: Prolusiones Academicæ. Rome, 1617.

—a work of much literary skill and ingenuity. Strada is quite specific, however, in his instructions. Two flat, smooth disks are to be provided, marked around their circumferential edges with the alphabet. Iron needles are pivoted at their centres and energized by one and the same lodestone. "Let your friend," he says, "about to depart, carry this disk with him, and let it be agreed beforehand at what time or at what days he shall observe whether the dial pin (needle) trembles, or what it marks on the indicator. These things so disposed, if you desire to address your friend secretly, whom a part of the earth separates far from you, bring your hand to the disk; take hold of the movable iron; here you observe the letters arranged round the whole margin with stops, of which there is no need for words; hither direct the iron and touch with the point the separate letters, now this one and now the other, whilst by turning the iron round again and again throughout these you may distinctly express all the sentiments of your mind. Strange, but true, the friend who is far distant sees the movable iron tremble without the touch of any one and to traverse now in one, now in another direction: he stands attentive and observes the leading of the iron and follows by collecting the letters from each direction, with which, being formed into words, he perceives what may be intended, and learns from the iron as his interpreter. Moreover, when he sees the dial pin stop, he, in his turn, if he thinks of any things to answer in the same manner by the letters being touched separately, writes back to his friend."

Addison[1] copied from Strada this conceit and talked about it charmingly in the Spectator and Guardian, nearly a century later, and Hakewill[2] and Akenside[3] allude to it.

[1] Addison: Spectator, 241, 1711; Guardian, 119, 1713.

[2] Hakewill: An Apologie or Declaration of the Power and Providence of God. London and Oxford, 1630.

[3] Akenside: The Pleasures of the Imagination. Bk Ill., v. 325-7. London, 1744.

But it was doomed to be ridiculed. Cabæus[1] although giving to it an air of reality by actually depicting the disk with the alphabet around it, denounces it as an absurd error and an instance of the outrageous things which heretics are willing to credit, although they reject the miracles of the faith. Galileo dealt with it in a way which has served ever since as an example to be followed by the skeptical capitalist besieged by the sanguine inventor.

"You remind me," he makes Sagredo say in one of the famous dialogues,[2] "of a man who wanted to sell me a secret of communication through the sympathy of magnetized needles, so that it would be possible to converse over a distance of three thousand miles. I told him that I would willingly purchase it, provided he would show me an experiment, and that it would suffice if I remained in one room while he went in another. He replied that the distance was too short to exhibit the operation of the invention properly; so I dismissed him, saying that it was not convenient for me to travel just then to Cairo or Moscow to test the matter, but that if he would go there himself I would remain in Venice and do the rest."

Yet, despite all the contradictions and ridicule, the conception that people far separated might find a way of communicating with one another, perhaps, through the magnet, or through some means depending upon the magnet or magnetic relations, persisted. There was Cardan's old notion of the magnetism of flesh, which became expanded by the Rosicrucians into the conception that if pieces of muscle cut from the arms of two persons were mutually transplanted, there would be such a community of feeling between the parties that if the alphabet were tattooed on the foreign flesh in the arm of each it would be simply necessary for one individual to prick with a needle the appropriate letter on his own arm to cause a similar sensa-

[1] Cabæus: Philosophia Magnetica. Ferrara, 1629, pp. 30⁻–6.
[2] Galileo: Dialogo Intorno ai Due Massimi Systemi del Mondo, etc.. 1632.

tion at the corresponding letter in the arm of the other person.[1] Van Helmont's story of the artificial nose belongs to the same category.

For such conceits as the flesh magnet and the sympathies attributed thereto, Browne has no stomach. He regards them all as "of that monstrosity that they refute themselves in their recitements." But the two needles and their alphabetical dials he evidently thinks are not to be disposed of with mere expression of disbelief or even contempt. There is a concreteness about that apparatus which makes strongly for its toleration. It is very simple, and the needle of the compass certainly does obey the earth from a long distance, and go to certain marked indications on a card, all of which are conditions closely allied to those in the sympathetic dials. If Browne had lived before Gilbert, he would have written a dissertation, very subtle and very ingenious, no doubt, which would have demonstrated that, from the nature of things and the canons of Aristotle and the names commonly bestowed on the phenomena and substances involved, the whole alleged effect could not be. But Gilbert had lived and passed away, and this thing which "was whispered through the world with some attention, credulous and vulgar auditors readily believing it, and more judicious and distinctive heads not altogether rejecting it," could not be satisfactorily dismissed in any such manner. And therefore Browne, for the first time, tested the sympathetic dials by actual experiment.[2]

"Having expressly framed two circles of wood," he says, "and, according to the number of Latin letters, divided

[1] Fahie. A History of Elec. Telegy. to the year 1837. Lond., 1884, p. 19. An excellent bibliography of the early works on the subject is given here on p. 20. See also

Bertelli: Di un supposto sistema Telegrafico Magnetico . . . dei secoli xvi. e xvii. Rome, 1868.

[2] Browne: Pseudodoxia Epidemica, cit. sup.

each into twenty-three parts, placing therein two stiles or needles composed of the same steel, touched with the same lodestone and at the same point; of these two, whensoever I removed the one although but the distance of half a span, the other would stand like Hercules' pillars, and (if the earth stand still) have surely no motion at all. Now, as it is not possible that any body should have no boundaries or sphere of its activity, so it is improbable it should effect that at a distance which nearer at hand it cannot at all perform."

That gave its quietus to Porta's ingenious conjecture, but still not to the idea which was the life and soul of it. "Now, though this desirable effect possibly may not yet answer the expectation of inquisitive experiment," says Glanvil twenty years later, "yet 'tis no despicable item that by some other such way of magnetick efficiency it may hereafter with success be attempted . . . to confer at the distance of the Indies by sympathetic contrivances may be as usual to future times as to us in a literary correspondence."[1] And again and again in after years this persistent, all-pervading world-notion, which, perhaps, begins with the Scriptural, "Canst thou send lightnings that they may go and say unto thee, here we are," reappeared. "Whatever the way or the manner or the means of it may be," says Beal,[2] writing to Boyle in 1670, in words which sound like those of a seer, "we are sure that we have a perception at great distance, and otherwise than by our known senses, and sometimes a secret anticipation of things future, which cannot be without correspondence with some causative. Whether aerial, more refinedly ethereal, intelligent or astral, whether by any one or other, or all of these strange expedients, we are sure of the great and strange effects; and when we see how quickly the sunbeams do pass to the borders of this vertex, we may well imagine that our spirits may hold an intercourse at

[1] Glanvil: Scepsis Scientifica. Lond., 1665, chaps. xix. and xxi.
[2] Boyle: Works, cit. sup.

like distance with equal dispatch in mental and spiritual affairs." So went on germination of the telegraph; but the plant was of slow growth, and nearly a century had yet to elapse before it began to expand and fructify.

Meanwhile, in the quaint old town of Magdeburg, in Germany, the Herr Burgomaster has been laboring with things stranger than any that the alchemists knew; and Balthasar de Monconys,[1] Lieutenant of Police from Lyons, having set out for the east to discover vestiges of the philosophy of Trismegistus and Zoroaster, has heard of these doings and has gone far out of his way to learn of them. And thus, in October of 1663, Monconys is told of a "globe made of nine minerals" which plays with feathers "continually and without end," and which shows why the moon always looks at us with the same face. Nor was he surprised thereat; for what could not be done by the man who had put two empty hemispheres of copper, not two feet in diameter, face to face, and then proved to the Emperor Ferdinand and all the princes sitting in the Diet at Ratisbon that thirty horses, (fifteen attached to each hemisphere, and the two huge teams tugging in opposite directions), could not pull them apart? Verily, he was a wonderful wizard—the Herr Burgomaster. The Magdeburgers said that he had a devil's contrivance which told him when the storms were coming, and that while his prophecies were always right it was dangerous to live near him—for the thunder one day fell on his house and broke a lot of his infernal toys, and heaven might serve him worse next time for tampering with the spirits of the air, which he shut up and tortured in his tubes and globes.[2]

But the Burgomaster knew his Magdeburgers and they him, and there was little danger that the fellow-officials with whom he dined and smoked and joked would hale him before their courts for sorcery. Besides, he was councillor to his most serene and potent Highness, the Elector

[1] Monconys: Voyages. Lyons, 1665.
[2] Phil. Trans. Abridgt, vol. ii., 29.

From the Frontispiece Portrait
of the Experimenta Nova.

of Brandenburg, and he had other titles bespeaking great consideration. But, at the present time, a couple of centuries later, the moths have eaten all these dignities and we know Otto von Guericke best as one of the first and greatest of the electrical discoverers.

Now, we have to find out what Monconys saw.

Otto von Guericke[1] was Burgomaster of Magdeburg for thirty-five years. He was a many-sided man. He had studied law at Leipsic, Helmstadt and Jena, and mathematics at Leyden, and had travelled through France and England. He had established himself as an engineer at Erfurt, when the attractions of an official career proved more potent than those of his profession, and he entered political life in 1627 as an alderman of his native town. But he could not divorce himself from his interest in physical science; and so, throughout his long public service, he made work in his laboratory his relaxation and his play: just as President Jefferson found pleasure in experimenting upon the conduction of heat through fabrics amid the engrossing cares of the White House, or as Charles II., discovered in physical experiments conducted in his closet at Whitehall, a welcome relief from the feverish excitements and frivolity of his court.

In the history of pneumatics, von Guericke stands in the highest place. He invented the air-pump in 1650, and discovered that in a vacuum animals cannot exist, and all bodies fall with equal rapidity. He recognized that gases have weight, and by means of the "Magdeburg hemispheres," already alluded to, he showed how great the force due to pressure exerted by the air, by comparing it with the contrary pulling strain of horses. He invented the air-balance and the anemoscope, and, by such means, weighing the air, he was enabled to make his astonishing

[1] Hoffman: Otto von Guericke. Magdebourg, 1874. Paschius: De Inventis, vii., § 29. Fontenelle: Eloges Hist. des Acad., vol. ii.

predictions of the weather. Besides all this, he made the remarkable electrical experiments and discoveries now to be described.

Most of the long and arduous researches whereby physical science has become established have been undertaken, not for the purpose of ascertaining results previously unknown, but in the hope that their outcome would afford support for some preconceived and favorite hypothesis. In this way, as I have already pointed out, Gilbert was led to his magnetic and electric investigations, trusting to find in them corroboration for his cosmical theory. So Cabæus undertook similar studies in the hope of eliciting evidence which would break down, not only Gilbert's conception, but the Copernican doctrine generally. So observation of the magnetic spectrum and its phenomena resulted in the mentally conceived spirals of Descartes and their percolation through and grouping about the magnet. So, in another field, the alchemists established the science of chemistry through their futile experiments in search of the transmutation of metals.

In dealing with the ancient and mediæval philosophers we have seen that they seldom narrowed their observation down to specific matters. Their treatises, as a rule, were on the Nature of Things—De Natura Rerum—from the days of Lucretius onward, and there was a time when a writer, such as St. Isidore, might reasonably compress all that was known about everything in natural philosophy, both celestial and terrestial, into a very moderate-sized tome. As sublunary things, however, became familiar and commonplace, philosophical speculations began to change, and finally, during the sixteenth century, the fundamental hypothesis was one pertaining, not to the nature of things in general, but to the nature of the extra-mundane regions and of the worlds moving therein. Hence to the three great theories of Ptolemy, Copernicus, and Tycho Brahe were added sometimes new hypotheses, sometimes new supporting arguments, just in proportion as new knowledge, based on physical experimenting, gave new basis for one or the other. Therefore when a man had made novel discoveries, instead of contenting himself with stating simply what he had done and how he had done it, and leaving other people to make and find useful applications of the new-found information, he was far more likely to begin his dissertation either with a new cosmical hypothesis or a re-statement of his favorite old one, and then to adduce the discoveries as establishing the new notion or as affording additional proof to the preferred doctrine. The suppression, by the way, of this discursiveness, as broad as the universe itself, and the limitation of scientific treatises to matters strictly germane and relevant to their subjects, is one of the great achievements of the Royal Society and of the various philosophical bodies modeled after it.

Consequently, as might be expected, when von Guericke gave the results of his pneumatic and electrical experiments to the world, he did it in a treatise on Vacuous Space "in quo totum Mundi Systemi consistit."[1] His first book deals with the universe generally, and his second with interstellar space. His own discoveries occupy the third and fourth books, and then he gets back to vast conceptions again, and, in successive divisions of his work, considers the earth and moon, comets (whereof it may be noted in passing he first pointed out the periodicity), the planets and the fixed stars. We need not here occupy ourselves with his astronomical or cosmical notions, and the detailed history of his beautiful discoveries in pneumatics belongs to a different field in physics from that in which we are now wandering. Hence the matters which interest us, and those which astonished Monconys more than two centuries ago, are also set forth in the book, wherein are treated "the mundane virtues and other things thereupon depending."

[1] Von Guericke; Experimenta Nova (ut vocantur) Magdeburgica de Vacuo Spatio. Amsterdam, 1672.

A few words of preface may be granted, because von Guericke has ideas of his own about these virtues, which, together with life and matter, enter into the constitution of all bodies. They are effluvia—sometimes corporeal, such as the air—sometimes incorporeal, or more properly highly diffused, such as those which emanate from a body, and surrounding it, form its orb or sphere of virtue. Of the incorporeal virtues there are many—not all perceptible, because of defects in our senses; but those which we are best able to recognize come from the earth or from the sun. Thus from the earth arise "impulsive virtue," "directive virtue," "turning virtue," "sounding virtue," and so on; while the sun yields "light and coloring virtue," and the moon "frost-making virtue." Then there are other virtues derived from the planets, which the astrologers call "influences."

All of these virtues are alike in that they can act at a distance. They join themselves to neighboring bodies, which simultaneously recognize them; but if any virtue meets a body not suited to it, it is repelled or reflected, and the repercussions may continue until the virtue is exhausted and ceases. The more solid the body, the more virtue it is capable of receiving. Certain virtues accord with certain bodies and there is mutual suitability. They are excited therein by attrition, collision, touch, vibration, and so on.

Von Guericke's "impulsive virtue," so called, appears to be simply momentum. "Conservative virtue" is gravity. "Directing virtue" is Gilbert's verticity, the magnetic force which he thought adjusted the earth's axis in space and prevented its nutation. "Turning virtue" is any impressed rotary motion; von Guericke gives as an example a man revolving on his heel. "Sounding virtue" is that which causes the sensation of sound, and is produced "by the friction of bodies." "Heating virtue" is heat due to subterranean fire or "friction of the sun's virtue." "Lighting virtue" is the sensation of light and color. It is clear that von Guericke perceives that there are certain resemblances between the phenomena due to the play of natural forces, and has, perhaps, a hazy notion of some correlation between them—of the development of one phenomenon from another, as the production of heat by friction of the sun's virtue. Hence he is seeking, in the language of the arithmetics, to reduce all these factors to a common denominator. This he finds in the idea of "virtues." They are all different, these natural happenings—light and sound and magnetism and heat and gravity—but none the less they are all "virtues" emanating from a physical body, such as the earth or the sun, and as such they are as necessary a concomitant of that body as the matter whereof it is composed.

This is von Guericke's main hypothesis, corresponding to the magnetic theory of Gilbert. But he differs squarely from Gilbert in the belief that the earth is a great magnet. The globe, he says, is moved by the rays of the sun and its own intrinsic turning virtue; by two forces, and hence it would naturally be controlled unevenly. Therefore it is given by nature a directive virtue (whereof its poles are merely termini), "so that it does not sway this way and that in its position, not even on account of the rubbing of the rays of the sun, and so that it does not wabble or nutate in its own daily rotation, and change the times of the year, length of days," etc. This directive virtue is not inherent to the earth itself, but is imparted to it by nature, which does nothing in vain, and for the express purpose of preventing wabbling. Consequently he concludes that Gilbert is wrong in regarding the globe as intrinsically a big magnet.

There is, however, still another virtue, but which is of higher import than all of the others. Von Guericke goes back to Aristarchus, and says that he, believing the earth to be animate, revised the opinion of those who thought that it had both an attractive and a repelling faculty. "This," says von Guericke, "appears harmonious with reason, for if the earth has the power of attracting those things which are agreeable to it, it will likewise have the power of repelling those things which injure it and do not please it;" and specifically, if one planet "impresses its contrary influence upon another, that other resists the same by its own repelling virtue." For a modern philosopher and a skillful experimentalist, this notion of an expulsive virtue or repelling force actually existing in the earth was a new one. Clearly, moreover, he is not evolving that supposition out of his inner consciousness, but because he has some tangible physical reason for it. Observe first, that he has imputed to the earth not one, but many different virtues; second, that he has denied that it is by substance a magnet; and, third, that he gives it this new repelling effect. Now why is he doing this?

Let me recall here a peculiarity in mode of reasoning common to all of these old philosophers—save perhaps Gilbert—and that is, they always present their theory first, and then detail some physical phenomena, usually new ones, which they think sustain it—the idea being to lead the student to suppose that the working of the superior brain alone has produced the conception, the truth of which detected nature is afterwards compelled to admit. This is not only putting the cart before the horse, but also causing it to appear that the vehicle tows the animal. In that way Descartes puts his spiral-ribbon theory first, and his observation of the iron filings in the magnetic spectrum last; Cabæus his rebounding-effluvia hypothesis before the repelled-chaff experiment.

Von Guericke is evidently following the same course; or in other words, he has found some strange and novel effects, and all this new theory about the virtues, etc., rests on that basis; while, like his predecessors, he fails to perceive that the credit and the honor due to him who is gifted with eyes to see and ears to hear what the laws of nature are revealing cannot be enhanced by an effort to make the world believe that the true source of it all is not nature, but himself.

Von Guericke's position is that if the earth is by substance a magnet, it must have all the magnetic properties—not only directive virtue or verticity, but also the attractive magnetic power. But in comparing what he calls the conservative virtue of the earth with its directive virtue, he points out that "the former attracts all bodies not only in the regions of the poles but everywhere. The bodies attracted are not changed but are held by a conservative force."

In other words, von Guericke believes that the magnetic quality of the earth simply adjusts and holds its axis in space and exerts no attractive force on exterior bodies. The attractive force which the earth does manifest is gravity, and that is owing to the "conservative virtue." And the "conservative virtue" is the same thing as electrical attraction, which he says is exerted not like magnetic attraction merely at the poles but at all points of the electric. So, according to him, gravity is not correlated to the earth's magnetism, but is simply the electrical attraction exerted by the earth upon exterior objects, and he believes it to be due to the rubbing of the globe by the sun's rays.

Note the difference between this conception and Gilbert's dictum that the "matter of the earth's globe is brought together and held together electrically." Von Guericke's idea is that the earth, as a mass, electrically attracts, not only its own matter, but also outside matter.

Now what Monconys saw was the experiment which illustrated the possession by the earth of conservative virtue—that is, electrical attraction—and also of the capacity of not only attracting, but repelling other bodies. Von Guericke had copied Gilbert's idea of the earth-kin—the terrella—but had made his miniature globe, not of a magnet, but of an electric. And Monconys saw von Guericke rotate that electric globe to imitate the rotating earth, meanwhile rubbing it; and then he also saw, and describes fairly well, the extraordinary phenomena revealed by this first of all electrical machines.

Monconys says that Guericke told him, among other things, that the smooth yellow sphere which was exhibited was made of nine different minerals, but either Monconys' memory was at fault or else Guericke thought a little misdirection justifiable in the circumstances. Guericke himself tells how to make the globe, as follows: "Take a sphere of glass (called a vial) of about the size of the head of an infant; put in it sulphur that has been pulverized in a mortar, and sufficiently liquefied by being placed near a fire. When this has become cool, break the sphere, take out the globe, and keep it in a place that is dry." Afterwards it is to be perforated, so that it can be rotated on an iron axis, and that, perhaps, is a reason why he did not rub the glass vial itself instead of the sulphur cast in it. It would be difficult to make a smooth round hole in the imperfect glass of those days, for the insertion of the axis.

The sulphur globe being described, Guericke proceeds to show how it possesses the different virtues and the results thereof.

"It has, first, the impulsive virtue (momentum), because, being heavy, it can be hurled by the hand further than if it were made of wood or lighter material." So also it has the conservative virtue, to exhibit which the axis of the globe is placed in two supports a hand's breadth above the supporting base or platform, upon which and beneath the globe are to be strewn "all sorts of little fragments, like leaves of gold, silver, paper, shavings," etc. The directions are to "stroke the globe with the dry palm, so that it may be rubbed or submitted to friction thus twice or thrice. Then it will attract the fragments, and when turned on its axis will take them along with itself. In this manner is placed before the eye the terrestrial globe, as it were, which by attracting all animals and other things which are on its surface, sustains them and takes them around with itself in its diurnal motion in twenty-four hours.

"Thus this globe when brought rather near drops of water causes them to swell and puff up. It likewise attracts air, smoke, etc.

" From these experiments it must be seen that there exists in the earth for the preservation of itself a virtue of this sort, which also can be excited in an especially suitable body, namely, this globe, so that it acts more in it than in the earth itself (for whatever this globe attracts, it snatches it, as it were, or draws it away from the earth)."

Now follows the first positive recognition of electric repulsion—which is none other than von Guericke's expulsive virtue. Cabæus had seen the chaff leap back from the electric, but he had not interpreted the phenomenon itself, although he had tried to concoct a theory in conformity with it. Not so von Guericke. Hear him:

"Even expulsive virtue is to be seen in this globe (namely when it is taken from the apparatus to the hand and is rubbed or stroked in said manner with the dry hand), for it not only attracts but also *repels* again from itself little bodies of this sort (in proportion to their temper), nor does it receive them until they have touched something else."

There is also the first suggestion of the discharge of the electrification of the attracted body on contact with an object other than the electric, and its conscquent re-attraction by the latter.

But note his experiment. He takes the globe out of its supports and holds it in his hands with its axis vertical. Then, after exciting the globe and causing it to repel a feather, he carries it around the room, so that it drives the feather, floating in the air, before it. His feather is a bit of down, which he says "extends itself and in some way shows itself alive"—its individual electrified filaments of course mutually repelling. He observes that when it is thus chased around the room it prefers to approach " the points of any object whatsoever before it, and it is possible to bring it where it may cling to the nose of any one." Here he is anticipating Franklin in recognizing the effect of pointed conductors in drawing off the electric charge.

"But," he continues, "if one places a lighted candle upon the table and drives the feather at a distance of about a hand-breadth from the candle up to the globe, the feather suddenly recedes and flies to the globe as a sort of guard;" and thus he observes the dissipation of the charge on the

VON GUERICKE'S ELECTRICAL MACHINE AND SULPHUR GLOBE.[1]

feather by the hot air, so that it becomes no longer repelled but once more attracted by the rubbed globe.

Now follows a number of other curious observations of the electrified feather. He finds that the same part or face of the feather by which the feather has been once caught up by the globe and then repelled is kept unchanged in the orb of virtue; so that if any one puts the

[1] Reproduced in reduced fac simile from von Guericke's Experimenta Nova Magdeburgica. On the right appears the first electrical machine; on the left, the sulphur globe on the end of the staff held by the figure is represented as repelling the floating feather (*a*).

THE DISCOVERY OF ELECTRICAL CONDUCTION. 399

globe above the feather, the latter "inverts itself in the air and views the globe always with the same face." Von Guericke stops here to suggest that "it is from the same cause that the moon always turns the same face toward the earth, and doubtless in the orb of the earth's virtue is thus repelled by it and there held." Then continuing: "if the feather begins to unfold its pinnules on the globe and you extend your finger or something else to it, it will fly to it and recede toward the globe, and repeat this several times; but if you present a linen thread to the feather, all its pinnules are straightway attached to the globe, and thus attached lie for quite a while as if dead, until they again erect and extend themselves. In the same manner this feather shows the fire to such an extent that if it thus unfolds itself and the flame of the candle is moved up to it, the feather throws itself back upon the globe."

"If the globe is suspended on its axis in the apparatus in such a manner that it can turn, and be excited by the palm in the accustomed manner, and a rather soft feather is placed beneath, the globe will then attract the feather many times and drive it around away from itself into the nearest place underneath itself, and continue this for some hours." Thus the feather is alternately charged and discharged and so kept in vibration. Von Guericke finds in this proof of the animate nature of the globe. "When it does not want to attract," he says, "it does not attract." Nor does it "allow the feather to approach until it has cast it against something else, perhaps in order that it may acquire something therefrom."

Now comes the announcement of a discovery of the highest import. Gilbert had seen a rod, rendered magnetic at one end, become magnetic at the other; but no one had observed any transference of the supposed electric effluvia except from the surface of the electric to the limits of the orb or sphere of virtue. Von Guericke now, for the first time, makes known electrical conduction—the transference of electrification from an electrified body to one not electrified, and the appearance of the electrification in a long conductor at the end opposite to that at which it is produced—or, in other words, the apparent instantaneous transfer of electricity from end to end of the line. He had noticed that "if you let down almost to the globe a linen thread suspended from above and try to touch it with the finger or something else, the thread recedes and will not allow the finger to meet it." So he fastens a similar thread, an ell in length, to the end of a sharp stick attached to a table, and allows it to hang vertically and so that its extremities will be situated "a thumb breadth distance from some other body"—the nature of which is not material. Now he excites his globe and brings it up to the stick which supports the thread. And then he sees the lower extremity of the thread move up to the adjacent body.

"By this," he says in one place, "it is demonstrated to the eye that the virtue extends itself in the linen thread even to the lowest parts where it either attracts or is itself drawn"—and in another "This experiment ocularly shows that the sulphur globe, having been previously excited by rubbing, can exercise likewise its virtue through a linen thread an ell or more long, and there attract something."

Here—and not in mythical sympathies of widely-separated magnets—was the true beginning of the harnessing of the lightning to compass the annihilation of distance and time. The first telegraph, the first conductor for the transmission of energy by electricity, were there in von Guericke's "linen thread an ell or more long;" and its quivering extremity, swinging to the juxtaposed body, indicated the approach of the excited globe to the distant supporting rod as certainly and by means of the same medium as does the equally swinging spot of light in the receiving station show the varying electrification of the great cable controlled on the other side of the Atlantic.

Note, moreover, that von Guericke attaches his little

linen line to a "pointed stick." He had before stated that points, even a person's nose, best attracted the floating electrified feather. Having found out the discharging advantage of the point, he thus applies it as the best means of causing the virtue to pass upon the linen thread.

This is one of the most remarkable examples of thoughtful invention which the history of electricity affords. He conceived the idea that the electrical virtue could be made to pass over a line; that the charge could be imparted to a thin thread conductor by connecting the latter to the sharpened extremity of a fixed support; that the support could be electrified by bringing the rubbed globe into proximity with it; that the end of the hanging thread would, when electrified, move toward and from an adjacent fixed body, and that therefore a movement to and fro of the excited globe at one end of the line would instantly cause a like vibratory motion of the other end.

And yet von Guericke had never heard of "Maxwell's laws," or "surface density," or "ether strains and stresses;" but he lived in the seventeenth century, and therefore it is conceivable that he may have made the discoveries above outlined. Had he lived in the nineteenth, plenty of people would be ready to argue the opposite, for to these doubting minds no man can now be presumed to have discovered anything if, after the event, it is objected that he was ignorant of the laws which higher intelligences think they would have followed had they made the discovery themselves. Besides, there would not be wanting other keen spirits to recognize a complete anticipation of his revelation of electrical conduction in Bacon's allegation of half a century before, that "it is an ancient tradition everywhere alleged, for example, of secret properties and influxes, that the torpedo marina, if it be touched with a *long stick*, doth stupefy the hand of him that toucheth it."[1]

[1] Bacon: Nat. Hist. Cent., x., No. 993.

But a still more wonderful discovery is yet to be recorded. Von Guericke says to rub the sulphur globe with the dry palm of the hand. Then if "you take the globe with you into a dark room and rub it, especially at night, light will result, as when sugar is beaten."

And that is the first announcement of the electric light.

He had seen a brush discharge between the electrified globe and his hand, and although he does not connect the two phenomena, he had also heard the snaps and crackling incident thereto. "There is likewise a virtue of sound," he says, "in this globe, for when it is carried in the hand or is held in a warm hand and thus brought to the ear, roarings and crashings are heard in it."

And thus von Guericke established, to his own satisfaction, that the sulphur globe is endowed with many virtues. When thrown by the hand it had momentum or impulsive virtue because of its weight. It drew light bodies to it and expelled others from it, and hence had both the conservative and expulsive virtues. It had also the sounding, lighting and heating (by friction) virtues, but not the turning and directing virtues. But in that it had the conservative and expulsive virtues it was like the earth. It was an electric instead of a magnetic terrella—showing that our globe is not a mass of primordial terrene Matter drawing things to itself while directing its own axis by its inherent magnetic quality, but a great electric mass having for its chief characteristic its conservative or attractive virtue, and endowed from outside with a capacity whereby its axis is prevented from wabbling; a vast electric machine rotated by the hand of the Almighty and excited by the friction of the solar rays.[1]

---

[1] The first record in the annals of the Royal Society which has any relation to electricity is a review of von Guericke's treatise. It was quickly recognized that his sulphur globe was an electric terrella—"by which experiment," it is added, "he thinks may be represented the chief virtues he enumerates of our earth," and that "the impulsive, attractive, expulsive and other virtues of the earth, as he calls them, may be ocularly exerted." Phil. Trans., 1672, No. 88, p. 5103.

Such was the conception which Guericke sought to establish. The effort resulted in the discovery of electrical conduction, of electrical polarity, of the transmission of electrification over an elongated conductor, of electric light, of sound produced by electricity, of the discharging capacity of points, of the dissipation of charge by hot air, and of the vibration of a freely-movable body due to its charge and discharge; the first recognition of electrical repulsion as such, a direct suggestion of the identity of electrical attraction and gravity, and the construction and successful use of the first machine for the production of electricity.[1]

And all these accomplishments remained practically unnoticed until the days of Dufay. Well might that generous discoverer detect in them the cause of subsequent progress, and express his astonishment that they had remained so long forgotten.

[1] On peut voir dans le recit abrégé de ces experiences, la base et la principe de toutes celles qui ont été faites depuis avec le tube et le globe de verre: et on ne peut s'empecher d'être surpris quelles ayant demeuré si longtemps dans l'oubli, ou du moins qu'on ne se soit pas avisé de les repeter et de tacher de les porter plus loin. Mem. de l'Acad. Roy. des Sci., 1733, 25.

CHAPTER XIII.

THE Invisible College in England continued to hold its meetings in Oxford and in Gresham College, but in 1659, upon the fall of Richard Cromwell, the members were scattered, and their gathering-place converted into barracks. The advent of King Charles, however, gave them new courage; and in 1660, twenty-one persons, including, among others, Sir Kenelm Digby, Dr. Wilkins and Mr. John Evelyn, regularly organized themselves into a society for the promotion of all kinds of experimental philosophy.[1]

The prospects of the new society were not flattering. At best it might plant a few seeds of sound knowledge of which chance might favor the growth, or maintain a cult which now and then might attract a disciple. But for the great body of the English people, exhausted after twenty years of incessant strife, and still in the turmoil and excitement of the Restoration, physical science probably possessed no more immediate living interest, than it had for the troublesome savages in the Irish bogs. So the new philosophy had little to expect by way of speedy advancement; nor had it the inherent motive power capable of diffusing it through the vast and sodden mass of popular ignorance and indifference; still less the more potent impetus required to effect the substitution of new learning for old, in minds which the latter had saturated and there become stagnant.

Nevertheless, it numbered among its members such men as John Wallis, the mathematician; John Wilkins, afterwards Bishop of Chester; Seth Ward, later Bishop of Salisbury; Jonathan Goddard, warden of Merton; Sir William Petty, and most eminent of all, Robert Boyle. And per-

[1] Thomson: Hist. Roy. Society. London, 1812.

haps because of its fostering care, we hear now and then of a new conceit; such as Hartlib's discovery of the ink which gives a dozen copies on a moist sheet of paper applied to the writing; or Colonel Blount's new plows, or Neale's telescopes, or Greatrex's fire engine, or Petty's double-bottomed ship.[1] Evelyn dines with Wilkins in 1654, and admires his ingenious apiaries, so made that the honey can be taken without disturbing the bees, his way-wiser, thermometer and monstrous magnet.[2] A year later, he records seeing a "pretty terrella, described with all the circles and showing all the magnetic deviations."[3]

Gilbert had said that the earth is a magnet and does rotate. The Jesuits, contrariwise, and with characteristic casuistry, had said that the earth is not a big magnet, but merely a magnetical body, and that it does not rotate. Others had admitted that the earth rotates, while denying its inherent magnetic quality. Now comes Father Grandamicus,[4] from the Jesuit College at Flêche, in France, with an effort to reconcile all difficulties on the new and original basis that the earth is a big magnet, and for that very reason does not rotate, because, like the magnet, it has poles, and no magnet has ever been seen, by its own inherent magnetism, to turn itself around its own poles. But Dr. Power[5] was ready with a "confutation," and, to the credit of the College, he talks of the corporeal effluviums of the magnet and the electric about as well as anybody had done before him, and sets the Frenchman right with all the emphasis peculiar to a semi-theological disputation of the times.

The great impulse which was to start anew the progress

[1] Knight: Hist. England, iv., 174.

[2] Evelyn's Diary. 13 July, 1654.

[3] Ibid., 3 July, 1655.

[4] Grandamicus: Nova Demonstratio Immobilitatis Terrae petita ex virtute magnetica. Flexiae, 1645.

[5] Power: A confutation of Grandamicus, his magnetical tractate de Immobilitate Terrae. London, 1663.

of English experimental science had not yet, however, been felt. It was to come, if at all, from without, and from without it came, and from a quarter least of all to be anticipated. For the first time in the history of mankind Fashion and Science joined hands. All the benefits which the stern goddess had offered had been as nothing; all her struggles to stir the inertia of the load had been futile; but now a beckon and a nod from the fickle and laughing dame, a touch of the finger, and the mountain moved.

The Society applied to Charles for a charter. There was no reason why so devoted a band of Royalists should not thus be rewarded, especially as doing so involved no settlement of old pecuniary scores for aid and comfort. So the King not only converted it into the Royal Society, but gave to it, what was far more immediately valuable than the charter, the light of his kingly countenance. The result upon the fortunes of the new philosophy was magical.[1] The Court, in lieu of baiting Puritans, place jobbing, flirting and gambling, fell to discussing the pneumatic engine, the ponderation of the air, blood transfusion, and the variation of the compass. My Lord Keeper Guilford thriftily had barometers constructed for sale in London, and united with my Lord Chief Justice Hale in making suitors wait pending the production of obiter dicta on hydrostatics. Prince Rupert invented mezzo-tinto engraving, and set the willingly admiring courtiers to breaking off the tails of the wonderful little drops of glass which he had brought into England, to see them fly to pieces. Even Buckingham found time, amid the pressing claims of wine, women, the gaming table and the stage, to dabble in chemistry. If one dropped in at Will's it was to find men of fashion discussing telescopes and the Vacuo Boyleano. My lady, in her boudoir, chattered of the shining phosphorus from Germany, or went in her coach and six to visit the Gresham curiosities, and "broke forth into cries of delight at finding that a magnet really attracted a needle,

[1] Macaulay: Hist. of England, Chap. iii.

and that a microscope really made a fly look as large as a sparrow." Did not that "mighty pretender to learning, poetry and philosophy," the Duchess of Newcastle (and with her the Ferabosco, with "good little black eyes"), visit the new Society to witness experiments "upon colors, lodestones, microscopes and liquors?" And did not the Lord President receive her (together with the Ferabosco) and escort her to her seat with the mace solemnly borne before?—and, let us hope, with a properly straight visage. And her Grace was indeed edified, for "after they had shown her many experiments," records Mr. Pepys,[1] "she cried still she was full of admiration, and departed," Mr. Evelyn being in waiting to hand her to her coach.

And Mr. Pepys likewise undertakes a little magnetic experimenting on his own account. "This day" (Nov. 2d, 1663), he records, "I received a letter from Mr. Barlow with a terrella which I had hoped he had sent me, but to my trouble I find it is to present from him to my Lord Sandwich; but I will make a little use of it first, and then give it to him." He kept it nearly a month before delivering to Sandwich, who, he says, received it with great pleasure.

And as for the king, he set up a laboratory at Whitehall and worked in it. He went to the Society's rooms and looked at experiments on the new liquid for staunching the flow of blood.[2] And, when the men of quality came to chat with him of a morning during the portentous ceremonies of tying his cravat or combing his wig, they found his Majesty with far less appetite for court gossip than for weather observations.[3] Even at the Council Board, the royal thoughts were apt to wander from the doings of the Dutch abroad and his last idea for extorting taxes at home, to the new baroscope with which he and his chaplain Beal amused themselves.

[1] Pepys' Diary, May 30, 1667. See also Evelyn's Diary, same year.
[2] Phil. Trans., 1673, No. 96, p. 6078.
[3] Thorpe: Essays on Hist. Chemistry. London, 1894 (Robert Boyle).

Hence it came about—because Iris had lent Minerva her wings—because Folly had put her shoulder to the load which Reason could not move—that a great progress in philosophical thought was made. The Aristotelian physics and the moribund relics of scholasticism expired; the newer vagaries of the Rosicrucians faded into thinner air than even their most refined spirits could breathe. The "sure arguments and demonstrated experiments" for which Gilbert had so strongly pleaded were hereafter to be the only foundation for physical knowledge. And all this, because the touch of that singularly wise, pure and good Charles had made experimental science the mode.

But however much people betook themselves to the new philosophy because it was fashionable, this was far from being the reason which influenced the king himself. His taste for science was no craving for new diversion, nor did he soon tire of his fancy. His inclination to physical study and experiment was natural. He would have been a good chemist or physicist had he not been king. Sprat, writing five years after the establishment of the Society, tells us that he constantly spurred the members onward to fresh exertion and "provok'd them to unwearied activity in their Experiments by the most effectual means of his Royal example:" that "the noise of Mechanick Instruments is heard in Whitehall itself, and the King has under his own roof found place for Chymical Operators." It is the king who "has endowed the College of London with new Priviledges . . planted a Physick Garden under his own eye" and "made Plantations enough, even almost to repair the ruines of a Civil War"—the king who offered rewards to "those that shall discover the Meridian," the king who, "acknowledged to be the best Judge amongst Seamen and Shipwrights," set the Society studying the problems of navigation and ship-building. That he was especially interested in magnetism is shown by his presentation of a terrella to the Society—a stone which the members examined twenty-five years later to see whether its poles had changed in position.[1]

"He has frequently committed many things to their search," says the future Bishop, beginning a succession of sentences which insist upon irrelevantly recalling the arraignment of George III. in the immortal Declaration—"he has referred many foreign rarities to their inspection: he has recommended many domestick improvements to their care: he has demanded the result of their trials in many appearances of Nature: he has been present and assisted with his own hands at the performing of many of their Experiments, in his Gardens, his Parks and on the River . . he has sometimes reproved them for the slowness of their proceedings."

Nor did he fail to recognize the democracy of science—for when the young Society demurred at admitting into its fold John Graunt, citizen of London, the judicious author of the Observations on the Bills of Mortality (the first great work on its subject) because he was a tradesman, it was speedily brought to its senses by a sharp message of disapproval from his Majesty and a curt order "that if they found any more such tradesmen, they should be sure to admit them all without any more ado."

The importance of the part which the Royal Society played in the development of the new philosophy, and later in that of the new science of electricity, can not be overrated. Indeed, it may be said that at the very beginning of its career, the sturdy blows which it dealt to witchcraft, sorcery and demonology, by shattering popular belief in these delusions, did much to emancipate electrical knowledge from the errors with which it was encumbered. But the example, the stimulus, the encouragement, the immediate help, without which its efforts might well have proved fruitless, it owes in no small measure to the king himself. Therefore in estimating the conditions of the philosophical renaissance now under review, it is necessary to remember

[1] Phil. Trans., No. 388, p. 344, 1687.

not the dissolute, prevaricating, pleasure-loving monarch, whose reign reduced England to the lowest political level she has touched in modern times, but rather the eager student who vied in making experiments with the other members of the Society, and who directed the influence of his great position toward the promotion of knowledge and research with a vigorous enthusiasm such as the world had never seen before, and very seldom since, in the occupant of a throne.

Meanwhile, events happened which, although disastrous to the community in general, tended to advance the new Society and create for it augmented popular interest. The frightful epidemic of plague of 1665, in London, followed by the great fire of 1666, caused all classes to turn to the Royal Society for advice looking to the prevention of such scourges and the rebuilding of the devastated town: this time seriously and earnestly, and not à la mode. And the Society rose to the occasion, and investigated building materials and new modes of construction, roadmaking and the laying out of streets, together with ways and means of destroying infection, and specifics against the dread disease.

Its enthusiasm matched that of its royal patron. "The Fellows set to work to prove all things that they might hold fast that which was good,"[1] remarks Professor De Morgan, satirically, forgetting that this was the first institution in which the idea of progress was distinctly embodied.[2] True, they considered whether sprats were young herrings; whether a spider would stay within a circle of powdered unicorn's horn (which it would not); whether barnacles turned into geese; whether diamonds grew in their beds like oysters; and if one should choose to select further absurdities, it would not be difficult to make their proceedings appear grotesque. But this is not only de-

---

[1] De Morgan: A Budget of Paradoxes. London, 1872.
[2] Buckle: Hist. of Civilization. N. Y., 1877, i. 269.

liberately to disregard a long list of experiments which are useful and valuable, but to ignore the famous announcement made by Robert Hooke, which is at once a declaration of independence of the old philosophy, and a tolling of its knell.[1]

Although Sir Kenelm Digby was of the council (and then in high favor at court, being named in the king's charter as "chancellor to his dearest mother Queen Mary") the Society, even before its regular organization, demolished his magnetic nostrum and apparently did not even think it worth while to consider the report of the "curators of the proposal of tormenting a man with the sympathetic powder"—a committee which it appointed in June, 1661.

Among the members was Dr. (afterwards Sir) Christopher Wren, who appears on a different eminence from that which he occupies as the great architect of St. Paul's cathedral in London, and of the graceful spire which throws its shadow across the busiest part of Broadway. He invented the first registering and recording apparatus —a weather-gage and clock combined, actuating a pencil over a record surface so that "the observer by the traces of the pencil on the paper might certainly conclude, what winds had blown in his absence over twelve hours space;" the registering thermometer, the pluviometer, balances for determining weight of air, besides many improvements in astronomical instruments; but more interesting to us is his arrangement of a huge terrella in an opening in a flat

[1] "This Society will not own any hypothesis, system or doctrine of the principles of natural philosophy, proposed or mentioned by any philosopher, ancient or modern, nor the explication of any phenomena, where recourse must be had to original causes (as not being explicable by heat, cold, weight, figure and the like, as effects produced thereby), nor dogmatically define nor fix axioms of scientifical things, but will question and canvass all opinions, adopting nor adhering to none, till by mature debate and clear arguments, chiefly such as are deduced from legitimate experiments, the truth of such experiment be demonstrated invincibly." Weld: Hist. R. S., i. 146.

board "till it be like a globe with the poles in the horizon." This board he dusted over with steel filings "equally from a sieve" and then studied the curves of the filings as they delineated the magnetic spectrum. Sprat tells us that he found that "the lines of the directive virtue of the lodestone" are "oval" and that appears to have been another recognition of "lines" of directive virtue—a conception curiously similar to Faraday's lines of magnetic force.

Among the other experiments which Sprat records, made prior to 1665, when the Society began the publication of its transactions, are essays "to manifest those lines of direction by the help of needles; to discover those lines of direction when the influence of many lodestones is compounded; to find what those lines are in compassing a spherical lodestone, what about a square, and what about a regular figure; to bore through the axis of a lodestone and fill it up with a cylindrical steel." Experiments also were made on lodestones "having many poles and yet the stones seeming uniform;" "on the directive virtue of the lodestone under water," and "to examine the force of the attractive power through several mediums."

No reform sought by the Society proved of higher moment to the progress of science than that which put an end to the De Natura Rerum treatise. If any one had anything to communicate, it compelled him to do so relevantly and briefly. It ruthlessly rejected dissertations starting from the time of Adam, introductory to a physical fact observed yesterday. It "exacted from all its members a close, naked, natural way of speaking, positive expressions, clear senses, a native easiness, bringing all things as near the mathematical plainness as they can, and preferring the language of artisans, countrymen and merchants before that of wits or scholars." Thence sprang that requirement which enters into all highly-developed modern systems of Patent Law, that a specification shall not be addressed to the erudite and learned, but shall be written in such full, clear and exact terms that any person skilled in the art to which it nearest relates shall be able to understand it and put it in practice. In a word, the Royal Society completely revolutionized didactic and technical writing and the mode of expressing scientific thought, and thereby did enough, had it immediately afterwards gone out of existence, to earn for itself the perpetual gratitude of mankind.

Yet the glowing language of the ode which Cowley addresses to the young Society, in which he compares it to Gideon's band picked out by divine design to do "noble wonders," and predicts its discovery of "New Scenes of Heaven" and "Crowds of Golden Worlds on High," not to mention numerous new countries on earth, by no means commanded universal assent. In fact, the poet especially desires that

> "Mischief and tru Dishonour fall on those
> Who would to laughter or to scorn expose
> So Virtuous and So Noble a Design"—

which referred, with direct indirection, to Butler, who lampooned, and to Hobbes who both sneered and thundered at the new repository of all wisdom, and to the many others who detested the Baconian method as subversive of religion, civil law, reason and true learning. Stubbe, writing to Robert Boyle, beseeches him to consider "the mischief it hath occasioned in this once flourishing kingdom," and warns him, that unless he seasonably relinquishes "these impertinents" "all the inconveniences that have befallen the land, all the debauchery of the gentry . . . will be charged on your account."[1] Imagine the most pious and amiable of English philosophers held to responsibility for the eccentricities of Lady Castlemaine and Mistress Eleanor Gwynne!

There was a deal of appropriateness in Mr. Stubbe's solicitude that Boyle should abandon the Society. He

[1] Thorpe: Essays on Hist. Chemistry, cit. sup.

was by all odds the most able man in it—its leading spirit; and, at the time of its establishment, easily the most eminent experimental philosopher in the land—for Bacon was dead and Newton yet a "sober, silent thinking lad."

In the year 1658, Gaspar Schott,[1] a German Jesuit and a pupil of Kircher, published a voluminous treatise on Universal Magic, in which he described, for the first time, von Guericke's air-pump and discovery of the weight of the air—facts which he had learned from von Guericke himself. Schott's work came into Boyle's hands, and he at once saw something which von Guericke apparently had overlooked—namely, that important results should follow the study of rarefied air. Thereupon, with the assistance of Robert Hooke, he devised a new and more effective form of air-pump and demonstrated the elasticity or spring of the air, and the law of the relation between gas volume and pressure, which has ever since borne his name. He was the first scientific chemist[2]—the first to teach that chemistry was independent of other arts and not a mere adjunct; and the publication of his Skeptical Chemist, in 1661, marked the overthrow of both the Aristotelian and the Paracelsan doctrines of the elements. With him began the new era in scientific research, when its highest aim became the simple advancement of natural knowledge.[3]

In Boyle's treatise[4] touching the spring of the air (1659), we find him experimenting upon the lodestone and observing that a vacuum does not prevent the passage of its

[1] Schott: Magiæ Universalis, Naturæ et Artis, Pars III. et IV. Herbipolis. 1658.

[2] And the object of Sir Boyle Roche's famous Hibernicism "The father of modern chemistry and cousin to the Earl of Cork!"

[3] Roscoe & Schorlemmer: Treatise on Chemistry. New York, 1883, i., 10.

[4] See The Works of the Honourable Robert Boyle, London, 1744. Edited by Thomas Birch.

effluvia. Later, in 1663, he rubs a diamond in the dark "upon my clothes, as is usual for the exciting of amber, wax, and other electrical bodies," when it did "manifestly shine like rotten wood."[1] He believed, as we shall now see, fully in the corporeal nature of the electric effluvium, regarding it as a part of the substance of the electric, and so material a thing that, as he averred, he could smell it. ("Many electrical bodies may, by the very nostrils, be discovered.") This seems to have been the first recognition of the peculiar odor of ozone, long subsequently observed by Van Marum, although the substance itself was not discovered until 1840 by Schönbein.

Boyle's conception of the nature of magnetic and electric attraction was by no means an arbitrary hypothesis framed to meet some special physical conditions. Here he differed from Cabæus, Descartes, and even from Von Guericke. He formulated the corpuscular or mechanical philosophy, wherein he neither agreed with the Plenists, as Hobbes and the Cartesians were called, nor with the older Vacuists, who denied the plenitude of the world. His primary concepts were matter and motion—matter, apparently, in one primordial substance. By variously determined motion, he believed matter to be divided into parts of differing sizes and shapes, and set moving in different ways. Natural bodies of several kinds, according "to the plenty of the matter and the various compositions and decompositions of the principles, are thus formed; and these, by virtue of their motion, rest, and other mechanical affections which fit them to act on and suffer from one another become endowed with several kinds of qualities," which, acting on the senses, result in perceptions, and, on the soul, in sensations. The summing up of his philosophy is in the following passage, which certainly, for its time, is wonderfully close to modern thought:

"I plead only for such philosophy as reaches but to

[1] Birch : The Life of the Hon. Robert Boyle. London, 1744.

things purely corporeal, and, distinguishing between the first original of things and the subsequent course of nature, teaches concerning the former, not only that God gave motion to matter, but, in the beginning, He so guided the motions of the various parts of it, as to contrive them into the world He designed they should compose . . . and established those rules of motion, and that order among things corporeal, which we are wont to call the laws of nature. . . . The laws of motion being settled and all upheld by His incessant concourse and general providence, the phenomena of the world, thus constituted, are physically produced by the mechanical affections of the parts of matter and what they operate upon one another, according to mechanical laws."

Surely it cannot be said that Boyle had not perceived that it was the province of science to concern herself not with matter, but with the changes in matter. "I am apt to think," he avers, "that men will never be able to explain the phenomena of nature while they endeavor to deduce them only from the presence and proportions of such and such material ingredients, and consider such ingredients or elements as bodies in a state of rest; whereas, indeed, the greatest part of the affections of matter, and consequently of the phenomena of nature, seem to depend upon the *motion and contrivance of the small parts of bodies.*"

By this corpuscular or mechanical philosophy Boyle explains such things as he regards as natural phenomena—such as heat and cold, tastes, corrosiveness, fixedness, volatility, chemical precipitation, and finally, magnetism and electricity. Thus, heat, he says, is "that mechanical affection of matter we call local motion, mechanically modified" in three ways: first, by the vehement agitation of the parts; second, that the motions be very various in direction; and third, that the agitated particles, or at least the greatest number of them, be so minute as to be singly insensible.

It is singular how the mechanical theory—or, as we now term it, the dynamical theory, as applied to heat—impressed itself upon the philosophers of the seventeenth century. Bacon defines heat as "a motion acting in its strife upon the smaller particles of bodies." Boyle saw clearly that when heat is generated by mechanical means, new heat is called into existence, and believed that the production of heat and electricity were somehow correlated. Locke, in his Essay on the Human Understanding, says that "what in our sensation is heat, in the object is nothing but motion." Hooke plainly perceived heat as a vibration, and denies the existence of anything without motion, and hence perfectly cold. Yet it was the Material, and not the Mechanical Theory, which prevailed and which held the beliefs of the world up to our own time.

When Boyle turns to the study of magnetism, his hypothesis grows obscure. He denies, in the beginning, Gilbert's conception that magnetic qualities flow from the substantial Form of the lodestone, and, on the basis of experiments showing the reversal of the poles and the destruction of magnetism by heat, he concludes that changes in the "pores, or some other mechanical alterations or inward disposition, either of the excited iron or of the lodestone itself," renders it capable or incapable of acting magnetically.

His subsequent experiments, such as cooling and hammering iron rods held north and south, are all old, and are interesting simply as leading him to the more definite dictum that "the change in magnetism communicated to iron may be produced in good part by mechanical operations procuring some change in the texture in the iron."[1]

He is not in nearly so much doubt, however, concerning the mechanical production of electricity.[2] Here he has

[1] Boyle: Experiments and Notes about the Mechanical Production of Magnetism. London, 1676.

[2] Boyle: Experiments and Notes about the Mechanical Origine or Production of Electricity. London, 1675.

made many experiments—so many indeed that he has learned that their event is "not always so certain as that of many others, being sometimes much varied by seemingly slight circumstances, and now and then by some that are altogether overlooked"—which is by no means out of harmony with modern conclusions. Besides, he has the backing of all the preceding philosophers—von Guericke alone excepted. There was Gilbert with his effluvia, "like material rods;" Cabæus with his "shrinking steams;" Descartes with his "ribbons shooting from the pores of the glass;" Digby and Browne with "unctuous filaments" contracting in the cold air; and Gassendi, whom I have hitherto not mentioned, but who imagined emanations which not only entered the pores of the chaff, but became crossed therein, and thus getting a better hold on it, pulled it back with greater force in retracting: every one of these philosophers finding the electrical effects due, not to a mere quality in Form, but to substantial emanations from the attracting body; and thus all seeking to solve the problem in a mechanical way. Heat, says Boyle, building on this foundation, agitates the parts of the body and makes it emit effluvia. Rubbing modifies the motions of the internal parts and gives the body a texture which disposes it to become vigorously electrical. And so it continues even after the exciting cause is removed, because some of the heat still remains. On a warm day, he was able to move a pivoted steel needle with an electric no larger than a pea, three minutes after the rubbing had ceased.

Then he remarks something altogether unaccountable; (although the discovery was not original with him, for it had been observed years before by the Florentine Academy del Cimento) namely, that an electric can apparently be moved by its own steams—as he observed by suspending a piece of amber, rubbing it and then causing it to swing so as to follow the rubbing cloth moved before it. He is not at all sure as to what this portends, and in fact is somewhat troubled about it. "His nature," says Humboldt, "was cautious and doubting." "Whether from such experiment one may argue," he says thoughtfully, "that it is but, as it were, by accident that amber attracts another body and not this the amber; and whether these ought to make us question if electrics may with so much propriety, as has been hitherto generally supposed, be said to attract, are doubts that my design does not here oblige me to examine."

So Boyle went on, and added some more things to the list of electrics—turpentine gum, and white sapphires, and English amethysts, and emerald (which Gilbert said would not attract), and carnelian and various other substances; which merely swell the list, and are of no importance. He comes back for a final blow at the Form theory, by distilling amber to a caput mortuum and showing that, as the attractive quality is still present, it cannot be due to the "substantial Form of amber" which has here thoroughly disappeared.

Boyle's idea of electric attraction having much in common with the hypothesis of corporeal emanations, which we have traced through different theories, it follows, of course, that he did not agree with von Guericke in the incorporeal nature of the expulsive force, but, on the contrary, refers to electric repulsion very much as Cabæus did long before, in proof of the fact that the briskly-moving steams from the electric physically drive away the attracted bodies. But, unlike Cabæus, he recognized the specific fact of the repulsion, indeed had to do so to reach the idea that the electric operated to "discharge and shoot out the attracting corpuscles" which carried away the chaff, although he finds it difficult to coördinate this action with the attractive effect, and admits that it happens only "at a certain nick of time."[1]

[1] Boyle: Of the Great Efficacy of Effluviums. 1673. Cap. iv. Works: Birch. Lond., 1744. Vol. iii., 323.

On the basis of a paragraph, which appeared in 1673, Boyle is very

Boyle closed the series of experiments recorded in his little treatise of 1675[1]—the first book entirely on electricity in the English language—with the doubts raised by the swinging amber unsettled, and, indeed, intensified, for he had encountered the same problem in other and even more enigmatic shapes. He tells us that once when he approached his finger to a down feather which had been attracted by a large piece of amber, the pinnules of the feather applied themselves to the finger "as it had been an electrical body." This was very obscure to him. First he thought that warm "steams" from his person might somehow have caused this attraction, but when he presented to the feather a rod of silver, an iron key, and a cold piece of black marble, the pinnules "did so readily and strongly fasten themselves to these extraneous and unexcited bodies that I have been able (though not easily) to make one of them draw the feather from the amber itself." But this, he is careful to note, happens only while the amber is sufficiently excited to make it sustain the feather, otherwise "neither the approach of my finger nor that of the other bodies would make the downy feathers change their posture. Yet as soon as ever the amber was by a light affriction excited over again," the finger attracted the feather.

He repeated this experiment over and over again, with "years of interval," he says; tried innumerable feathers and substituted brimstone for amber as the electric—always

frequently credited with the original discovery of electrical repulsion. In the same connection he speaks, however, of the observation as one which he "made many years ago, and which I have been lately informed to have long been since made by the very learned Fabri." This was published a year after the appearance of von Guericke's treatise, and hence ten years after the visit of Monconys to Magdeburg. Honoré Fabri, the French mathematician, to whom Boyle alludes, did not issue his treatise on Physics untill 1669, so that not only is the actual date of Boyle's reference to the phenomena long after that of von Guericke's discovery, but there is nothing inconsistent with the priority of von Guericke in Boyle's assertions that the fact had long been known to himself and Fabri.

[1] Cit. sup.

with the same result; always the same insoluble problem. He had no more conception of bodies becoming electrified by induction, when brought into the field of an excited electric, than von Guericke had; although both clearly saw the resulting phenomena, and both knew the essential conditions, that the electric must be excited and that the body must be brought within a certain distance of it. Of course, Boyle's finger became electrified by induction oppositely to the amber, and, hence, easily attracted the light pinnules on the down; but that knowledge was in the far future.

It is not difficult to imagine the host of puzzling questions which forced themselves upon Boyle. So far as he knew, only certain things (the so-called electrics), when rubbed, would attract the feather. Most things would not. Yet here it seemed that after the electric had once seized the down, all sorts of things would attract it, whether electric or otherwise. There was his own finger. He might rub the very skin off of it, and yet it would not attract; but put it near the feather on the amber, and at once it exhibits this astonishing capacity. Was he an electric? If so, why at one time and not at another? If he and the silver rod, and the marble, and the iron key, were all in fact electrics, why would not rubbing arouse the attractive capacity in any of them? and what sort of electrics were they which would attract without being rubbed? How could rubbing a totally distinct and separate body, such as that lump of amber or brimstone, convert a man's finger into an electric?

To the ad hominem argument of his own finger became added another, ad feminam, which deepened the mystery. Those were the days of colossal headdresses, when the men encased their craniums in huge full-bottomed wigs; while above every woman of quality arose a complicated structure of curled hair, wire, ribbons, artificial flowers and miscellaneous trinkets. The curling of the hair in wigs naturally made it dry and stiff; and especially so in these feminine towers, because the presence of the millinery in the edifice precluded the use of lubricants.

Now, the fine ladies, as I have stated, got into the habit of visiting the Royal Society and witnessing experiments because it was fashionable to do so, and perhaps there were wandering spirits of inquiry pervading the air about that grave institution which sought more attractive lodging-places than existed under the scrubby head-coverings of the philosophers. At all events, one of them made its abode beneath a more than usually alluring head-dress, the owner whereof came to Boyle and told him that her "knotted and combined locks" persisted in flying to her cheeks and sticking there, and demanded to know the why and wherefore of it. Boyle says that he "turned it into a Complemental Raillery, as suspecting there might be some trick in it." Being quickly disabused of that notion, with the characteristic brutality of his sex, he insinuated "sticky paint," but retreated at once before the instantly-ensuing flash of deepened color. Then he attacked the subject philosophically—for it was troublesome. The apparent electricity of his finger was surprising enough, but to find it in women's cheeks—and this time without the intervention of any rubbed amber or brimstone at all—was incomprehensible. So he experimented further upon his fair inquisitor. "She is no ordinary virtuosa," he says, doubtless feeling the full conviction of the expression, "and she very ingeniously removed my suspicions (that there was some trick involved), and, as I requested, gave me leave to satisfy myself further by desiring her to hold her warm hand at a convenient distance from one of those locks off and held in the air."

It remains to the lasting discredit of Boyle that he failed to transmit to fame the name of probably the first woman who thus sacrificed her finery in the cause of electrical science; but to continue:

"For as soon as she did this, the lower end of the lock, which was free, applied itself presently to her hand, which seemed the more strange because so great a multitude of hair would not have been easily attracted by an ordinary electrical body that had not been considerably large or extraordinarily vigorous. This repeated observation put me upon inquiring of some other young ladies whether they had observed any such like thing; but I found little satisfaction to my question, except from one of them eminent for being ingenious, who told me that sometimes she had met with these troublesome locks, but that all she could tell me of the circumstances which I would have been inform'd about was that they seem'd to her to flye most to her cheeks when they had been put into a somewhat stiff curle, and when the weather was frosty."

And her observation was right. The stiff curling of the hair had electrified it, and for this the frosty weather offered the best of conditions. History was repeating itself. Ages before, an unknown Phœnician woman had seen her whirling amber spindle pick up the leaves and chaff from the ground. Now an unknown Englishwoman saw the same strange attraction, excited by her own light locks, move the hair. And the learned philosopher of the 17th century to whom she told it first doubted it, and ultimately did not understand what he saw any better than did perhaps the Phœnician wise men five thousand years before. In fact, man has never put proper faith and credit in woman's discoveries since he accepted Eve's apple.

As Boyle, with all his ingenuity, could make nothing of the problem, he took refuge in the decrepit mediæval theory that the occurrence was due to the "effects of unheeded and, as it were, fortuitous causes," which, of course, as an explanation, is exceeded in logical absurdity only by that which attempts to elucidate an unknown matter by giving an entirely new name to it. One is apt to wonder why he did not attack the difficulty with something of the same enthusiasm and experimental skill which he brought to bear upon his chemical researches. Perhaps the reason is that, what with chemistry and pneumatics, and especially theology, he had abundant work to exhaust his energies. If it were not plain that he took a genuine pleasure in sermonizing, and often in evolving homilies concerning the most trivial topics ("on the paring of a summer apple;" "on drinking water out of the brim of his hat;" "on his horse stumbling," savagely burlesqued by Dean Swift in "Pious Meditations on a Broomstick"), we might be tempted to regard such efforts as misdirected. But no man, having contributed so much to the progress of his age, ever satisfied himself with so harmless an amusement.

Leave out Boyle's sermons, and the contents of his treatises again and again suggest Faraday. The descriptions of his experiments often have the same ladder-like quality. To such investigations as he did devote himself he brought a most untiring persistence. "Never," says Evelyn, who had known him for forty years, "did stubborn Nature come under his inquisition, but he extorted a confession of all that lay in her most intimate recesses; and what he did he as faithfully registered and frankly communicated." "Glasses, pots, chemical and mathematical instruments, books and bundles of papers did so fill and crowd his bedchamber, that there was just room for a few chairs—a small library, as learning more from men, real experiments and in his laboratory, than from books," continues the diarist. Some one who asked to inspect his library, he conducted to a room where he was dissecting a calf.

Among Americans, Boyle has especial claim to remembrance, for perhaps to him, more than to any one else, is due the first implanting and encouragement of scientific thought in the struggling colonies. He was the friend of John Winthrop, who joined the Royal Society as a founder, when he came to England in 1662 for the charter of Connecticut. And Winthrop seems to have been our first scientist. Bancroft says of him that he took delight in "the study of nature according to Bacon," in which way he studied Indian corn and told the Royal Society all about it. Later we find Winthrop writing to Boyle from Boston, to inquire whether lightning could kill fish, as the Indians had told him; and Leonard Hoar from Cambridge, about the Indian canoe, anent which "if you lay your tongue on one side of your mouth it may overset;" and William Penn from Philadelphia, telling him of the valuable resources of the tracts newly bought from the natives.

Through this correspondence, there came to be recorded an observation which mightily disturbed the reverend John Clayton, who had settled in James City in Virginia. He had taken great interest in Boyle's experiments, and had sent him accounts of the great luminosity of the American fire-flies; but the new occurence was far more surprising. It had been communicated to Clayton through the following epistle:

"MARYLAND, ANNO 1653.

"There happened about the month of November to one Mrs. Susanna Sewall, wife of Major Nicholas Sewall, of the province aforesaid, a strange flashing of sparks (seem'd to be of fire), in all the wearing apparel she put on, and so continued to Candlemas; and in the company of several, viz., Captain John Harris, Mr. Edward Braines, Captain Edward Poneson, etc., the said Susanna did send several of her wearing apparel, and when they were shaken it would fly out in sparks and make a noise much like unto bay leaves when flung into the fire; and one spark lit on Major Sewall's thumb-nail, and there continued at least a minute before it went out, without any heat; all which happened in the company of

"WILLIAM DIGGES."

Clayton transmitted this to Boyle with the following annotation:

"My Lady Baltimore, his mother-in-law, for some time before the death of her son, Cæcilius Calvert, had the like happened to her, which has made Madam Sewall much troubled as to what has happened to her. They caused Mrs. Susanna Sewall, one day, to put on her sister Digges' petticoat, which they had tried beforehand and would not sparkle; but at night when Madam Sewall put it off, it would sparkle as the rest of her garments did."

The astonishing behavior of Madam Sewall's garments, or even of the petticoat of her sister Digges when worn by her, was outrivaled by the performances of Madam Digges herself, concerning which Clayton writes to Boyle in 1684. Whether it was because Madam Sewall had lately departed for England that Madam Digges felt it incumbent upon herself to surpass the first-mentioned lady in luminous manifestations, is unknown; but Clayton says that she developed crepitations and shining flames about her person, "and," adds the good parson, "how it should transpire through the pores, and not be inflamed by the joint motion and heat of the body, and afterwards so suddenly to be actuated into sparks by the shaking or brushing of the coats, raises much my wonder."

Such was the first electrical observation in the New World.

Whether a man of unusual inventive genius is a product or a factor of the circumstances about him is always a debatable question. We may believe, with Emerson, that souls out of time, extraordinary, prophetic, are born who are rather related to the system of the world than to their particular age or locality,[1] or, contrariwise, with Froude, that even the greatness of a Shakespeare is never more than the highest degree of excellence, which prevails widely, and in fact forms the environment.[2] We may regard all invention as inspiration, or maintain that the presence of the divine afflatus is not to be presumed, and that upward progress is, on the whole, more commonly made by way of the beanstalk which springs from the ground than by way of the chariot descending from the skies.

A just apportionment of honor among men of the same

[1] Worship. [2] Science of History.

time who labor to the same end is the more difficult problem. True, the reward rightfully belongs to him who advances to the goal, and not to the finger-posts which stand still, though pointing the way; that it is the last step which counts, and should count, if it happens to carry one over the border of the promised land. But, on the other hand, the steep path of discovery is never occupied only by finger-posts and a single inspired wayfarer continually shouting "Excelsior." It is always a ladder crowded with a struggling throng, sometimes pushing, sometimes carrying one another upward; and the prize is often grasped by the fortunate climber who, from the vantage of other men's shoulders, first perceives it to be within his reach. For each "mute inglorious Milton" the world has held scores of mute inglorious Gilberts and Galileos, with the difference that the unsung songs never helped to the singing of those that were sung, while the stooping backs of such lifelong plodders as penurious, embittered, disease-racked Robert Hooke[1] have over and over again made sturdy treads upon which others of far less merit have scrambled upward to fame and fortune.

I know of no prototype for Hooke, unless it be Leonardo da Vinci, and the similarity here exists only in the remarkable fecundity of invention which each displayed, due regard being had to the differences in the epochs in which they lived. Hooke illustrates the dictum of Froude as perfectly as Isaac Newton does that of Emerson. To Hooke no one would concede inspiration; to Newton few would deny it. Hooke was the natural complement of Boyle. Matters ethical concerned him not at all; of spirituality he had none, and deductive reasoning had little place in his mind; but he devised and made the air-pump with which Boyle discovered his law. He began to con-

---

[1] See Waller: The Posthumous Works of Robert Hooke.. London, 1705. Durham: Phil. Exp'ts. and Obs'ns. of the late Dr. R. Hooke. London, 1726. Bib. Britannica, article, Hooke. Also Hooke's papers in Phil. Trans.

trive when a little child; he stopped contriving when old, blind, bed-ridden and dying, his shaking hand refused to feel out to the end the scrawl descriptive of a new instrument. He invented as he breathed, because he could not help doing so, and he ceased inventing when he ceased breathing. He cared for nothing else. After he had passed away, when his few earthly belongings were sifted by his fewer friends, they found a large iron chest which had been locked down with the key in it, with a date of the time, by which it appeared to have been shut up for more than thirty years, and the contents amounted in value to many thousands of pounds in gold and silver. In the prime of life, at thirty-seven, when the world looked bright before him, he had conceived a great project —that same great project which every inventor has lurking somewhere in his brain to come forth in his benevolent moments—of founding a vast laboratory and museum and library containing everything to help every one who needed to be helped in the study of science, every one who had felt the needs which he had felt; but when Hooke found others not only climbing his ladder, but making a ladder of him, he locked up the hoard and waited for the millennium when the inventor and his kind shall dwell in perfect peace, mutual love and harmony, and all their competitions be squared by the Golden Rule; a result even the dim outline of which in the blue of the furthest horizon, it is needless to add, we are still as unable as ever to discern.

Robert Hooke was indeed the typical inventor. To say that his inventions are numbered by hundreds conveys little in these days of inventive attenuations. Their diversity, however, was extraordinary, and now and then an idea flashes out which, in the light of after discovery, is surprising. From Boyle's air-pump he turns to the flying machine, tries to construct "artificial muscles," and then a contrivance to raise a man by "horizontal vanes a little aslope to the wind," toward which last there is a tendency now to come back. He experimented in about every known branch of physics, but he shone most as a contriver of measuring instruments. In this respect no predecessor approaches him in ingenuity and skill; very few have since equaled him. He was the first curator of the Royal Society, and whenever anything was to be investigated, it was Hooke who evolved the mechanical devices for doing so. Clocks and chronometers, astronomical apparatus in great variety, instruments for measuring specific weight, refraction, velocity of falling bodies, freezing and boiling points, strength of gunpowder, vibrations of dense bodies, degrees on the earth, magnetism, and so on through such a variety and multiplicity of mechanisms that it may be imagined that of the many 'scopes, 'graphs, and meters which now sharpen the senses of modern physicists, few exist in which something originally emanating from Hooke's tireless brain cannot be found. His volute spring, opposing and counterbalancing the motion of a rotary arbor in all positions, which made the pendulum clock into the portable chronometer, now weighs the pressure of the electric current, and in the same way has rendered electrical instruments portable.[1] And as for the foreshadowing of modern achievements, we are told by Richard Waller, secretary of the Royal Society, his biographer and immediate personal friend, that "he shewed a way of making musical and other sounds by the striking of the teeth of several brass wheels proportionally cut as to their numbers and turned very fast round, in which it was observable that the equal or proportional strokes of the teeth, that is 2 to 1, 4 to 3, etc., made the musical notes, but the unequal strokes of the teeth *more answered to the sound of the voice in speaking*"—which is remark-

[1] Butler in Hudibras alludes to Hooke's spiral spring and its effects in the lines:

"And did not doubt to bring the wretches
To serve for pendulums to watches
Which, modern virtuosi say,
Incline for hanging every way."

ably suggestive of the principle and operation of the phonograph.

Hooke's essay on the method of improving natural philosophy is a forecast of possibilities replete with suggestion. "There may be a possibility," he says, "that by otocousticons[1] many sounds very far distant (I had almost said as far off as some planets) may be made sensible. I cannot, I confess, myself so far throw off prejudice as not to look upon it as a very extravagant conjecture; but yet methinks I should have the same thoughts of a conjecture to find out a help for the eye to see the smaller parts and rocks of the moon and to discover their height and shadow, before I had seen or known the excellent contrivance of telescopes." So, perhaps, he might not have thought the telephone or even the photophone—or the hearing of explosions in the sun—quite so marvelous as did those who came after him.

He affirmed that electric light is due to the same cause as heat—that is, internal motion in the parts of the body. To produce electric luminosity, however, it is not enough, he says, merely to cause this internal motion, but certain bodies—such as diamonds, sugar, black silk, clean warmed linen, or a cat's back—must be rubbed and agitated up to a certain degree, and then "the more you rub it the more it shines, and any little stroke upon it with the nail of one's finger when it so shines, will make it seem to flash." That was written in 1680, and it appears to have

[1] Otocousticons were probably speaking tubes, an invention as old as the Egyptians, and then newly coming into vogue. Burton (Anat. Mel., part 2, § 2, mem. 4.), speaks of them as serving to aid hearing, as telescopes do sight. Evelyn (Diary, 13 July, 1654), notes a hollow statue contrived by Bishop Wilkins, "which gave a voice and uttered words by a long, concealed pipe which went to its mouth, whilst one speaks through it at a good distance," something after the fashion of the talking head of Albertus Magnus. The aroused interest in the transmission of sound resulted in the invention of the speaking trumpet by Sir Samuel Morland in 1671. Butler's lines in Hudibras

"And speaks through hollow empty soul
As through a trunk or whispering hole—

allude to this.

been the first recognition of the electric flash—the spark—in contradistinction to the glow. This was also the first attempt to explain light electrically caused, as being in common with all light a "peculiar kind of internal motion of the particles of a body," but specifically due to the nature of the body itself, the mode of exciting the motion (rubbing) and the degree of excitation produced. Moreover, he went further and asserted that there is "an internal vibrative motion of the parts of the electric bodies, and so soon as ever that motion ceases, the electricity also ceases"—so that, not only did he find the particular manifestation of electricity as light due to vibration, but ascribed the entire electrical phenomenon, even when appearing as attraction, to the same cause.

Hooke's theory of light, following substantially that of Descartes, and involving the assumption that space is filled with something that transmits light instantaneously, was overthrown by Roemer's observations of Jupiter's moons in 1676, resulting in a determination of the velocity of light.[1] Then came Newton's emission theory, which yielded to the now-accepted undulatory hypothesis of Young; but, none the less, such concepts of the electrical phenomena as Hooke made were a long way ahead of the "unctuous steams" and "rebounding effluvia" which had preceded them.

So far as is known, Hooke made no electrical or magnetic discovery of major importance. The catholicity of his work was against his doing so. It is seldom that the inventor who expends his energy in an infinitude of details ever leaves behind him any one great monumental achievement. There is an apparent gap between the endless mechanical refinements of Hooke's multitudinous instruments and his dynamical theories of heat, light and electricity, which it seems should be filled by tangible accomplishments of a higher order than the former. If he did so, he concealed them; and again revealed another one

[1] Tyndall: On Light. London, 1875, 45.

of those strange characteristics which often drive the inventor to hiding his work from the world, as a bird hides her eggs from the serpent.

"He was in the beginning of his being made known to the learned," says Waller, "very communicative of his philosophical discoveries and inventions, till some accidents made him, to a crime, close and reserved. He laid the cause upon some persons challenging his discoveries for their own, taking occasion from his hints to perfect what he had not; which made him say he would suggest nothing till he had time to perfect it himself, which had been the reason that many things are lost which he affirmed he knew."

The Royal Society was not the first of the institutions for the promotion of experimental science, the organization of which followed as a consequence of the renewed interest in physical discovery which, in the last half of the century, spread throughout Europe. In 1657 the Florentine Academy del Cimento was established under the immediate patronage of Prince Leopold, of Tuscany, (a potentate whose interest in natural science rivaled that of Charles), and attracted to itself many of the most eminent Italian philosophers. Its transactions were not published, however, until 1667, when it went out of existence; so that the exact dates when the experiments recorded were made, cannot be assigned. Although the researches included the first demonstration of the incompressibility of water, and mainly related to air pressure, physical conditions in vacuo and effects of high and low temperatures, those on electricity and magnetism which are interspersed show notable insight and skill.

In magnetism, many attempts were made to find a substance which would cut off the influence of the lodestone, but without avail; and the Academy records that the virtue is neither barred nor impaired by any interposed body, solid or fluid, except a plate of iron and steel. Of the electrical experiments one is of great importance, for it is the same which so much puzzled Robert Boyle. The Academy, however, was by no means mystified, nor need Boyle have been so had he read the clear description of it which was already in the archives of the Royal Society, to which the transactions of the Florentine Academy had been solemnly presented by emissaries from Leopold in 1667. How important this experiment was, will now soon appear; meanwhile, note how clear the Florentine philosophers' perception of it, in contrast to Boyle's obscurity.

"It is commonly believed," they say,[1] "that amber attracts the little bodies to itself; but the action is indeed mutual, not more properly belonging to the amber than to the bodies moved, by which also itself is attracted; or rather it applies itself to them. Of this we made the experiment, and found that the amber being hung at liberty by a thread in the air, or counterpoised upon a point like a magnetical needle, when it was rubbed and heated, made a stoop to those little bodies which likewise *proportionally* presented themselves thereto and readily obeyed its call." Such was the first announcement of the mutual attraction of electrified bodies—corresponding to the mutual attraction of magnet and iron which Gilbert had recognized.

Let us now recall some facts which, to the intelligent student of physics attending the meetings of the Royal Society toward the end of the 17th century, might seem as fairly well established.

Standing apart by themselves, he perceives four things, each able to control mechanically other things even at a distance and without apparent means of communication. These are first, the sun which controls the earth; second,

---

[1] Saggi di Naturali Esperienze fatte nell'Accad. del Cimento. Florence, 2d ed., 1691. Waller: Essayes of Nat. Exp'ts made in the Academie del Cimento. London, 1684, 128.

the earth which controls the moon and all sublunary bodies; third, the electric which controls all light objects; and fourth, the magnet which controls iron or steel. The attraction between sun and earth, or earth and moon, is sufficiently accounted for to most people of the time as a creative act. The fall of a stone to the ground, our student might consider to be the return of a part to its origin, source or reservoir. But electrics and the lodestone he knows to be physical outlaws. True, there is a choice of several theories wherefrom to select, but on the whole, no law seems exactly to reach them, and one is quite safe in holding that they act for the same reason that the dogs in good Dr. Watts' verse (if it had then been written), delight to bark and bite; "for 'tis their nature to." But what is actually seen to be true, concerning either lodestone and iron or electric and its objects? This; that when the two bodies (as stone and iron) are placed one in proximity to the other, although separated by a considerable interval, not only will the stone influence the iron, but the iron will influence the stone. Gilbert had already described what he called the mutual concourse of lodestone and iron, and Boyle had as plainly seen the swinging amber, in its turn, attracted by the rubbing cloth. Thus both had observed, and others were now observing, the two ends, so to speak, of what happened, the inherent attractive power of the magnet or electric at one extremity and the movement of the attracted body at the other.

Still another fact is also perceived, namely, that around the electric there is a certain space or field in which light bodies are either attracted or repelled, and similarly that around the magnet there is also a certain space or field within which iron is attracted, like effects not appearing upon bodies located outside of these fields. That the power of magnet and electric is inherent to and resides in the substance of each, is commonly believed. How that power became exerted was, as we have seen, the subject of many speculations, all of which, generically considered, had now become reducible to two—that there are physical emanations from stone or electric which come into contact with the attracted body and so move it, or that there are no emanations, no material linkage of any sort, but that either magnet or electric has the capacity of "action at a distance," a term which obviously merely describes without explaining. As measuring instruments had increased in numbers, and experimental tests had become more rigorous, so the emanation doctrine had lost ground, for the simple reason that the imagined effluviums refused to reveal themselves, and correspondingly the "action at a distance" notion had gained in favor. Hence it is not uncommon at this time, when Mr. Isaac Newton announces his great discovery of universal gravitation, to consider the magnet and electric as possessed of a certain occult capacity for moving far-off objects. In other words, people had begun to realize that they did not know anything about the matter, in which circumstances a little mystery has, in all ages, been regarded as quite human, and not unconducive to the preservation of a proper self-respect.

When Newton fell into the famous reverie in his garden, the magnetical-cosmical theory which Gilbert had proposed had been for some time moribund. The modification of it which Kepler had adopted had preserved its vitality somewhat, although Kepler had used it for little else than a scaffolding. It had not served Gilbert's purposes in lending any material support to the Copernican doctrine now firmly established, and in fact had acted rather to divert attention from the experiments on which it was founded, and so to obscure rather than enhance its author's fame. There was also a strong inclination among English philosophers, never stronger than just before Newton's advent, to reject all explanations of the movements of the planets based on analogies and guesses, and in place thereof to regard their motions and relations as consequent upon physical laws, and capable of mathematical determination. Hooke was so far in the van of this thought that when

Newton's discovery was announced he claimed it as already his own, though without sufficient grounds.

While the Gilbertian theory would probably have soon succumbed to the changed conditions, the Newtonian conception more directly led to its disappearance, not by refuting so much as by displacing it. Why, is best shown by tracing the contrast between the two theories, and at the same time this will bring us by the shortest route to the vantage ground whence Newton's remarkable part in the development of electricity can be most clearly discerned.

"The force which emanates from the moon," says Gilbert, "reaches to the earth, and in like manner the magnetic virtue of the earth pervades the region of the moon; both correspond and conspire by the joint action of both according to a proportion and conformity of motions."[1] Newton says that the earth draws the moon and the moon the earth.

"The earth," continues Gilbert, "has more effect because of its superior mass." "The motion which the moon receives from the earth bears to the motion which the earth receives from the moon the same proportion as the mass of the earth bears to the mass of the moon," says Newton, with mathematical brevity.

[1] Gilbert: Physiologia Nova. Amsterdam, 1651.

Bishop Wilkins, writing in 1638, says:

"This great Globe of Earth and Water hath been proved by many Observations to participate of Magnetical Properties. And as the Loadstone does cast forth its own Vigour round about its Body, in a Magnetical Compass, so likewise does our Earth. The difference is, that it is another kind of Affection which causes the Union betwixt the Iron and Loadstone from that which makes Bodies move unto the Earth. The former is some kind of nearness and similitude in their Natures, for which Philosophy, as yet, has not found a particular Name. The latter does not arise from that peculiar Quality whereby the Earth is properly distinguish'd from the other Elements, which is its Condensity. Of which the more any thing does participate, by so much the stronger will be the desire of union to it. So Gold and other Metals which are most close in their Composition are likewise most swift in their Motion of Descent."—*The Discovery of a New World.*

And thus both Gilbert and Newton agree that earth and moon attract one another, and in proportion to the quantity of matter in each. So much for similarities which are certainly striking enough.

But Gilbert regarded the earth as emitting a magnetic virtue, and the moon (which he does not suppose to be a magnet) also as emitting a virtue, but of a different nature. Here Newton differs and moves ahead. The attractive power in the members of the solar system, he declares, is no different, but of the same nature in all, for it acts in each in the same proportion to the distance and in the same manner upon every particle of matter.

Not even is this power new or unfamiliar. It is "one no different from that existing on earth which we call gravity." With what was then called gravity, Gilbert was well acquainted, for he tells how the earth not only attracts magnetic bodies but also "all others in which the primary force is absent *by reason of material.*" "And this inclination," he adds, "in terrene substances is commonly called 'gravity.'" It must not be forgotten, however, that Gilbert had never assumed that the gravity of the earth could control aught but earthly things. It could make a stone fall to the ground to "the source, the mother where all (parts of the earth) are united and safely kept." The idea that mother earth could govern by her gravity attraction "th' inconstant moon" never entered his head.

The magnetic attraction of that great magnet, the earth, on the other hand, was to him a different attribute altogether; and it was not at all difficult to imagine the colossal enclosing sphere of magnetic virtue as sufficiently enormous to "pervade the regions of the moon."

But that was imagination, which rigorous proof pushed aside as a great steamer displaces fog. Then it was gravity which became colossal, and, under the mighty conception of Newton, grew into an attraction as broad as the universe itself—existing between all masses, all sorts of matter, always, everywhere; between worlds as well as between sand grains. As a cause it explained all of the observed effects, and "more causes," said Newton sententiously, "are not to be received into philosophy than are sufficient to explain the appearances of nature." The magnetic attraction of the earth, in any event, he regarded as "very small and unknown."[1] To argue whether our little globe governs its littler satellite by magnetism or electricity or any other power, virtue, agency, or force, became needless when the mode in which the Almighty had ordered the mechanism of all worlds stood definitely revealed.

So vanished Gilbert's theory. The sun's

> "Magnetic beam, that gently warms
> The universe and to each inward part,
> With gentle penetration, though unseen
> Shoots invisible virtue ev'n to the deep,"[2]

gilded the pages of the great epic and then disappeared, to return only in later days when light and electricity and magnetism began to be as one.

So Newton proved that whatever other influence electricity and magnetism may exert as cosmical forces, it is not necessary to assume the action of either to account for the motions of the planets. And as no one had hitherto seen, for either, any other useful purpose except the ordering of the heavenly bodies, it followed that this left them with their "occupation gone;" a mere nebulous cloud of facts and fancies gathered about the nucleus which Gilbert had segregated from the pre-existing chaos. But then, just as such a body—a vagrant new-born world perhaps—finding itself within the control of a greater orb, becomes under Newton's precepts a satellite, forever after pursuing its orderly round in the celestial mechanism; so, at his bidding, this unrelated mass of knowledge fell into its appointed place and became obedient to the reign of

[1] Principia, B. III., Prop xxxvii.
[2] Paradise Lost, Book III.

law. There is no event in this history more significant, more epoch-making, than this.

It will be remembered, that among the fundamental principles of physics are the three laws of motion which Newton formulated; the first stating the effect of force upon a body left to itself; the second defining the relation of the change of motion of the body to the force impressed, and the third—that perennial stumbling-block to all the perpetual-motion seekers of the past and most "new motor" contrivers of the present—that action and reaction are equal and in contrary directions. This last obviously defines the effect of the action of two bodies one upon the other—that of the first upon the second being equaled by the contrary reaction of the second upon the first; or, to borrow Newton's own illustration, "If you press a stone with your finger, the finger is also pressed by the stone. If a horse draws a stone tied to a rope, the horse will be equally drawn back toward the stone; for the distended rope, by the same endeavor to relax or unbend itself, will draw the horse as much toward the stone, as it does the stone toward the horse, and will obstruct the progress of the one as much as it advances that of the other."[1]

Under this law, Newton makes the first close linkage of gravity, electricity and magnetism. If the sun draws a planet, so that planet draws the sun; if the amber draws chaff, so that chaff draws the amber; if the lodestone draws iron, so the iron draws the stone. The law is the same for all. It is the law of stress.[2]

But the bond is closer than this. He mentions the common habit of referring the reacting forces to that body of

[1] Principia, Axioms or Laws of Motion.

[2] "Every force, in fact, is one of a pair of equal opposite ones—one component, that is of a stress—either like the stress exerted by a piece of stretched elastic, which pulls the two things to which it is attached with equal force in opposite directions and which is called a *tension;* or like the stress of compressed railway buffers, or of a piece of squeezed india rubber, which exerts an equal push each way and is called a pressure." (Lodge.)

the two which is least moved; as when we call the attraction of sun and planet the attractive power of the sun. Yet more correctly, he says, we should regard the force as acting between the sun and earth, between the sun and Jupiter, between the earth and moon, for both bodies are moved by it, in the same manner as when tied together by a rope, which shrinks on becoming wet, and so draws them each one to the other. Equally true is this—another link forged—of electrical and magnetic attractions; for although as to the nature of this he has no hypothesis to offer ("Hypotheses non fingo," is his motto everywhere), yet concerning it he says, if we would speak more correctly, and not extend the sense of our expressions beyond what we see, we can only say that the *neighborhood* of a lodestone and a piece of iron is attended with a power, whereby the lodestone and the iron are drawn toward each other;[1] and the rubbing of electrical bodies gives rise to a power whereby those bodies and other substances are mutually attracted. Thus, we would also understand in the power of gravity, that the two bodies are mutually made to approach each other by the action of that power.[2]

Such was the first suggestion that the seat of electric and magnetic forces is not in the electric, or the substance attracted by it, or the magnet, or the iron, but in the intervening medium; whatever the last may be.[3]

[1] "I made the experiment on the lodestone and iron. If these placed apart in proper vessels are made to float by one another in standing water, neither of them will propel the other; but by being equally attracted, they will sustain each other's pressure and rest at last in an equilibrium."—*Principia cor.* vi.

[2] Pemberton: A view of Sir Isaac Newton's Philosophy. London, 1728, 254.

[3] "We may conceive the physical relation between the electrified bodies, either as the result of the state of the intervening medium, or as the result of a direct action between the electrified bodies at a distance. If we adopt the latter conception, we may determine the law of the action, but we can go no further in speaking on its cause. If, on the other hand, we adopt the conception of action through a medium, we are led to inquire into the nature of that action in each part of the medium. . . . If we now proceed to investigate the mechanical state of the medium

But he does not stop here. To follow him further we must look backward to find the ladder he is climbing; for Newton has a way of not leaving his ladders readily available, and sometimes he is charged with pulling them up after him.

When Peregrinus placed his bit of iron in different positions on the lodestone globe, he saw it stand upright at the poles, and at various inclinations between poles and equator. Gilbert, three centuries afterward, observed the same thing; but neither perceived that a line drawn lengthwise through the needle in all its positions would be curved and extend between the poles. Porta, multiplying the piece of iron many times in the form of filings sprinkled about the stone, saw them branch out from the poles like hairs, but not in continuous curves; while to Cabæus they seemed to fall into lines more plainly curved, but still not arching from pole to pole. Then came Descartes, who found what all had missed, namely, that not only did the filings fall into regular curved lines from pole to pole, but that their arrangement in such lines in that intervening space must be the effect of some force there existing and acting on them. This Christopher Wren had also seen, and Sprat, in recording his experiment, even refers to the "lines of directive force." Not only did Descartes note these lines arching between opposite poles of the same magnet, but as extending between the poles of two magnets and seemingly connecting them. These curves, which the filings traced for Descartes, occupy the magnetic field or Gilbert's orb of virtue, and, when so rendered visible, map it. And

---

on the hypothesis that the mechanical action observed between electrified bodies is exerted through and by means of the medium, as in the familiar instances of the action of one body on another, *by means of the tension of a rope* or the pressure of a rod, we find that the medium must be in a state of mechanical stress. . . .

The nature of this stress is, as Faraday pointed out, a tension along the lines of force combined with an equal pressure in all directions at right angles to these lines." Maxwell: A Treatise on Electricity and Magnetism. 3d ed. London, 1892, vol. I., 63.

as a piece of iron placed in the lodestone's field becomes itself a magnet by induction, these lines extend through the intervening space between the stone and the iron.

Now turn to Newton, remembering that it was the action-at-a-distance theory which confronted him as the current explanation of attraction. He says:

"That gravity should be innate, inherent, essential to matter, so that one body may act upon another at a distance through a vacuum, without the mediation of anything else by and through which their action and force may be conveyed from one to another, is to me so great an absurdity that I believe no man who has in philosophical matters a competent faculty of thinking can ever fall into it. Gravity must be caused by an agent acting constantly according to certain laws; but whether this agent be material or immaterial, I have left to the consideration of my readers."[1]

Again and again Faraday quotes this passage. As Tyndall says,[2] he loved to do so.

He found from it, to use his own words, that Newton was "an unhesitating believer in physical lines of gravitating force."[3] But in his co-ordination of electricity, magnetism and gravity under the law of action and reaction, Newton makes himself even clearer as to this, than in the passage which Faraday selects. For what is the imaginary rope connecting the two bodies and contracting to draw them together but the direct expression of a physical line, not only of gravitating, but of electric and magnetic force? He not only sustains the last indirectly, as Faraday seems to intimate, but directly.[4]

---

[1] Third Letter to Dr. Bentley. Horsley: Opera. London, 1782, vol. iv., p. 438.

[2] Tyndall: Faraday as a Discoverer. N. Y., 1873.

[3] Exp'l. Researches, 3305. Dec., 1854. Jan., 1853, vol. iii., 507.

[4] "The attractive virtue (of magnetic bodies) is terminated nearly in bodies of their own kind that are next them. The virtue of a magnet is contracted by the interposition of an iron plate and is almost terminated

The law of action and reaction is true of electric, magnetic, as well as of gravitating attraction. The seat of the attracting power is in the interval between the bodies, whether electric, magnetic, or gravitating; and it is exerted in every case along lines of physical force. Such was Newton's discovery.

It was reserved for Faraday to direct renewed attention to the part taken by the medium, or as he called it, the dielectric, existing between electrified bodies, and to point out the nature and properties of the lines of force extending between these bodies and indicating the state of strain existing in this intervening space. Amplifying upon Newton, he inferred the existence of both magnetic and electric lines of force "from the dual nature of the powers (electricity and magnetism), and the necessity at all times of a relation and dependence between the polarities of the magnet and the positive and negative electrical surfaces.[1]

To pass beyond Newton's conception, in his time, was to struggle against the limits of the human intellect. So Faraday, in his epoch, dashed against the same barriers, only to recoil baffled, but never disheartened. The effects of the physical lines of force could be observed and dealt with experimentally; but their intimate nature remained, and still remains, unknown. That electricity and gravity and magnetism might be but manifestations of but one great controlling power pervading all matter was Newton's conception. For this power, throughout his whole life, Faraday searched. In this quest he made all his great discoveries.[2] Again and again, he exhausts the matchless powers of his imagination and his consummate experimental skill upon the problem, only to fail. The genius

---

at it; for bodies further off are not attracted by the magnet so much as by the iron plate." Principia, b. iii., prop. xxiii. Bence Jones: Life and Letters of Faraday. London, 1870, ii., 279.

[1] Faraday: Observations on the Magnetic Force. Proc. R. Inst., Jan. 21, 1853. Expl. Researches, vol. iii., 506.

[2] Bence Jones: Life and Letters of Faraday, London, 1870, vol. ii, 484.

which established the interconvertibility of electricity and magnetism could not identify gravity with either electricity or heat; and yet he felt this identity to exist, despite the negative experimental results. And so he left the world, even as Newton had left it, richer by vast accomplishments, challenging posterity to the grandest achievement to which the human intellect can aspire—the revelation of the unity of all natural force.[1]

"Electricity is often called wonderful, beautiful; but it is so only in common with the other forces of nature," writes Faraday, among his lecture notes. "The beauty of electricity, or of any other force, is not that the power is mysterious and unexpected, but that it is under *law*, and that the taught intellect can even now govern it largely. The human mind is placed above and not beneath it."[2] And the first mind which brought it into subjection to law was that of Isaac Newton.

The medium pervading space, Newton regarded as an ether; filling the universe "adequately without leaving any pores, and, by consequence, much denser than quicksilver and gold,"[3] yet offering an inconsiderable resistance to planetary motion. As it was questioned how such a medium could at the same time be both subtle and dense, he refers the critic to the electric and the magnet. "Let him also tell me," says Newton, "how an electric body can, by friction, emit an exhalation, so rare and subtle, and yet so potent, as by its emission to cause no sensible diminution in the weight of the electric body, and so be expanded through a sphere whose diameter is above two feet, and yet to be able to agitate and carry up leaf copper or leaf gold at the distance of above a foot from the electric body;" and as for the magnet, he points out that its emanations are capable of passing through glass without meeting apparent resistance or losing force.

[1] Bence Jones: Mrs. Somerville to Faraday, ii, 424.
[2] Ibid.: The Life and Letters of Faraday, London, 1870, vol. ii, 404.
[3] Newton: Optics. Qy. 22.

The recorded electrical experiments made by Newton are few, and are separated by long intervals of time. The earliest one was made in 1675,[1] when he found that a telescope glass, a couple of inches in diameter, mounted in a ring so as to be held about a third of an inch above the table on which it was placed flatwise, would, when rubbed on its upper side, attract bits of paper, etc., lying beneath it; and that the paper would vibrate up and down between glass and table for some time after the rubbing ceased. The Royal Society, to which this was communicated, tried to repeat the experiment and failed. Newton then discovered that not only were better results secured by using a larger glass disposed barely a sixth of an inch distant from the table, but that the nature of the substance with which the glass was rubbed appeared to influence its excitation. This last seems to have impressed him, as well it might, for it was an entirely new observation. He says that he obtained twice as much excitement of the glass when he rubbed it with his gown as he got on rubbing it with a napkin; and he advises the Society not to use linen or soft woolen, but "stuff whose threads may rake the surface of the glass." The Society, curiously enough, obtained the best results by employing a "scrubbing brush made of short hogs' bristles," "the haft of a whalebone knife," and finally resorted to merely scraping the glass with the finger-nails. This experiment of Newton appears to be the first suggestion of the different effects attending the rubbing of the electric with dissimilar bodies, a subject which became of great importance through the subsequent brilliant research of Dufay.

The principal discovery in magnetism resulting from actual experiment which belongs to the early days of the Royal Society, is the first production of artificial magnets

[1] Horsley: Isaaci Newtoni, Opera. London, 1782, vol. iv., 373.

by Sellers in 1667.[1] It was of course old to magnetize iron needles by rubbing them with the lodestone; and that even a succession or chain of armatures could be rendered magnetic by induction from a single stone, both by actual contact and through simple location in the field, had been known for ages. Sellers, however, had been rubbing needles on the stone to find out the conditions under which they would become most strongly magnetized; and he made up his mind that the needle's strength or direction did not depend so much upon "fainter or stronger touches on the stone nor the multiplicity of strokes" as upon "the nature of the steel whereof the needle is made, and the temper that is given thereunto." So he tried all sorts of steel, and finding the magnetism apparently permanent in his needles, easily made the succeeding step—which was to regard the magnetized steel itself in the same light as the lodestone; or, in other words, as an artificial magnet which "shall take up a piece of iron of two ounces weight or more; and give also to a needle the virtue of conforming to the magnetic meridian without the help of a lodestone or anything else that has received virtue therefrom."

As the century drew to its close, the growing commerce of England created an urgent demand for more definite knowledge concerning the variation of the compass. In 1580, William Burrowes determined the variation in London to be 11° 15' to the East. Edmund Gunter, the inventor of the scale and rule which bears his name, found that, in 1622, it had diminished some five degrees. Gellibrand, Gunter's successor in the Chair of Astronomy at Gresham College, observed that it had become reduced some two degrees more. In 1640, Henry Bond, a teacher of navigation in London, published his Seaman's Calendar, showing the progressive nature of this secular variation, and in 1668 issued a table predicting, though inaccurately, its changes in London for the next forty-eight

[1] Phil. Trans., No. 26, 478, 1667. Abridg., vol. i., 166.

years. But who actually discovered the secular variation is not certainly known. Bond attributes the honor to John Mair—other contemporary authority to Gellibrand, who at least has the preponderance of assent in his favor.[1]

The whole subject of compass variation, however, was thoroughly studied by Dr. Edmund Halley,[2] a mathematician and astronomer of great ability, who proposed the odd theory to account for it, that the earth has four magnetical poles, two near each geographical pole, and that the needle is governed by the pole to which it happens to be nearest. Unfortunately, however, the observed changes in the variation itself over certain periods of time interfered so greatly with this doctrine that it became evident to Halley that the notion of four fixed poles would not meet the observed conditions. Thereupon he evolved a still more striking supposition, to the effect that the earth really consists of two concentric magnetic shells, each having poles differently placed and not coincident with the geographical poles. Then as the poles on the inner shell "by a gradual and slow motion change their place in respect to the external, we may give a reasonable account of the four magnetic poles, as also of the changes in the needle's variations."

It is hard to believe that the imagination could exercise such control in the days of Newton. Yet the theory attracted considerable attention and had even great vitality, for in 1698, thirteen years after he had proposed it, Halley induced William III. to appoint him a captain in the Navy and give him command of a ship, in order to make long voyages for the express purpose of establishing the truth of his supposition. He made two voyages to various parts

[1] Dr. Wallis (Phil. Trans., 1702, No. 278, 1106), says that "at about the beginning of the reign of Charles I., Gellibrand caused the great concave dial in the Privy Garden at Whitehall, which is still remaining, to be erected in order to fix a true meridian line.

[2] Phil. Trans., No. 28, p. 525, 1667; No. 148, p. 208, 1683; No. 195, p. 563, 1692.

448   THE INTELLECTUAL RISE IN ELECTRICITY.

of the Atlantic and Pacific Oceans, and came back not with the desired proof exactly, but with a useful chart exhibiting the variation of the needle in many parts of the world, and the general law of its phenomena[1].

[1] Brewster: Treatise on Magnetism. Edinburgh, 1836, p. 13.

DALANCE'S TITLE PAGE.

NOTE.—A curious illustration of the mixture of old and new ideas concerning magnetism which existed at the end of the seventeenth century is found in the title page of Dalance's "Traitté de l'Aiman," published in 1687, which is here reproduced in fac simile.

The lodestone, disposed in a bowl after the mode suggested by Neckam and Peregrinus, and marked with a longitudinal directing line, appears floating in front of the vessel, which the mariner, holding a rudder in one hand and a compass in the other, is about to board. The goddess, who appears to be advising him, points to the Great Bear, represented by the actual animal in the heavens, with the Pole Star situated at his tail, and also to a compass and a dipping needle, while in her left hand she has a sounding line. The idea evidently intended is that the divinity is advising the sailor to avail himself of all these means of guidance. There is also shown on the left a suspended armed lodestone, supporting at one pole a series of keys, and at the other a number of iron plates, this being possibly designed to indicate in some way the strength and consequent trustworthiness of the magnet.

CHAPTER XIV.

FOUR years after the foundation of the English Royal Society, Colbert, the astute and far-seeing minister of Louis XIV, perceived in the gatherings of philosophers which were still held at the houses of Thevenot and others, the possible nucleus of a great national institution, capable of advancing science and the industries of France. The Royal Academy of Sciences was therefore duly established by royal command in 1666, and with princely generosity, intended to be in marked contrast with what English Charles did not do, Louis endowed the new body with ample funds for its future experiments, and added pensions and rewards for deserving members. Thus equipped, the philosophers had nothing to do but startle the world with the magnitude and originality of their discoveries, to the making of which they might now devote themselves without troubling as to cost.

At first they proceeded slowly. The original members were chiefly mathematicians, and experiments can hardly be said to have begun until the physicists were admitted. Then they went at it with a will. They experimented in concert, with results fully equal to such as might reasonably be expected to follow the production of Shakespeare's tragedy with a chorus of simultaneous—if not concordant —Hamlets. There was no gathering in a room and reading one another asleep with interminable papers, suitable only for the phlegmatic plodding English. The sessions were held in the laboratory. Nature should be made to yield up her secrets by the combined efforts of several brains attacking her stronghold simultaneously, like the concentrated fire of a battery. They needed no Charles to suggest subjects and spur them on. Indeed, when Louis the Magnificent and Monsieur the Dauphin and le Grand Condé, attended by a gorgeous retinue, came in state to visit them, it was the king himself who, after intrepidly withstanding several chemical lectures, remarked that he had "no need to exhort them to work, for they were doing it enough for themselves."

So they kept on experimenting manfully, and quarrelling fiercely; and their activity was prodigious. The results of these practical labors appeared principally in the shape of dissertations on abstract mathematics, and they fill ten volumes of "Anciens Memoirs." Still, as long as Colbert lived, the philosophers were protected, and experimental science—as they viewed it—flourished.

But when Louvois became Minister, matters took a new turn. If the work of the Academy thus far was properly defined as experimental, then Louvois soon showed the most opposite, and hence theoretical, disposition. When the public-spirited king decided to improve the landscape at Versailles with more indispensable cascades and the erection of a much-needed additional mountain, it was Louvois who told the members that they were paid to work, and set them at such theoretical tasks as aqueduct building, pipe laying and surveying. He made La Hire and Picard supervise the building and engineering, Thevenot plan watercourses, and Mariotte attack the problems of water supply. When there was not sufficient of this sort of theorizing to do at Versailles, Condé invited them to theorize in the same fashion at Chantilly.

Besides, the *haut monde* of Paris had heard of the new fashion at Whitehall, and how all the great English milords and miladies were besieging the Royal Society. Should the Court of the Grand Monarque be distanced in a matter of la mode? Immediately were the mathematicians invited to calculate the chances in every gambling game in vogue, in "quinque nova," in "le hoca" and "le lansquenet." Sauveur, however, who too complacently evolved a surely winning system adapted to "la barsette" in his capacity of "mathematician to the Court"—found himself abruptly invited into the closet of his irate sovereign, and given distinctly to understand that the royal prerogative included secrets of that sort, and that kings were not to be left subject to the run of luck ordained to common people. "What was the Royal Academy for, if" etc., etc.?

Every one knows how Louis went to the wars, dragging poor Racine from his theatre to write history as he made it, and Perrault and Roemer and Mariotte and Blondel, regardless of the fact that some were mathematicians and others astronomers, to study bombs and ballistics. It was sufficient for Louis that they were all scientific persons. About the only philosopher of eminence whom he let alone was Cassini, and that because the astronomical observations in progress were useful for the Navy. It is perhaps not altogether surprising that in these circumstances the Academy, as one of its historians remarks, "lost its lustre and fell into a languor." There it remained until De Ponchartrain reorganized it in 1699, mainly after the bureaucratic system, so dear to the Gallic heart, and with such singular astuteness that it at once provided a variety of new offices for hangers-on of the Court. Thus inspired with new life, it proceeded to dispute the Newtonian theories for the next half century, and patriotically stuck to Descartes and his vortices long after they had become abandoned by Holland, Germany and St. Petersburg.[1]

All of this accounts for the fact that one may turn over the pages of the ten volumes of Anciens Memoirs before noted—yes, and those of many of the later tomes of the Histoire de l'Académie Royale—and find little or nothing to show that French philosophy had ever heard of the discoveries of Boyle or Hooke, or even of the German, Von Guericke. Yet in that (to us) dreary waste of antiquated natural history, anatomy and mathematics, there may be

[1] Maury: L'Ancienne Académie des Sciences. Paris, 1864.

found a short note, barely filling a printed page, which contains the suggestion which was the original cause which started the whole scientific world to puzzling over the wonders of the electric light.

The terrestrial measurements which enabled Newton to correct his calculations concerning the moon and to verify his belief in the effect of the earth's gravity thereon, were made by Jean Picard, a priest and an astronomer of remarkable ability. It was Picard who informed the Royal Academy of a curious effect which he had observed in the barometer which he employed in the Paris Observatory. The instrument of that time was merely a glass tube closed above, open below, exhausted of air and inserted, open end downwards, in a cup of mercury: the metal, of course, rising in the tube under the atmospheric pressure. Picard observed that when the instrument itself was moved so as to cause the mercury to vibrate in the tube, a light appeared in the empty portion of the latter, clearly visible in the dark. It is said that he first saw it while carrying the apparatus in his hands from one part of the observatory to another after nightfall. At all events, there was no mistaking the luminosity—which was a sort of broken glow above the quicksilver, and which appeared best when the mercury descended quickly. The note, which bears the date of 1675, adds that efforts had been made (combined experiments, probably) to find other barometers which would behave similarly, but not one had been encountered; that it had been resolved to examine the matter in every possible way, and that the future discoveries would be set forth in detail.[1] The same cheerful confidence which the king had shown concerning coming developments in general, is here reflected with regard to what was going to be found out about this singular light.

[1] Mem. de l'Acad. Roy. des Sciences. Paris, 1730, vol. x., p. 556.

But the years went by, and if the discoveries were made nobody mentioned them, and the strange light which Picard had seen in the barometer was as little remembered as the glow which Guericke had obtained years before from his sulphur ball.

There had been known, since the beginning of the century, a mineral, sometimes termed the Bologna stone, sometimes the Bononian stone, from the place of its discovery, which would become luminous in the dark.[1] It had been accidentally found by one Casciorolus, a shoemaker who had deserted his trade for alchemy, and who gave it the name of "lapis solaris," because, from its illuminating properties, he conceived it especially suitable for the transmutation of silver into gold—the alchemical *sol*. As the Italian chemists seem to have agreed in this opinion, the stone soon became in great demand and brought fabulous prices, which were maintained despite the claim of Potier, a French chemist, that he could produce it artificially. In 1666, the English Royal Society records the death of a clergyman who was said to have exclusively possessed the art, without communicating it to any one.

The value placed upon the substance—which was barium sulphide, frequently used now as a basis for the so-called luminous paint—incited the chemists to endeavor to imitate it; with the result that, at about the time of Picard's observation of the light in his barometer, Brand, of Germany, produced a light-giving substance from animal excretions, and sold the secret of its manufacture to Krafft. Krafft named it "phosphorus" and took it to England, where it was exhibited to the king, and, as we have already seen, it constituted one of the most interesting of the Gresham College curiosities. In Germany, Kunkel, who learned of it from Krafft, published, in 1678, a pamphlet describing it, and the interest excited in Eng-

---

[1] Beckmann: A History of Inv'ns and Discoveries. 3d ed., 1817, vol. iv., 419. Roscoe and Schorlemmer: A Treatise on Chemistry. N. Y., 1883, vol. i., 457.

land spread rapidly over the continent. It was termed "phosphorus mirabilis," "phosphorus igneus," and sometimes "light magnet"—although the last name is often also applied to the Bologna stone.

The effect of this discovery was to draw especial attention to all substances which appeared to be naturally luminous, and decaying fish, sea-water and glow-worms, sparks produced by abrasion, the heating of metals to redness by friction or impact, were all studied as allied effects, because all of them gave light. Boyle made the subject one of special research; and in aid thereof Clayton sent him huge fire-flies from Virginia, and told him about the sparks which flashed from Madam Sewall's petticoats.

It is curious to observe how frequently the accidental acquirement of a book precedes the making of a train of discoveries. A little tract on barometers, which happened to have in it an account of Picard's observation, fell into the hands of John Bernouilli, who was then professor of mathematics at Gröningen.[1] Bernouilli made up his mind that here was a way of producing light naturally, without the aid of any chemical phosphorus at all; but as the word "phosphorus" was then applied to any substance which became luminous without combustion, he called Picard's phenomenon the "mercurial phosphorus," and, in June, 1700, gives the results of his own experiments on the subject in a letter to Varignon, then a member of the French Academy. The ensuing consequences are a warning against hasty deductions, and besides exhibit the wisdom of the profound remark of Mr. Diedrich Knickerbocker, that "it is a mortifying circumstance which greatly perplexes many a painstaking philosopher that nature often refuses to second his most profound and elaborate

---

[1] See Martin and Chambers: The Phil. Histy. and Memoirs of the R. Acad. of Sci., Paris. London, 1742.
Histoire de l'Acad. R. des Sci., from 1666 to 1699. Paris, 1733.
Histoire de l'Acad. R. des Sci., for years 1700 to 1707. Paris, 1701 to 1708. With accompanying memoirs. Bernouilli's letters are here published in full.

efforts; so that, after having invented one of the most ingenious and natural theories imaginable, she will have the perversity to act directly in the teeth of his system, and flatly contradict his most favorite positions."[1]

Bernouilli gave not only an elaborate explanation of the effect in accordance with the Cartesian theory, by assuming different matters respectively entering the vacuum through the glass from without and arising from the mercury within and then clashing together (in which he was quite safe, seeing that he was communicating with the Cartesian stronghold); but also laid down numerous precautions, which he said it was indispensable to observe in order to reproduce the effect. This last rather surprised the Frenchmen, because Cassini for one had been getting light from his barometer for the last six years without troubling himself with any precautions at all. And Picard's old instrument had been taken to pieces by De la Hire and set up over again, and sometimes it had given light and sometimes refused to do it, from apparent sheer wilfulness. In fact Cassini and De la Hire had compared notes, and even thought they found differences in the sort of light which their respective barometers yielded. However, it was thought best to follow Bernouilli's directions, with the unexpected sequel that the apparatus so made refused to glow at all—while more people began to produce instruments which behaved beautifully.

Bernouilli, on being informed, calmly modified his requirements, insisting, however, upon absolutely pure mercury and total exclusion of air. But old barometers obviously containing air bubbles still persisted in glowing. Then Bernouilli himself discovered that the vacuum was not needed, and that mercury shaken in an ordinary vial shone finely. The French Academy seems to have been unable to reproduce this, and Bernouilli investigated the matter far enough to reach firm ground. He found that so long as the mercury was fairly pure he could get lumi-

---

[1] Irving: Knickerbocker History of New York.

nosity with certainty in the vial; and stranger still, that when the vial contained air, the light appeared like sparks "which arise simultaneously and perish almost at the same time;" but when the vial was exhausted of air "the light is like a continuous flame which lasts incessantly while the quicksilver is in agitation." The least humidity, even the perspiration of the hand, would put the light out.

Bernouilli's discovery was hailed in Germany with enthusiasm. It was supposed that he had invented a new mode of mechanical illumination which might perhaps render candles and lamps things of the past. And he probably so believed himself, for he seems then to have had no conception of the real cause of the glow.

Before long the news reached the Royal Society. Hooke was then incapacitated for arduous work by both age and illness, and Francis Hauksbee,[1] who held the office of curator of experiments, undertook to investigate the matter. Little is known concerning Hauksbee further than that he had already achieved reputation as an experimentalist. His first recorded researches bear date 1705, and he seems to have been a persistent student until he died, some seven years later. That he was a man of unusual genius in original research is abundantly shown. His mind was philosophical, and but little influenced by the prevalent hypotheses which to many seemed axiomatic. To him is due not merely the recognition of the effect of Newton's reduction of electric phenomena under general law, but the almost instant perception that the next logical step was the seeking of "the Nature and Laws of Electrical Attractions" which "have not yet been much considered by any." He invented a form of air-pump that is still known by his name; but his fame ought to rest, and deservedly, upon his extraordinary electrical experiments now to be recounted.

[1] Hauksbee: Physico Mechanical Experiments on Various Subjects. London, 1709. See also his communications to the Royal Society in years 1705 to 1712 inclusive.

His starting-point is the strange light seen in the mercurial barometer, the cause of which it is his task to discover. Like Bernouilli, he calls it the "mercurial phosphorus." In common with others, he believes the radiance to be due to some quality of the mercury, brought into action by the peculiar conditions of vacuum, or agitation, or both. The Cartesian theory had few adherents among the English philosophers of the time, and certainly Hauksbee was not among them.

From the moment he begins his experiments (1705) the results astonish him. It must be borne in mind that, at the outset, he had no suspicion that the mercury light had anything to do with electricity. As I have already stated, these odd luminosities, which did not appear to be the immediate consequence of actual burning, were all grouped together, and the effort was often made to refer them to some common origin. Even Newton[1] held this belief. "Do not all bodies," he asks, "which abound with terrestrial parts, and especially with sulphurous ones, emit light as often as those parts are sufficiently agitated; whether that agitation be made by heat, or by friction, or percussion, or putrefaction, or by any vital motion on any other cause? As, for instance, sea-water in a raging storm; quicksilver agitated in vacuo; the back of a cat or neck of a horse, obliquely struck or rubbed, in a dark place; wood, flesh and fish, while they putrefy; vapors arising from putrefied waters, usually called Ignes Fatui; stacking of moist hay or corn growing hot by fermentation; glow-worms and the eyes of some animals by vital motions; the vulgar phosphorus, agitated by the attrition of any body or by the particles of the air; amber and some diamonds, by striking, or pressing, or rubbing them; scrapings of steel, struck off with a flint; iron hammered very nimbly till it become so hot as to kindle sulphur thrown upon it." Obviously there was no more reason why Hauksbee, in the beginning, should have supposed the barometer light to be

[1] Optics. Q. 8.

kindred to the amber light or cat's-back light, than to the light due to the striking of flint and steel. In fact, as will be apparent further on, his impressions evidently were that the last-named alliance was the most probable.

The question which had been most debated bore upon the need of a vacuum existing in the vessel which contained the mercury, and to that he first directs attention. He proves almost immediately that, by allowing air to rush through quicksilver in an exhausted receiver, he can convert the liquid metal into a jet dashing in drops in every direction against the sides of the vessel, and looking, as he says, "like one Great Flaming Masse." Then he permits mercury to flow downward into an exhausted receiver so as to strike a rounded glass surface therein and so become spread. A shower of fire appears; luminous, however, only (his observation is very quick) "where it strikes the glass in its fall." Now he lets in three pounds of mercury at once in a cascade, and then "the light darted thick from the crown of the included Glass like Flashes of Lightning."

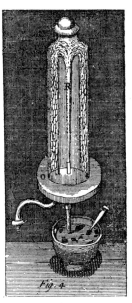

HAUKSBEE'S LUMINOUS MERCURIAL FOUNTAIN.[1]

They *were* flashes of lightning, and this was the first suggestion of that great identity by one who was building far better than he ever knew.

The behavior of the light is curious. When the mercury falls into a vacuum, there is a gentle, uniform glow; but when it pours into air, the sparks dance between the glistening drops. What are the sparks? Certainly,

[1] Reproduced in fac simile from s'Gravesande's Elements of Natural Philosophy. 4th ed. 1731.

concludes Hauksbee, whose fascination with his work shows itself now in every line of his description, that sort of light does not resemble the little bluish radiance in the barometer! What is it?

He undertakes to find out by noting the conditions under which he can produce a similar light. He arranges a piece of amber so that he can revolve it swiftly in contact with a pad of woolen cloth within his exhausted glass vessel. The light appears; he can see it at a distance of three or four feet. It is brighter as the vacuum increases; but then the amber begins to burn and the woolen scorches. Did the heat so produced make the light?

Try flint and steel, and see, his active brain answers. A steel ring is made to revolve in the glass vessel and a bit of flint is pressed against it. Before the air is exhausted the sparks fly in showers, but as the air-pump draws out the reluctant atmosphere they fade and finally disappear, and only a just perceptible luminous ring where the stone touches the whirling metal at last remains. No; it is not the flint and steel light which needs the air—this unearthly glow, which thrives best when the air is gone.

Singular, that this light, so like the lightning, should have been produced in an exhausted glass bulb, and almost two hundred years ago!

Hauksbee now determines that the mercury light on the whole is more like the amber glow than like the coruscations flying from the steel; but as amber is resinous and inflammable he substitutes glass as the material to be rubbed, and makes a new discovery. The light in the exhausted receiver becomes purple; but, as the air is let into the vessel, it fades, turns reddish, and then gray—very feeble when the vessel is full of air. It is odd how the color changes as more or less air is admitted; odder still that there should be flashes and no longer a glow when the woolen rubber is soaked with a saltpeter solution. He rubs glass on glass, glass on oyster shells, oyster shells on woolen, sometimes in vacuo, sometimes in air; puzzling continually over the varied results, for it is difficult to tell when the light comes from the high heating of these substances, due to friction, and when not.

At last a novel idea strikes him. Why rub glass in a glass vessel exhausted of air? Why not rub the exhausted glass vessel itself? At once he mounts a glass globe in a

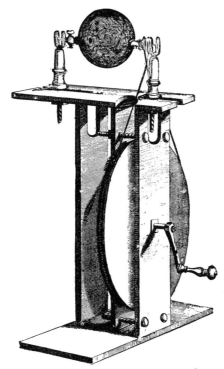

HAUKSBEE'S ELECTRIC MACHINE.[1]

sort of lathe, sets it whirling, and holds his hand to the surface. The results, in point of brilliancy, overtop those of all predecessors.

[1] Reproduced in reduced fac simile from Hauksbee's "Physico-Mechanical Experiments on various subjects containing An Account of several Surprizing Phenomena touching Light and Electricity." London: 1709. The wavy lines on the globe are evidently intended to represent the play of light therein.

The light, still purple, is "so great that large print, without much difficulty, could be read by it, and at the same time the room, which was large and wide, became sensibly enlightened, and the wall was visible at the remotest distance, which was at least ten feet." As he lets air into the globe the radiance diminishes; but something of the meaning of what he sees begins to dawn upon him. He notices a similarity between the mercury light and this glow of the glass—the difference is as great "between the light in the globe exhausted and the light produced when the globe was empty of air, as between the lights produced from mercury when the experiment was made in vacuo and in the open air."

It was fortunate for Hauksbee that he was experimenting in the days of good Queen Anne, and showing all these marvelous things to the Royal Society, which had done more to overthrow superstition and especially belief in witchcraft and sorcery in England than perhaps any other of the great civilizing forces. Conceive of Scotch James hearing with complacency of a man who makes spots of light appear under his fingers as he touches a glass bottle; who says, "Nay, while my hand continued upon the glass—the glass being in motion—if any person approached his fingers toward any part of it in the same horizontal plane with my hand, a light would be seen to stick to 'em at the distance of about an inch or thereabouts without their touching the glass at all." A light, a corpse-light, clinging to the very hands of the foolhardy wretch who ventured near the infernal apparatus of this prince of wizards, might well be the royal conclusion, followed by a disposition of Mr. Hauksbee which would have left the world ignorant that he had ever thus exhibited the electrification of the human body by induction from his glass globe.

But, as I have said, these were the days of Queen Anne, when Marlborough was returning in triumph from Ramillies, and England and Scotland were uniting, and other great political events happening, all proving how greatly the times had changed; and as to this last it is significant that Hauksbee's treatise and the Tatler newspaper—the first real Anglo-Saxon newspaper which did not get its home news by way of the Dutch—appeared in the same year. One not unnaturally follows such a chronicle of physical discovery as this, tracing the struggles of men to wrest from unwilling nature her secrets, often forgetting that the achievements or the failures are correlated to other and widely different events peculiar to especial ages and times. True, such research merely reveals natural laws which are the same yesterday, to-day and forever; and whether this is done a hundred years earlier or later, or brings to the discoverer fame or a fagot, cannot alter the ultimate supremacy of the truth. Yet there is a great world living and moving outside the walls of the laboratory and influencing in his every act the man that is within it, sometimes to encourage, oftener to dishearten. It has had a great deal to do with the rise of electrical knowledge, mainly in the way of prevention; but never before have its ignorance and credulity and superstition strewn fewer obstacles in the pathway. Mr. Hauksbee's hands may glow and his fingers may sparkle with the ineffectual fires of the excited glass, without fear of a change to the flames of Smithfield. Perhaps his future associate in the Royal Society, the Reverend Cotton Mather, resident in New England, might feel moved to offer him the joys of martyrdom were his lights flashing in Boston instead of in London; but in Old England, the England of Steele and Addison and Swift—of Isaac Bickerstaffe and Sir Roger de Coverley and Gulliver—even Mr. Hauksbee's neck-cloth may become as "fiery" as it likes without provoking the grewsome summons of the witch-finder. Besides, his "Physico-Mechanical Experiments," and the first volume of Mr. Addison's Spectator own the same noble patron, John, Lord Somers, sometime President of the Royal Society and Lord Chancellor of England; a good and stalwart bulwark at home, even if across the Atlantic, in Cotton Mather's land, that growing settlement, New York, is eyeing him suspiciously as an accomplice of her most picturesque pirate, Captain Kidd.

Mr. Hauksbee, however, bending over his globe, is uttering new, and even more fervid expressions of amazement and admiration. His lights are becoming fantastic, branching here and there, dashing against the crystal walls, while his notions are being turned so completely around, that he is beginning to believe that this illumination and that given by the mercury are, after all, very much alike. Certainly both seem to come from glass, and, as he says, "one might conjecture with some probability that the

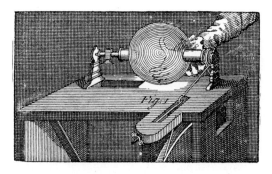

HAUKSBEE'S ELECTRIC GLOW.[1]

light produced proceeds from some quality in the glass (upon such as friction or motion given it), and not from the mercury, upon any other account than only as does a proper body, which by beating or rubbing on the glass, produces the light."

Observe how easy it is after the event to foresee conclu-

[1] Reproduced in fac simile from s'Gravesande's Elements of Natural Philosophy, 4th ed., 1731. There is no picture of the electric glow given in either of the two editions of Hauksbee's treatise. s'Gravesande's work from which Dr. Desaguiliers made the translation above noted, appears to have been published not long after Hauksbee's second edition, so that the present illustration is a fairly near contemporary representation of the phenomenon.

sions. Hauksbee has found that the rubbed glass glows, mercury rubs glass, glass is an electric excited by rubbing; *ergo*, says the Keen Intelligence, glancing at this page and bounding unerringly to the inevitable sequel, he has discovered the mercurial light to be electric. That, however, is what the Keen Intelligence would have done in Hauksbee's place; but it should be remembered that minds differ, and Hauksbee's was not of the superior nineteenth century, but of the inferior eighteenth century variety; and hence, unable as yet, despite all that has happened, to harbor the notion that electricity has anything to do with the matter at all. So we must follow him a little further in his gropings.

Serious physical discoveries, untinged by any trace of levity, have a way of getting into that stage in which Charles Lamb records the cooking of roast pig to have long remained before the important fact was revealed that it was not necessary to burn down a whole house in order to roast that succulent animal.

The imagination always recoils from abstractions, and insensibly links an idea with the particular thing in which it happens first to be embodied, or through which it first came to be known. Consequently when Hauksbee desires to test his explanation of the light as due to the friction of mercury on glass, he goes back to the barometer, although that instrument, as a barometer, had nothing to do with the effect; just as people all over Europe, for a considerable time, depopulated the frog ponds, under the notion that Galvani's discovery could not be made manifest except through the actual frogs' legs. He rubs the empty tube above the mercury with his fingers, and then again he sees the light, which follows his fingers without any motion of the quicksilver at all. That brings him to Newton's experiment—the attraction of the bits of leaf brass and paper by rubbed glass—although he does not recognize it as Newton's, because he has reached it by his own independent reasoning, and, in fact, has re-invented it. Then he begins to forge the link between the rubbed glass giving light and the rubbed glass attracting scraps of paper, and suspects that both phenomena are electrical.

But now new questions crowd upon him. The light appears and the attraction is exercised outside of the glass. The supposed emanations producing both can be cut off by moist air or even by fine muslin, not because of any inherent property in either air or cloth, but—and here he uses a well-known word in modern electrical language—because there is an interposition of something which acts as a "resistance."

The belief that all these effects are governed by law is uppermost in his mind, and so he says that, "the effluvia, how subtle soever they can be imagined to be, are yet body and matter, and must therefore be liable to the common laws of bodies, which is to be resisted in some proportion to the strength and density of the medium." At once he seeks to "find in what manner such a motion is propagated and in what figure or sort of track it went along." A chance observation spurs him on. He has held the rubbed tube to his face and felt, with amazement, the electric wind from the dense charge at its extremity, making "very nearly such sort of strokes upon the skin as a number of fine limber hairs pushing against it might be supposed to do."

As he rubs his tube, the light breaks forth and crackles like green leaves in the fire. The substitution of a solid glass rod for the tube, makes little difference in the effect. Rubbing the tube by hand is awkward, so he arranges a glass cylinder in his lathe and revolves it, noticing now not only the purple light, but again the sensation of a current or wind striking his finger held near "with some force, being easily felt by a kind of gentle pressure, though the moving body was not touched with it by near half an inch." The object lesson is plain. If whatever that is which seems to come from the glass is so powerful that it can be felt, it ought to be able to influence bodies which

ELECTRIC LINES OF FORCE.   467

are placed in it—perhaps just as the wind moves the weather-cock, or causes a flag to stand out in its current. He places a semicircle of wire having a number of woolen threads hanging from it, transversely over his glass cylinder. The threads at first are perpendicular. Then as the cylinder is rotated, no pressure of the hand being exerted upon it, the threads are blown aside all in one direction by the wind or eddy caused by the revolving glass; but, as soon as he places his hand on the cylinder to rub it, the threads immediately straighten, and every one of them assumes a radial position pointing to the axis of the cylinder, while the light and the cracklings are simultaneously seen and heard. He changes the position of the threads, sometimes fastening the semicircle of wire below the cylinder, and then the threads are compelled to stand up and point to the axis; and sometimes he places the cylinder vertically with the semicircle horizontal, and then the threads stand out horizontally, thus proving that the force in the space about the cylinder is strong enough to direct the threads in straight lines despite the tendency of the air to swing them aside. The extension of the threads certainly depends "upon the action of some matter whose direction is in straight lines toward the glass." There was a recognition, clearly and plainly, of the lines of electric force—for he says that when a body is interposed between the threads and the glass "they lose their regular extension and hang as their own weight causes them."

Now follows a discovery of capital importance, but which to Hauksbee is a complete puzzle. He disposes two glass globes within an inch of each other, but mounted in separate lathes so that they can be rotated independently. He exhausts the air in one and applies his hand to the unexhausted globe. Then he sees the light appear, not only on the globe that is rubbed, but on the exhausted globe which is not rubbed. But he soon finds that motion of both globes is not necessary, and that he has only to bring near to the excited globe a vacuum tube to see the light at once flash therein. This he supposes to be due to the frictional action of the effluvia emanating from the rubbed electric, but in fact it was a demonstration of electric induction.

After many more experiments—variations of one kind and another—he is confirmed in his belief of the effects happening in the field of force, and states it in the following explicit terms: It is (the italics are his own)—

"Not only a *Communication*, but a *Continuity* of the Matter which occasions the Motion of the Threads. The Progress of it seems to be in a straight and direct track; in which the Matter is push'd by the *shortest Course* from the *Approach'd Body* to the Threads that are shaken by it. And if the Threads are mov'd by influence of any Matter emitted from the Glass, it appears to be impossible to explain how they should be so, and *at such distances*, without a Continuity. So that the Case seems to be thus: That the Effluvia pass along, as it were, in so many *Physical Lines* or *Rays;* and all the Parts that compose them, adhere and joyn to one-another, in such manner, that when any of 'em are push'd, all in the same Line are affected by the Impulse given to others."

It is not necessary to review his concluding experiments in detail, although some of them, such as the outlining of his hand in fire on the inside of a globe partly lined with sealing-wax, and the movements of threads electrified within the cylinder following that of his finger outside of it, are striking enough. The research extended over four years, interrupted at times by other investigations, mainly in pneumatics. It was a brilliant piece of work, and probably the first thoroughly scientific investigation of electrical phenomena.

Hauksbee's achievements attracted great attention. Newton, after the publication of his Optics in 1704, experimented on a glass globe for himself, and the results appear in the second edition of that work, which was published in 1717. He also felt the electric wind dashing against paper, and saw the sheet "become lurid like a glow worm."[1]

In 1708, Dr. Wall,[2] who evidently disagreed with Hauksbee's conclusions as to the electric nature of the barometer light, evolved a hypothesis concerning the amber about as odd as that which Father Grandamicus had proposed to account for the earth's rotation. Grandamicus said that the earth does not rotate because it is a magnet, and Wall asserts that the amber attracts, not because of its electrical quality, but because it is "a natural phosphorus, a mineral *oleosum* coagulated with a mineral acid of spirit of salt." Wall, however, attains immortality neither for his theories nor for his experiments, but for an expression. Hauksbee, long before, had heard the crackles and had likened the fires in his glass globe to flashes of lightning. Wall, rubbing a large piece of amber and seeing the sparks and hearing the noise, however, says: "Now I doubt not, but on using a longer and larger piece of amber, both the cracklings and light would be much greater, because I have never yet found any crackling from the head of my cane, though it is a pretty large one: and it seems in some degree to resemble Thunder and Lightning." It is a pity that Wall's far-fetched notion that the amber is phosphorus, and its light that of phosphorus, should cast a shadow on his title to being the first who saw in the electric spark and detonations the effects of Jove's armory in miniature.

Bernouilli, to whom Frederick of Prussia, on the recommendation of Leibnitz, then President of the Berlin Academy of Sciences, had presented a gold medal, worth forty ducats, as a reward for his discovery, denied Hauksbee's explanation of the mercury light. It is needless to review his contentions; they went the way of the learned arguments whereby the Italian ecclesiastics in Galileo's time sought to eliminate the moons of Jupiter.

The progress of electrical discovery had now reached one

---

[1] Optics. Q. 8.   [2] Phil. Trans., No. 314, p. 69, 1708.

of those temporary halting-places which are easily discernible in looking back over its path. The latest problem had apparently been solved. To many it no doubt appeared that all the capabilities of the rubbed electric had been revealed. It had given light, attracted and directed threads, yielded effluvia sensible to the touch and transmitted its virtue to other bodies near it so as to cause them to glow. No new possibilities were in sight. For nearly twenty years no one sought for any, and the very few experiments that are recorded merely thrash over old straw. The philosophical world was devoting all its energies to the digestion of the colossal intellectual banquet which Newton had spread before it.

"A masterpiece of English charity" is what old Fuller says of it—that ancient foundation of James I., in the chapel whereof the boys of Grey Friars school and the fourscore old pensioners of the Hospital used to assemble on Founder's Day listening to the prayers and psalms.

Who does not know Thackeray's description of the place? It is one of those old Charterhouse brethren whom I have now to call back; an old brother who sat on the same old benches in the ancient chapel, and who passed away and gave place to another old brother, and he to another, and another, long before Thackeray's time, and one whom, if we may credit what another philosopher high in favor in court said about him, was a testy and crusty old gentleman. But philosophers high at court, and philosophers who are poor brothers, rarely appreciate all one another's excellences; and besides, the young Cistercians had a much better opportunity of knowing this particular poor brother than the dignified gentry at Whitehall. For Stephen Gray never hung up his chief critic, the Reverend Joseph Desaguiliers, tutor to his Royal Highness, by the neck and heels and drew sparks from him—and that is what he did, besides many other astonishing things, to the

Grey Friars lads; and we may be quite sure, not without their entire consent and approbation.

There is no biographer to tell Gray's history, however curtly. His memorial is hidden in the early volumes of Philosophical Transactions[1]—the annals of the Royal Society which are seldom read except through some one's abridgments. He appears there first during the halt period in 1720, and evidently with Boyle's experiments on the feminine head-gear in mind, says that he has made leather and parchment and paper and hair and feathers and threads all electrical by rubbing them, so that it almost looks as if he had procured one of those towering structures of millinery and, after dissecting it, had electrified every bit of it in detail. Then he disappears for nine years, and we do not know what he was doing in that interval any more than before his sudden advent, although it is said that in his early days he devoted much attention to optics. When he returns to view in February, 1732, and recounts his discoveries of the preceding three years, he dates his letter to the Society from the Charterhouse, and the presumption follows that the world has shown him its seamy side, and that after fifty years of struggle, he welcomes the peaceful asylum and sober garb of the poor brethren. But it would be altogether wrong to suppose that he utters any note of repining. On the contrary, it is evident that he is now in possession of facilities for doing work in which he delights; and besides, he has two good friends, one a well-to-do country gentleman, the other resident in London and a member of the Society; and, better still, both cordially sympathetic in all his aims and endeavors. He spends his summers with them, and makes the house of one of them the scene of a great discovery, and worthy of a commemorative tablet, if it could be now identified.

[1] Gray's papers are as follows: Phil. Trans., 1720, vol. 31, p. 104; 1731, vol. 37, p. 18; 1732, vol. 37, pp. 285, 397; 1735, vol. 39, pp. 16, 166; 1736, vol. 39, p. 400.

A fine laboratory fitted with delicate and costly apparatus, skilled workmen at one's call, and unlimited capital to draw upon, did not fall to the lot of the electrical discoverer of Gray's time. There were no electrical shares quoted on the world's exchanges in those days, and whatever the magnetizers may have gained, no one had ever made a penny out of electricity, or even perceived channels whereby profitably to lead other people to lose pounds. Therefore, no one supplied Gray with means pecuniary or otherwise for the prosecution of his work. But that did not trouble him. There were his fishing-rods and his canes, the kitchen poker and cabbages and pieces of brick; hemp twine was cheap, and by getting along with these he could economize sufficiently to acquire the more expensive part of his apparatus, a little silk and a few glass tubes. If a suspended boy was wanted, no doubt there were plenty of the Grey Friars lads willing enough to undergo the astonishing experiences which the old brother contrived for them.

Up to this time no one (Von Guericke excepted, and he forgotten) had thought to inquire whether the electric virtue could be made to pass from one body to another. This Gray did, and came to do so through the idea suggesting itself that if Hauksbee's glass tube could communicate light to another object by its electric quality, why could it not communicate the quality itself?—in which case the body receiving the virtue would have the same property of attracting and repelling light bodies as the excited tube. It also struck him that if this could be done, "the attractive virtue might be carried to bodies that were many feet distant from the tube."

He procures a glass tube about a yard long and a little over an inch in diameter. To keep out the dust, he puts corks in the ends—an expedient which turns out to be the quickest possible means of revealing exactly what he was looking for. Now he rubs the tube in order to excite it electrically, and to his surprise he finds that feathers and pieces of foil fly as readily *to the cork* in the end of the tube as to the tube itself—and thus it was plain that the virtue had instantly passed from glass to cork.

He at once attacks the second part of his problem: how far will this virtue travel? Into the cork in the glass tube he inserts a wooden rod four inches long, having an ivory ball—which he "happened to have by him"—at its end. The ball attracts brass foil when the tube is rubbed. Gradually he increases the length of the rod, then substitutes for it a wire until the sagging of the latter makes it troublesome to handle, and then he hangs the ball from the tube by a long piece of hemp thread. Still no change in the attractive power, despite the distance between ball and tube.

All this is so far beyond his expectations that it seems to him that the effect must in some measure depend upon the nature of the ivory ball; so he takes it off and substitutes other objects. He has no store of rare chemicals to draw upon; but the street and courtyard yield him bits of brick and stone and tiles and chalk; and the garden, different vegetables and plants; and his purse a gold guinea, a silver shilling and a copper halfpenny. After he has tried all these things—always with the same result—he looks about his chamber and finds the fire shovel, and the tongs and the poker, and the tea-kettle (which works just the same whether full of water or empty), and his silver pint-pot. By this time he considers the question sufficiently settled, and gets out his fishing-rod to see if the virtue will go over even so long an object as that. But it does and over other rods fastened thereto; and how much further it might travel he cannot tell, because his little chamber is not large enough to let him use a series of rods over eighteen feet in length.

The month of May, 1729, has now come, and Gray is glad to exchange the bricks and mortar of London for the country fields. His "honored friend, John Godfrey, Esq.," of Norton Court, near Faversham, Kent, has in vited him thither, and, to the joys of a country life, Gray has the added felicity of plenty of room. The rod is lengthened to thirty-two feet, and gives way to thread, so that he can stand on Godfrey's balcony and swing the ivory ball attached to its lower end over the scraps of foil on the ground, thirty-four feet below. As he can find no higher elevation, he concludes to suspend his experiments until his return to London and try them in the dome of St. Paul's, where he could get just ten times the above altitude. He had thought of a horizontal line of thread, and had tried one looped to a beam in order to suspend the ball. But then the ball refused to attract, because, as he says, the electric virtue runs to the beam and not to the ball.

Instead of going back to the Charterhouse, he proceeds to Otterden Place, the residence of "Granvile Wheler, Esq., member R. S., with whom I have the honor to be lately acquainted," and takes with him a little glass tube "in order to give Mr. Wheler a specimen of my experiments."

But the moment Wheler sees what has been done, he wants much more than a specimen, for his interest is enthusiastic. In fact, he develops a desire more burning even than that of Gray himself to find out how far the electricity will travel. He insists upon a long horizontal line being put up immediately. Gray tells him that it will be useless, for the virtue will run off at the supports. Then says Gray, "he proposed a silk line to support the line by which the electric virtue was to pass. I told him it might be better upon the account of its smallness, so that there would be less virtue carried from the line of communication." Gray therefore had already found out that the conductivity of his line depended upon its "smallness," and that the smaller it was the less virtue it would carry.

There is a gallery eighty feet long in Wheler's house, and there Wheler, and all his servants helping him, speedily stretch a packthread line over taut silk threads. The virtue seemingly has no more trouble in traversing eighty feet than as many inches; and then the line is carried backward and forward to increase its length, until it measures over three hundred feet, when the silk threads break under its weight. However, that is easily repaired, thinks Gray, substituting metal wire for the silk; but now, to his dismay, the attraction of the ball disappears. No matter how vigorously they rub the tube, apparently no virtue from it goes upon the line, for the bits of brass foil under the ball at the far end remain motionless. Wheler's happy suggestion of the silk thread supports, now results in a great discovery. Why silk?

"We are now convinced," says Gray, "that the success we had before, depended upon the lines that supported the line of communication being silk, and not upon their being small."

More than a century before, Gilbert had cut off electric attraction by interposing silk or water between the electric and the attracted body; and this had been done by Hauksbee, and, in fact, all the later experimenters. So also the latter believed that substances were divided into electrics and non-electrics, although the list of the former was constantly increasing. But no one before had recognized the fact that the electric virtue would apparently refuse to pass over certain substances while freely traversing others, and this even when the first were short bodies and the second very long. In other words, Gray had discovered the difference between the electric conductivity of bodies depending on the substances composing them, and had found in silk threads this conductivity so small as to be inconsiderable. Some bodies evidently conveyed electricity and some did not, and those which did not could be used to prevent the electrical virtue escaping from those which did. Here began the world's practical and useful knowledge of electrical conduction and insulation.

Wheler's ingenuity rose to the occasion, and, by multiplying the silk threads, he managed to make the line outgrow the gallery. Then he erected the first aërial line of communication on poles, extending over his land for a distance of 650 feet. The weather was warm—it was July—and the experimenters were ablaze with enthusiasm. In fact, they exerted themselves so much in running from one end of the line to the other, Wheler now rubbing the tube and Gray watching the bits of foil, and then vice versa, that, suddenly, in the late afternoon, when the attraction ceased for the day, Gray naively remarks that it could not positively be said whether this was "caused by the dew falling or by my being very hot, but I rather impute it to the latter."

After discovering that the virtue could be made to travel from the tube over three lines simultaneously, to Mr. Wheler's "great parlor, little parlor and hall," Gray departs, leaving Wheler to expend his excitement in electrifying ' a hot poker, a live chicken, a large map and an umbrella."

In the fall of 1729 the discovery that the virtue will travel from tube to line and then over the latter without direct contact of the tube—or, in other words, by induction—is made. Then follow further researches into the electrification of different bodies. Gray charges a soap-bubble and makes it attractive. By means of hollow and solid suspended wooden cubes he demonstrates the important fact that the charge is resident on the surface of the electrified body, for "no part but the surface attracts." Then, in the spring of 1730, he suspends a boy, and finds that when the tube is rubbed and held to the boy's feet, the leaf brass is vigorously attracted by the boy's face, thus demonstrating the conductibility of the human body. It is, doubtless, not pleasant to the urchin to feel the fire pattering against his cheeks, but Gray encourages him to bear it manfully, because should he turn the back of his head the virtue, says the discoverer, would be greatly "cut off by the short hair."

In the fall of 1730 Wheler again appears with his unquenchable desire for longer lines, and one of 866 feet, is successfully used; but Gray has seemingly satisfied himself on this subject, for, after several months' silence, he reappears in June, 1731, with a host of new experiments, depending mainly upon his discovery that it is possible to insulate electrified bodies by placing them on cakes of resin. This gives more employment for the Charterhouse lads, who are hung up on hair lines and stood up on blocks, and electrified tubes are applied to them in all sorts of ways, which need not here be detailed.

A year later, 1732, Godfrey and Wheler are both pressed into service to aid him in making experiments to show induction; and these lead him to the conclusion that "the electric virtue may not only be carried from the tube by a rod or line to distant bodies, but that the same rod or line will communicate that virtue to another rod or line at a distance from it, and by that other rod or line the attractive force may be carried to other distant bodies." And thus was proved for the first time that an electrified line could induce a charge on another line; and, in fact, Gray found that this induction would take place over distances as great as a foot between the two lines.

Gray's experiments had now extended over three years, during which time, despite the attention which results so novel and unprecedented naturally excited, no one had appeared to rival him. Dr. Desaguiliers, writing some years after Gray's death, finds an explanation of this in Gray's irascible temperament and intolerance of opposition, and gives as an excuse for the long withholding of his own observations that their publication would probably have caused Gray to abandon the research. Nevertheless, when the field was entered, Gray welcomed the interloper, and, so far from relaxing his efforts, continued them to the end of his life with a pertinacity rivaling that of Hooke. At all events, if such solicitude as Desaguiliers manifests was sufficient to deter the English philosophers from independent investigation, it at least seems not to have extended across the channel; for, in the spring of 1733, Charles François de Cisternay Dufay[1] began his famous work.

Dufay was then thirty-five years of age, and perhaps as widely different from Gray as one man can be from another. To the broadest general culture and knowledge of the world he united a charming personality, a keen wit, and exquisite tact, the last never better exhibited than when, instead of antagonizing Gray, he managed to convert the sensitive philosopher into a cordial and communicative friend and colleague. He had been educated as a soldier, and was a lieutenant in the Picardy regiment at the age of fourteen; but his natural taste was for scientific study, and not at all for military life. He exchanged arms for diplomacy, and the latter for science. In his brief lifetime of forty-one years (he died in 1739) he made himself a chemist, an anatomist, a botanist, a geometrician, an astronomer, a mechanician, an antiquary, and an electrician, and in every one of these varied capacities shone with unusual brilliancy. The French Academy then recognized only six subjects as worthy of public discussion, namely, chemistry, anatomy, botany, geometry, astronomy, and physics. Dufay, says Fontenelle,[2] in his celebrated eulogy, was the only man of his time who contributed to the Academic annals investigations in every one of these branches. His early studies on the Bologna stone and phosphorus resulted in the discovery that all stones containing salts of lime become luminous on calcination; his essay on the magnet, published in 1728, the phenomena of which he regarded as in accordance with the Cartesian theory, epitomizes all existing knowledge

[1] See Dufay's eight original memoirs. Histoire de l'Académie Royal des Sciences, Paris, for years 1733, 1734 and 1737.

[2] Fontenelle: Eloge de M. Dufay. Hist. de l'Acad. Roy. des Sciences, 1739.

on the subject. In the spring of 1733 he learned, with absorbing interest, of the achievements of Gray and Wheler, and determined at once to prosecute them further and in entirely new directions.

At the very outset he makes a discovery which overthrows the distinction between electrics and non-electrics, and brings to an end the efforts to enlarge the list of the former, which had continued ever since the time of Gilbert. The number of different substances which he tests is legion—all sorts of woods and stones, especially all those materials which earlier investigators had been unable to electrify. Some he finds require more "chafing or heating" than others; some, such as the gums, he cannot so treat without rendering them viscid; while the electrification of the metals is so slight that he doubts whether he has really recognized it: but in the end he announces that all bodies (the metals and soft substances excepted) are endowed with the property which for ages was supposed to be peculiar to the amber, or, in other words, become electrics by themselves (électriques par eux-mêmes).

Then he turns to Gray's experiments on conduction and verifies them, but in so doing his attention becomes concentrated upon the supports for the electrified body—Gray's silk strings and cakes of resin. He varies the material of which these supports are made. Pieces of metal, or wood, or stone, on wooden or metal standards, he could not electrify by bringing the excited glass tube near to them, but when he substituted glass supports then he could do so. Immediately it dawns upon him that the possibility of electrifying a body does not depend upon the nature of the body itself so much as upon its being insulated, so that the virtue cannot escape from it. Again he collects a great variety of objects—woods and stones and amber and agate, even oranges and books and red-hot coals—and placing them, one after another, on the glass standards, brings the rubbed glass tube near to them, when every one of them becomes electrified; and what is still more curious, the very ones, the metals, which it was most difficult or apparently impossible to charge simply by rubbing them, now receive more electricity than all others on mere approximation of the excited tube. Here are two capital discoveries made at the very threshold of his labor.

This is not altogether unusual, as many a later investigator can testify. Long study and thought produces a sort of mental polarization that somehow dulls the perceptive faculties, or results in an intellectual inertia which renders it difficult for the mind to turn itself out of the path in which it has been moving. And, as a consequence, the power of original thought, of invention, is apt to weaken even in those most highly gifted with creative genius, unless the brain-work be differently directed for a time, or wholly intermitted for a period of rest. There seems to be no exhaustion of energy, for the thinking mechanism may continue its operation, although fruitlessly, with even greater assiduity than ever. It is rather a new condition of the apparatus which causes a change in the quality of its accomplishment. Therefore when a new mind—not polarized—attacks the problem, it is very apt not only to perceive solutions which evade the recognition of those which have long grappled with it, but to see the most prominent and general ones first. It is an incident of progress, and apparently a necessary one, that obstacles shall be attacked by a succession of new minds; and it constantly happens that a new mind without experience is often more potent in overcoming them than one rich with accumulated knowledge.

Gray had almost instantly discovered that electricity would pass from the excited body to one not excited: from the glass tube to the cork. Dufay also at once finds all bodies capable of electrification. Gray was halted by doubts as to the effect of the physical conditions of the body to which the charge is communicated. So Dufay similarly pauses because of misgivings as to the influence of color—these not

DUFAY'S EXPERIMENTS. 481

of his own suggestion, but because Gray had said that among electrified bodies physically alike, those which are red, orange, or yellow, attract very much more strongly than those which are blue, green or purple. Dufay saw in this not merely a possible cause of error in his future researches, but a suggestion that there might be a relation between electricity and light, if the former had a capacity for color selection. For both reasons, he proceeds.

His initial experiments seem to confirm Gray decisively. Of nine suspended ribbons (black, white and the rainbow colors), the rubbed glass tube attracts the black first and the red last. White gauze and black gauze intercept the electric virtue, while gauzes of the rainbow hues, the red especially, allow it to pass. Dufay presses on to the broader question, fully believing that he is on the track of a startling discovery.

If color alone exercises the effect, it can make no difference, he argues, whether the hue be natural or artificial: whether it appear on the rose-leaf or on a painting. So he tries the flowers—and the signs fail. The scarlet geranium responds to the attracting glass as readily as does the purple pansy—the green leaves as quickly as the white petals of the lily. Perhaps there is something in the inherent quality of these vegetable substances which interferes. Clearly the crucial test requires pure color, and that is only in the rainbow.

He directs a sunbeam through a prism, and spreads it out into its gorgeous spectrum, and distributes therein white ribbons, so that the sun paints one red, another orange, another yellow, and so on through nature's color box. But the ribbons act like the flowers. No one of them responds to the electric pull any more than does another. The notion that electric attraction could tear the sunbeam to pieces, and change it from white to red by drawing out the blue rays, was only a delusion.

Then Dufay went back to his colored ribbons and wet them—and their differences vanished. He heated his gauzes and the virtue went through black or red with equal facility. He had been misled by the dressing which the makers had put in the ribbons to give them body: that was all—the color exerted no influence.

Perhaps this left him in something of a questioning attitude toward Gray's other conclusions, for he begins to investigate long-distance transmission anew; and finally reaches the conclusion that the substances which are most difficult to electrify—such as metals or wet objects—best convey the virtue; while on the other hand, those easiest excited—amber or silk—can hardly be got to convey it at all. He puts up a packthread line 1256 feet long, and wets it; and the electricity traverses it with the same freedom with which it nowadays runs along wet telegraph poles, or escapes from the wires which touch the dripping foliage. For Gray's silk supports, he substitutes glass tubes and masses of Spanish wax, and thus, for the first time, uses solid insulators upon an electric line of communication. The new principle destroyed the non-electric and the electric as distinctive significations—it made non-electrics into "conductors," and electrics into "non-conductors."

Gray had shown how one line may electrify another placed near it. Dufay varies this by placing two short lines, respectively six and eight feet in length, end to end with an air-space intervening. When the gap is a foot wide he says that the attraction, despite the shortness of the lines, is as weak as if the virtue had traversed the continuous length of 1256 feet. Nevertheless it seems to him that the charge can escape from line to air, and therefore he says, coining the word, the necessity is apparent that the transmitting cord should be "insulated."[1]

He has meanwhile remarked that if he touches the ball hanging at the end of his electrified line, it refuses to attract; the electricity, he says, being dissipated through him to the floor. But suppose he touches it with a small

[1] "Que la corde dont on se sert pour transmettre au loin l'électricité soit *isolée*."

body, itself insulated. Then the ball loses only a part of its electricity, which goes to the last-named body. Consequently he says, the volume of the electrified ball must be considered. If too large, the virtue reaching it becomes too extended to act quickly; if not large enough, it will not take all that is brought to it by the cord. These were the first perceptions of the distribution of an electric charge on a conductor. Gray had found it resident on the surface.

Dufay now emulates the English philosopher in suspending people by silk lines and electrifying them; but he soon discards children and suspends himself. Then he compares the sensation caused by an electrified tube near his face to that of a spider-web drawn over it, and for the first time feels the pricks and burns of the electric sparks as they dart from his fingers. He believes them to be fire, and, as such, altogether different from the hitherto seen glow.

His is a nimble mind, and it leaps from one subject to another with marvelous rapidity. But this is necessary; for he is not only breaking a new path, but rebuilding the old one. As he meets a new problem he discovers that the vantage ground from which he must proceed is infirm. That necessitates re-examination of the foundation facts; and in this way he finds himself side by side with Von Guericke, contemplating the singular behavior of the feather which the sulphur globe drives away, and which, nevertheless, like the moon, always turns the same face. Dufay lets fall some gold-leaf upon his excited tube and sees it repelled in the same way, avoiding the tube as he chases the fragments around the room. But if, meanwhile, he rubs the tube, the leaf comes to it and goes away from it alternately, following the motion of the hand. When the leaf touches the tube, he says, it becomes electrified thereby by communication. Yet obviously it is repelled. Therefore all electrified bodies first attract bodies that are not electrified, communicate to them their own electricity, and that done, repel them. Nor will the latter be again attracted until, having touched some other body, the acquired electricity is lost. This, which Von Guericke saw, is now explained by Dufay.

But Dufay went a little further and imagined a whirl, a field of force, around the tube, and figured to himself the action going on there and not in the body of the tube. The attracted body, on touching the tube and becoming electrified, acquires a field of its own, the two fields repel, and so long as that of either body remains the same, the relative position of the two is unchanged. But if the field of the attracted feather, for example, is dissipated, the feather falls back to the tube; if the field of the tube is varied, as it is by the hand moving from one end of the tube to the other, then the feather swings to and fro, following the changes caused.

It is while examining the repulsive action of the glass tube that Dufay accidentally notes an effect which he says "disconcerted me prodigiously;" and well it might, for it seemed to be subversive of every conclusion which he had hitherto formed concerning the behavior of electrified bodies. He is watching a bit of gold-leaf float in the air under the repulsion of his excited glass tube. It occurs to him to see what it will do when subjected to the action of two electrified bodies; and therefore he rubs a piece of gum-copal and brings it to the leaf. To his utter astonishment the leaf, instead of retreating from the electrified gum, as it certainly did from the electrified glass, adheres to it. He tries the experiment again and again, but in every instance the leaf is drawn by the gum or by amber or by Spanish wax, while it is repelled by the glass tube. Yet a second glass tube or a piece of rock crystal brought near the leaf exercises the same repelling effect as the original tube.

This was Dufay's most important discovery. "I cannot doubt," he says, "that glass and crystal operate in exactly the opposite way to gum-copal and amber; so that a leaf repelled by the former because of the electricity which it contracted will be drawn by the latter. And this leads me to conclude that there are perhaps *two kinds of different electricities.*"

Further tests confirm the belief, and he announces that electrified glass repels electrified glass, or all bodies receiving electricity therefrom, and attracts electrified amber and all bodies to which its charge has been communicated. In other words, he had established the fundamental law that similarly-electrified bodies repel, while dissimilarly-electrified bodies attract one another.

He calls the electricity yielded by glass *vitreous*, and that derived from the rubbed gum *resinous;* because "glass and copal are the two substances which have led me to the discovery of the two different electricities."

Thus Dufay had found that all bodies may become electric either by direct communication or by induction; that the so-called electrics are the least suitable to convey the virtue; that the electric light may appear as fire or burning sparks, and that there are two different kinds of electricity, of which one attracts bodies repelled by the other; and that bodies, if similarly charged, repel, while attracting if dissimilarly electrified. These are only his more important conclusions; others, although ingenious and original, relate to details which need not be entered into here.

In December, 1733, Dufay wrote a brief synopsis[1] of the long memoirs which he had already published in the annals of the French Academy, and sent it to the Duke of Richmond and Lenox for presentation to the Royal Society and (with characteristic diplomacy) to Mr. Gray, "who works on this subject with so much application and success, and to whom I acknowledge myself indebted for the discoveries I have made, as well as for those I may possibly make hereafter, since it is from his writings that I took the resolution of applying myself to this kind of experiments." Whether in all the history of discovery there exists a more handsome recognition than this of the work

[1] Phil. Trans., No. 431, p. 258, 1733.

of a prior student may well be doubted. It is a custom which nowadays in the struggle for profit is too often forgotten. At all events Gray's heart was won. He ceremoniously salutes Monsieur Dufay and felicitates himself that his experiments should have been confirmed by so judicious a philosopher; and, no doubt, in the quietude of his little chamber at Grey Friars, wonders if it is really "poor brother" Gray, with his experiments with the tea-kettle and the pint-pot and the fishing-poles and threads, who is receiving these compliments from the distinguished French scientist through the Royal Society and his Grace of Richmond.

But he was invigorated—much invigorated. And besides, what Dufay had said about the burning sparks piqued his curiosity immensely. Out came the poker and the tongs, and the fire shovel, too, this time, to be hung up on silk threads and the crackling sparks produced, of which last a small boy was made to suffer the pain, even through his stockings. The next victim was a large white rooster, replaced by a sirloin of beef, and finally an iron rod astonished him beyond measure by exhibiting the true brush discharge, "rays of light diverging from the point," and hissing. Pewter plates, iron balls, dishes of water, were all pressed into service. The flames were real, and they burned and crackled and exploded. "The effects at present," says Gray, "are but in minimis, but in time there may be found out a Way to collect a greater Quantity of it, and consequently to increase the force of this Electric Fire, which by several of these experiments (si licet magnis componere parva) seems to be of the same Nature with that of Thunder and Lightning."

From that time on, Gray and Dufay maintained communication with a degree of friendliness which leads Fontenelle to wish that it might always typify the intercourse of the two great nations to which they severally belonged, and to add, with pardonable exaggeration, that "they enlightened and animated one another, and together made discoveries so strange and surprising that their respective beliefs in them perforce rested solely upon their mutual assurances." But, in fact, neither afterwards made any especially important discovery. It was not long before Gray died. He had wandered off into the old belief which von Guericke held, that somehow the planets were controlled by electrical influence, and he fancied he could make an apparatus in which a sphere would of its own accord revolve from west to east around an electrified body. But he was stricken unexpectedly, and he could tell Dr. Mortimer, the Secretary of the Royal Society, who attended his death-bed, only a few disjointed ideas, mingled with expressions of a hope "that God would spare his life a little longer, so that he should, from what these phenomena point out, bring his electrical experiments to greater perfection." But it was ordained otherwise, and he passed away on February 15th, 1736.

Dufay's last memoir is dated in 1737, and expresses his broadest view of the great phenomena which he had so well studied. "Electricity," he says, "is a quality universally expanded in all the matter we know, and which influences the mechanism of the universe far more than we think." He has left his monument in the magnificent Jardin des Plantes which he organized, and so made every student of Nature his debtor. His solicitude that the full meed of honor due to the poor brother of the Charter house should be yielded never failed; and when the world shall pay its tribute in enduring marble and brass to the memory of Stephen Gray, electrician, it will find no words more fitting to place upon it than those of his generous and brilliant rival:

"He was almost alone in England in pursuing his object. To him we owe the most remarkable discoveries pertaining to it; so all those who love Nature and her work must infinitely regret him."

Apart from the discoveries in which they resulted, the researches of both Gray and Dufay are remarkable for their inductive character and the absence of dogmatizing on the nature and cause of electricity. Concerning the last, opinions were undergoing radical change. Shortly after Hauksbee's experiments were published, Dr. s'Gravesande, Professor of Mathematics at Leyden, issued one of the earliest, if not the earliest, didactic work in which electricity is treated as a branch of physics, and there gives it as his ultimate conclusion, based on preceding experiments, that there is an atmosphere excited in rubbed glass by friction, which attracts and repels light bodies, and also that out of the glass, fire is forced ; but he does not regard either the atmosphere or the fire as electricity, which he defines as "that property of bodies by which (when they are heated by attrition) they attract and repel lighter bodies at a sensible distance."

The experiments of Gray and Dufay showed the light and the fire to be as much an electrical phenomenon as the attraction and repulsion; but Dufay's discovery of the dual nature of electricity had undermined the old conception of material emanations, while definitely establishing no new theory in its place.

After the death of Dufay appears Dr. Desaguiliers, a man of considerable prominence in the Royal Society. He had never found it expedient to discourse about electrical matters so long as either Gray, whom he seems to have disliked, or Dufay survived; but afterwards he contributes many papers to the Philosophical Transactions, in which he collects a great mass of experiments, chiefly in the nature of cumulative evidence. He invented the term "electrics per se," which, for a long time afterwards, was used to designate those bodies which could be made electric by rubbing them, although it was nothing but a polyglot translation of Dufay's term "électriques par eux-mêmes." He also first used the word "conductor," applying it to the string over which the electricity passes, and also was the first to electrify running water. Gilbert, of course, had made his rubbed amber attract a water-drop; the Florentine Academy, by like means, had drawn oil up into little viscous strings, and Gray had electrified soap-bubbles; but Desaguiliers found that, when he let water run in a stream out of a copper fountain, he could render the jet electric, so that it would attract thread, by merely holding the rubbed tube above the fountain, and when he applied the tube to the stream, he could draw it sidewise into a curve, or even cause it to fall outside of the vessel placed to receive it. He also appears to have been the first to conceive of atmospheric electricity, and to point out that a cloud or mass of vapor may be an electrified body. He had already recognized that air may be rendered electrical; and supposed it to be made up of electric particles constantly repelling one another. He imagined that the air-current which flows along the surface of the ocean is electrical in proportion to the heat of the weather, and that, as he had seen little particles of water leap up in spray to the excited tube, so he conceived the watery particles of the sea to rise to meet the excited air particles, and then, being of the same electricity, to be repelled by them, so that "a cubic inch of vapor is lighter than a cubic inch of air." In the recognition by Hauksbee and Wall and Gray of the similarity of the crackling electric spark to the thunder and lightning, and in this hazy conception of Desaguiliers of electrically-charged clouds and atmosphere, we can now begin to perceive the drift of thought leading toward Franklin's great discovery.

CHAPTER XV.

An assemblage of despotisms, big and little, engaged in constant bickerings and dissensions among themselves, and involved in foreign wars which drained every resource, formed the loosely-coherent German Empire of the eighteenth century. For the first forty years of this period, as might well be expected, German progress in physical science was far behind that of England, France or Italy. Learned societies had, however, been established, the most important of which was the Leopoldine or Collegium Naturale Curiosorum, modeled on the English Royal Society; but their existence was precarious, and their work little more than the gathering and glossing of the records of discoveries made abroad. The partial adoption of the Gregorian calendar by the Protestant States of Germany in 1700 is said to have led to the foundation in that year of the Berlin Royal Society of Sciences by Frederick I. of Prussia; but the real motive was that especially pompous king's desire to imitate and rival Louis XIV. of France.

It soon became apparent that to organize a philosophical society is one thing, and to find members of genius for it, another. The latter were manifestly wanting. Even the gigantic intellect of a Leibnitz in the Presidential chair could not leaven the entire mass. Hence its existence remained merely nominal until 1711, when a solemn opening of its proceedings was held; and it started on what might have been from that time a useful career. But a couple of years later, the sergeant king, who had less use for learned societies than for giant grenadiers, succeeded to the throne, and encouragement failed.

In 1715, Weidler of Wittenberg, and Leibknecht of Giessen, were still studying the mercurial phosphorus. The authority of Bernouilli remained potent against Hauksbee's plain demonstration of the electrical nature of the barometer light, although Leupold reconstructed Hauksbee's machine, and verified many of his conclusions. Little volumes of transactions in Latin printed at long intervals, became the sole sign of the continued animation of the Berlin Society. One electrical dissertation here appears written by Johan Jacob Schilling[1] in 1734, wherein he details experiments made with the rubbed tube; but they are of minor consequence, and merely go to show how prevalent was the belief that the electrical action resided in an atmosphere around the excited body, although Schilling's particular conception of his atmosphere involves its rarefaction by the heat due to the friction incident to rubbing the tube, and subsequent condensation on cooling.

Von Guericke was famous only for his pneumatic discoveries, fixed in the popular mind by his theatrical display of the Madgeburg hemispheres resisting the pull of many horses. His electrical discoveries, unimportant by contrast, and described in but a few terse paragraphs in his book, were forgotten or misunderstood in his own country; while the foreign philosophers (always excepting the liberal and cultured Dufay, whose appreciation of Von Guericke we have seen), regarded Germany very much as the British literati looked upon the United States seventy years ago—as a Nazareth whence little good might be expected to come.

The year 1742 probably marks the beginning of the singular and sudden interest in things electrical which arose in Germany, and which swiftly reached a stage of feverish enthusiasm.[2] It differed widely from the per-

[1] Schilling: Misc. Beroliniensia, Tome x., 3, 4.

[2] See Gralath : Geschichte der Elektrizität. Versuche und Abhandlungen, der Naturforschenden Gesellschaft in Dantzig. I. Theil. Dantzig, 1747.
Priestley : History of Electricity. London, 1767, and later editions.
Fischer : Geschichte der Physik. Band V. Göttingen, 1804. Hoppe's Geschichte der Elektrizität, 1884, and Poggendorff's Geschichte der Physik, 1879, follow these works.

functory craze which had taken possession of the English aristocracy at the behest of Charles. It had still less resemblance to the combined onslaught of the French philosophers which was designed to take all of Nature's secrets by storm. It was distinctively popular. It was the first instance—many times since repeated—of the intelligent portion of an entire community regarding with absorbing wonder the working of electric powers.

No unexpected desire for electrical knowledge in general had been born. The German naturalists were familiar with progress abroad during the last fifty years, but had shown no emulative spirit. The new motive force now came not from them, but from the people; and the people, in all times and in all ages, have never failed to respond to an appeal to their sense of the marvelous—to a conviction that something new has been found—something at once new and incomprehensible. The masses had cared little for Hauksbee's lights, and less for the vagrant virtue on Gray's lines, assuming that the knowledge of either percolated to them; but when it came to be noised about that the strange radiance which the English and French philosophers were exhibiting was fire,—fire which flamed in jets from the ends of rods, or, more wondrous still, leaped from the tips of men's fingers—that was a matter for every one's personal concern. For fire was then believed to be a material substance—phlogiston—and while perhaps it might exist in iron bars and inanimate things of that kind, and be forced visibly to come out of them by friction, as well as by heating, no one had ever supposed that it resided in the human body and could be compelled to escape, with an accompaniment of sparks and crackles, from one's person. It was the idea of a human being becoming such a torch that stirred the Teutonic mind to its profoundest depths. The impetus which electrical science had received from the fancy of a dissolute king was nearly spent: now progress was resumed with renewed vigor under that due to the astonishment and wonder which the latest electrical manifestations had created in the now thoroughly awakened Germans.

The activity of the German investigators is not reflected in the annals of the Berlin Academy, but in a host of individual treatises issued so closely together in point of time that it is impracticable to determine, from their often contradictory statements, the chronological sequence in which the recorded discoveries were made. It is even doubtful to whom is due the credit of accomplishing the work which began the new era; some contemporary writers according it to Christian August Hausen, others to George Matthias Bose. The achievement itself involved no new discovery; but, in the light of its consequences, its history is important.

Bose[1] was a teacher in Leipsic and master of an "experimental college." So slow was the diffusion of scientific knowledge at the time that the memoirs of the French Academy, containing the account of Dufay's experiments made in 1733-4, did not reach him until three years later. He had already studied electricity sufficiently to appreciate keenly the discoveries of the French scientist, and to be eager to repeat them. No glass tube of proper size was available in all Leipsic, and Bose's straitened means prevented his procuring one from Paris. There stood, however, in his laboratory a large distilling apparatus, the retort of which was of glass, and capable of holding six or seven gallons. Upon the nozzle of this vessel Bose's eye fell one day, and in an instant the sacrifice was made, and the long-desired tube was in his hands. It is singular that Dufay, with all his acumen, should not have perceived the disadvantages incident to the use of the tube, which required constantly renewed rubbing, and worked always with diminishing effects. Bose's fresher perceptions recognized them quickly, and his mind at once recurred to the rotary glass globe of Hauksbee and Newton as a much more convenient apparatus for generating electricity. But

[1] Bose, Tentamina Electrica. Wittenberg, 1744.

he had no globe, and saw no chance of obtaining one, until the old still suddenly revealed itself in a new light. There was the noseless glass retort; big, but all the better for that, for perhaps the effects would be stronger. Down came the vessel to be mounted lathe fashion, and the results, as I shall shortly relate, were amazing.

Meanwhile Hausen,[1] who was a professor of mathematics at the Leipsic Academy, and lectured there on electricity, while using the glass tube in one of his demonstrations, inveighed against its inconvenience, when a student reminded him of the Hauksbee globe. Hausen at once constructed such an apparatus, and, by means of a large crank-wheel and belt, made it possible to rotate the sphere very rapidly. Both Hausen and Bose now found, at about the same time, that not only could a practically continuous supply of electricity thus be obtained, but one of much greater strength than had hitherto been known. Hausen suspended a boy with his toes in proximity to the globe, and drew sparks from his fingers. Bose disposed twenty soldiers in line, with hands touching, and administered a shock to all of them at once. Hausen remarked the sulphurous odor of the electrical discharge, and distinguished three kinds of electric light—due respectively to the "spark," the "brush" and the "glow," as the phenomena are now termed; but he was before all a theorist. He announced that the electric field is formed of vortices of electric matter, caused by its being attracted and repelled in oppositely convex curves, that the vortex becomes a spiral around a rubbed tube, and that all electric action is due to the influences of vortices upon vortices, or vortices upon matter.

In the light of modern conceptions Hausen's hypothesis of the identity of his so-called electric matter with the ether of Newton and Huyghens is remarkable. He considers ether to be electric matter, because both glow as soon as the proper motion is impressed; and from this he

---

[1] Hausen: Novi Profectus in Historia Electricitatis. Leipsic, 1743.

advances to the assumption that solidity, fluidity, expansibility, electric and magnetic forces, density, light, sound, heat, etc., have all a common origin in ether or electric matter motion. The drawing of fire from the person shows the presence of this same matter, he maintains, in the blood; and hence it may be the seat of the soul, or at least exercise control of the sensory faculties. Hausen died in 1743, leaving his conceptions far from developed and his experimental researches unfinished.

Bose, on the other hand, was no theorist. His temperament unfitted him for abstract speculation, and he expressly avoids committing himself to any electrical theory, preferring merely to formulate questions for others to answer. But he was a genius. No one knew better the art of playing to the gallery; in fact, in the great electrical drama he created the part of the "modern wizard," and it is doubtful whether any one since has ever excelled him in it. He set jets of fire streaming from electrified objects, and exhibited them to the people who flocked to his laboratory. He invited guests to an elegant supper-table loaded with silver and glass and flowers and viands of every description, and, as they were about to regale themselves, caused them to stand transfixed with wonder at the sight of flames breaking forth from the dishes and the food and every object on the board. The table was insulated on pitch cakes, and received the discharge from the huge glass retort which was revolved in another room. He introduced his ardent pupils to a young woman of transcendent attractions, and as they advanced to press her fair hand, a spark shot from it accompanied by a shock which made them reel. Others, who had the boldness to accept his challenge to imprint a chaste salute upon the damsel's lips, received therefrom a discharge which Bose says "broke their teeth;" but Bose here either exaggerates more than usual, or else neglects to explain how the young lady bore her share of the injury.

Meanwhile he had become professor of physics at Wittenberg and an Imperial Count Palatine, so that his fortunes had evidently improved. He was now certainly producing the most powerful electrical discharges that had ever been seen, and popular excitement (and his own) concerning them was rapidly increasing. His constant desire was stronger effects, and with this object he sacrificed a large telescope in order to obtain its metal tube, some twenty-one feet in length. When he brought this close to his revolving globe the sparks leaped to it in great profusion, and finally, when it barely touched the glass, a ring of intense light appeared at the place of contact, while the discharge from the tube itself was powerful enough to knock a dollar from between his teeth, and cause a wound whenever it was allowed to strike the exposed skin.

He had now added to the electric machine, for the first time, the prime conductor. The tube was first held to the globe by hand, but afterwards suspended by silk cords. It collected the charge from the excited glass by a number of threads resting upon the revolving surface, performing the same functions as the numerous points of the collecting comb in the modern frictional machine.

It will be remembered that one of the discoveries which Dufay believed possible and desired to make, but in which he failed, as he conceived, because of the omission of some necessary precaution, was the proof of the identity of the electric spark with actual fire. Bose, in 1743, had reached sufficient faith in this to suggest the question anew, but then announced no proof. In January of 1744 the reorganized Academy of Sciences was formally opened in Berlin before an assembly of all the notabilities of the kingdom, and an address on electricity was delivered by Dr. Christian Friedrich Ludolff, in the course of which he exhibited the attractive effect of a rubbed glass tube upon water, and the apparent projection of the sparks from the tube to the liquid. While performing this experiment it occurred to him to substitute for the water some highly inflammable fluid, and see what effect the sparks flashing from the end of a metal rod would have upon it. Accordingly he brought to the rod a spoonful of previously-warmed sulphuric ether, which instantly, to the amazement of the entire assembly, burst into flame. There could now be no doubt that the electric spark and fire were the same. The resulting notion that the human body might thus be a miniature volcano created a profound impression, and popular excitement over the subject increased. Lectures on electricity were in great demand; exhibitions of electrical phenomena drew large audiences; even at the didactic discourses at the colleges the populace flocked to the halls and crowded the students out of their seats.

Daniel Gralath, writing at the time, records electrical experiments as in progress in the palaces of kings and princes and in the castles of the great. Meanwhile Ludolff continued his work, and ignited alcohol and turpentine in the same way, and is said even to have drawn the kindling sparks from pieces of ice and snow. It may here be noted that he turned his attention from this subject to that of the luminous barometer, and with his research ends even the German belief in the phosphorescent character of the mercury light, for he affirms it positively to be electrical.[1]

The news of Ludolff's exhibition drew immediately from Bose a claim to prior discovery, not only of the electric ignition of liquids, such as alcohol and ether, but of butter, resin, sealing-wax, sulphur, and a great variety of light and inflammable materials. These, being previously partly melted, he set on fire, not merely by the discharge from rods, but by the sparks from men's fingers. Then he turned to gunpowder, and succeeded in exploding it after getting it in a state so as not to be scattered by the discharge from the rod. Thus he made the first step toward the electric fuse, now a necessity in every mine, every quarry, every fort and every war-ship.

---

[1] Hist. de l'Acad. Roy. des Sciences et Belles Lettres. Berlin, 1746 and 1750.

In fact, with Bose the language of ordinary narrative seems to have become inadequate for the expression of his electrical achievements; and hence he followed the example of Leibnitz, who celebrated the discovery of phosphorus in pompous Latin verse, and became the author of the first of electrical poems.[1] Its opening canto is dedicated to Frederica Louise, Margravine of Brandenburg, and it epitomizes other people's discoveries; but the second part, under the frankly egotistical motto, "nunc mea sola cano," is entirely devoted to the celebration of his own. By judicious degrees he proceeds from things remarkable to things surprising, reaches the explosion of powder,[2] and then winds up with the announcement of an achievement calculated to throw the discoveries of every one else completely in the shade. He had found out, he said, how to reproduce, around any one's cranium, the halo or glory with which the old painters encircled the heads of saintly personages. It was necessary simply to place the individual on a cake of pitch and electrify him from a large globe, when a lambent flame, rising from the pitch, would first spread around his feet, and then gradually rise to his head, until the whole body would appear bathed in a heavenly glow; or if he were seated in a chair suspended by silken ropes, "a continual radiance or corona of light appears encircling his head."[3]

In Germany this astonishing claim was accepted, and Bose certainly exhibited people with flames about them. In England, however, where jealousy of both Gallic and

[1] Bose: Die Elektrizität, nach ihrer Entdeckung und Fortgang mit poetischer feder entworffen. Wittenberg, 1744.

[2] " Des Pulvers donnernd Schwartz wird auf zwölf Zoll belebt,
Das es dem Metall, und denen Fingern klebt.
Doch schmeltz es. Sieh dich für. Lass deine Funcken strahlen.
Es fängt, blitzt, donnert, zündt, und knallt zu tausend mahlen."

[3] " Wie man die Heiligen, ja selbst die Engel mahlt,
Wie das gemeine Volck von einem Irrwisch prahlt,
So steht mein Held alsdenn in einem Schimmer-Glantze,
In einem feurigen, fast fürchterlichen Krantze."

Teutonic achievements was now beginning to show itself, the electricians determined to test the matter, and to that end, as Priestley remarks, "went to a great deal of expense." Ultimately Dr. Watson, of whom there will be much to say hereafter, procured a huge cake of pitch, three feet high, mounted it himself, and submitted to vigorous electrification, with no better result than the cobweb sensation and a slight tingling of the scalp. A sharp correspondence followed between Watson and Bose, ending in the discomfiture of the latter and the admission that his boasted discovery was a mere trick; the beatification, as he called it, being produced by dressing the electrified person in a suit of concealed armor having many points, at which the brush discharge appeared. The older German historians either omit this episode in Bose's career, or else treat Bose's claims as mere "poetic license," on a par with his offer to shock an entire army if some one would furnish it.

Bose tried to increase the strength of the discharge by multiplying the number of rotating globes in his machine, and asserts that he obtained especially vivid sparks from an apparatus constructed of three globes, varying in diameter from ten to eighteen inches, and a beer glass. By like means he produced, in exhausted vessels, glow discharges which he says, "flowed, and turned, and wandered and flashed," so that "no name is so applicable to them as that of Northern Lights." This was the first suggestion of the electrical origin of the Aurora Borealis. So powerful, says Bose, were the discharges from the multiple globe machine, that the blood escaping from the opened vein of an electrified person appeared "lucid like phosphorus,"[1] and escaped faster, because of the electrification. In fact, water spouting from an electrified fountain flowed more freely than before. Thus came to light the principle now embodied in the Thomson siphon-recorder and other ap-

[1] Phil. Trans., No. 476, p. 419, 1745.

paratus, in which the flow of ink to a record surface is regulated by the electrifying charge.

It is a remarkable fact that, despite the progress which had been made in electrical knowledge since the time of Gilbert, no one had demonstrated any practical utilization of it. Of course, the discoveries made were the foundation of modern useful developments; but, at the period now under review, they had not been recognized as meeting any human need. Perhaps it was enough that they should have freed themselves from the ancient atmosphere of mysticism which surrounded all electrical effects, and had come to be clearly distinguished as purely natural happenings. From this, however, came the noteworthy sequel, that as popular familiarity with them increased, so far from its bringing with it indifference or sated curiosity, its accompaniment was augmented wonder. And this in turn led to the query, soon the demand, whether the new force could not be made to do its part in the world's work. Because those who ask it seldom have any conception how, or in what channels, such utilization is possible, this question, in the beginning of a new art, always takes the form of "cui bono;" and, moreover, as it often bears rather the aspect of belittling the importance or merit of the achievement than of evincing a desire that it shall be conclusively answered, the discoverer is as likely to retaliate with such counter demands as that the utility of mosquitoes or earthquakes shall first be explained, as he is to adopt Faraday's advice and silently proceed to "endeavor to make it useful;" or Franklin's genial philosophy summed up in the famous reply of "What is the use of a baby?"

So when the Germans had digested the feast of marvels which Bose and others spread before them, instead of glorifying the philosophers they manifested an inclination to taunt them with the uselessness of human fireworks, and such electrical shows generally. The man who answered these flings did it in a curious way. His name was Johann Gottlob Krüger, of Halle, a doctor and professor of philosphy and medicine; and his medium, an address delivered in the fall of 1743, to his students who had asked him to explain his views concerning possible utilizations of electricity. It is witty, prophetic, and pre-eminently the utterance of a sage, whose philosophy is indicated by his epigram that the philosopher's life consists in "trying to understand what you do not see, and not believing what you do." "What's the use of bugs, fleas and grasshoppers?" he demands, exemplifying the usual resentment of the closet student, yet in the next breath repeating, "God only knows what the ingenious heads of our time will get out of it all." "It is too early," he says, "even to try to venture explanations or predictions." But he believes—curious prescience—that the "Germans have laid the foundation, the English will erect the building, and the French will add the decorations." And as to what utilizations of electricity there may be in store, "if it must have some practical use, it is certain that none has been found for it in Theology or Jurisprudence, and therefore where else can the use be than in Medicine?"

Here begin the modern efforts to apply electricity to the curing of human ills. Not magnetism, for that, as we have seen, was used therapeutically at periods of remote antiquity; but with Krüger apparently starts the idea that electricity can be beneficially employed in the healing art.

It was one fraught with especial difficulties at the time, because of the imperfection of the electrical machine, which was then nothing more than a globe, or possibly two or three globes, of glass, seldom provided with Bose's prime conductor, and excited by the contact of the operator's dry palm. Nevertheless, Krüger urges his students to investigate. He has heard it rumored that certain electrified bodies will not decay because they attract only "balsamic vapors" from the air. The "true human body," he says, "is not electric of itself, nor can it be made so by rubbing; but only by the approach of the electric glass;" but consider, he urges, the immense value of such a discovery as that electricity will prevent wasting or decay of the human frame—"what reward would be too great for the discoverer?"

Not only, he says, does electricity make blisters on the skin, but it is apparently propagated through the entire body. Clearly, therefore, by means of electricity, changes can be caused in the most hidden parts of the frame. Lost health may perhaps be thus restored, or present health maintained, if the application be made at the proper time and in the proper way. Hence does it not follow that electrification is a new curative agent?

He conjectured that electrification of the body would augment the circulation of the blood, and cause contractions of the solid parts, and regretted that so little was known on the subject that no one could exactly predict what internal bodily changes would occur—a statement which can still be made with little qualification.

In the spring of 1744, Christian Gottlieb Kratzenstein,[1] of Halle, made the first experiments on the living body to determine the effects of electricity. He observed at once a marked increase in the pulse-beats, and the accelerated circulation predicted by Krüger, and also the contractile and irritating effect of the discharge upon the muscles. Sparks leaping from the blood running from the opened vein of an electrified man to a tin dish placed to receive the flow, added to the general conviction that electricity was a material substance in the body. Kratzenstein began to administer the discharge as a specific for all congestive ailments—rheumatism, malignant fevers and the plague—and claimed to have made remarkable cures of lameness and palsy, one woman with a lame or stiff finger being relieved in fifteen minutes. Lange,[2] who followed in

---

[1] Kratzenstein: Abhandlung von dem Nutzen der Elek in der Arzneiwissenschaft. (Gralath, cit. sup., 296.)

[2] Lange: Wochenliche Hallische Anzeigen, xxiv., An., 1744.

MYSTERIOUS SPARKLINGS.   503

Kratzenstein's path, in the same year announced that such fingers could be restored so completely as to fit them for the piano forte. Quelmalz[1] soon after evolved a theory that electric matter, nervous fluid and the Newtonian ether, are all of the same nature.

Meanwhile, the notion that fire exists in the human body, capable of being kindled or at least expelled by electrification, finding a support in the opinions of the German physicians, began to spread throughout Europe. In England, Dr. Henry Miles[2] at once associated with it the sparkling frock of Mrs. Susanna Sewall, concerning which Clayton had written to Boyle from Maryland in 1683, and exhumed other instances of mysterious bodily illuminations, notably the "Mulier Splendens," described by Bartholinus, of Copenhagen, and the remarkable lights which Dr. Simpson had recorded in 1675 as appearing on the combing of hair, the currying of a horse, or the rubbing of a cat's back—an effect which he ascribed to "fermentation." He might have added the "miracle" told by Bacon[3]—"that a few years since a girl's apron sparkled when a little shaken or rubbed," although Bacon himself attributed the light to the "alum or other salts with which the apron was imbued, and which, after having been stuck together and incrusted rather strongly, were broken by the friction." Miles connects such phenomena with Gray's mention of the great quantity of electric effluvia received by animals. It was reserved, however, for Paul Rolli,[4] another member of the Royal Society, to give the matter a new turn, well calculated to increase the already-aroused public apprehension.

An Italian treatise of 1733, written by Bianchini, Prebendary of Verona, contained an account of the spontaneous combustion of the Countess Cornelia Bandi, who,

---

[1] Quelmalz: Programma Solemnia Inaug. July, 1744.
[2] Phil. Trans., No. 476, p. 441, 1745.
[3] Novum Organum, ii., xii.
[4] Phil. Trans., No. 476, p. 447, 1745.

having retired one night in good health, was found in the morning a heap of ashes.[1] To this he added other equally gruesome instances, of a poor woman in Paris who, having drunk alcohol for years, "contracted a combustible disposition," and of a Polish gentleman who, over-indulging in brandy, exhaled flames and was consumed. As Miles had already linked together people who sparkled and glowed mysteriously and people who emitted fire when electrified, it remained simply for Rolli to suggest the connection between combustible people and mysteriously sparkling people; and of the latter, research in the ancient books reveals plenty of instances.

There is Eusebius Nierembergius telling how all the limbs of the father of the Emperor Theodoric exhibited lambent luminosity, and Bartholinus affirming the same of Carlo Gonzaga, Duke of Mantua; Licetus asserting that Antony Cianfio, a bookseller of Pisa, when he changed his garments "shone all over with great brightness;" Cardan relating that a friend of his, in like circumstances, "shot forth clear sparkles of fire;" Kircher describing a Roman grotto which possessed the capability of causing fire to "evaporate" from the heads of visitors; Father d'Ovale averring with equal recklessness the existence of mountains in Peru on the summits of which not only men, but beasts, became luminous; and Castro's story of the wonderful arms of a Veronese countess, which needed only the gentle friction of a cambric handkerchief to become resplendent.

"These flames," remarks the alarming Rolli, "seem harmless, but it is only for want of proper fuel;" and then he proceeds to relate how similar sparkles reduced to ashes the hair of a young man; depicts graphically the discomforts of a Spanish lady who perspired explosively, and crowns all with a quotation from Albertus Krantzius to the effect that in the time of the crusades "people were

---

[1] This is the story upon which Dickens bases the episode of the death of Krook in Bleak House.

burning of invisible fire in their entrails, and some had cut off a foot or a hand where the burning began that it should not go further."

Another member of the Society supplemented Rolli with an account of a carpenter who was set on fire by lightning and burned for three days. Still another presented a recent instance of a woman who ignited spontaneously because of the gin habit. And then came Dr. Cromwell Mortimer, directly suggesting the electrical fire as a cause of these automatic cremations. "The element of fire," he says,[1] "may . . . lie latent in fluid bodies ready to become active as soon as it meets with air, or even to kindle if it meets with sulphureous particles under proper conditions. . . . Animals appearing more susceptible of electric fire than other bodies greatly confirms these conjectures of the phosphoreal principles, and probably being rendered electric to any high degree might prove a dangerous experiment to a person habituated to the use of spirituous liquors or to embrocations with camphorated spirit of wine."

Thus a new factor was added to those which were gradually bringing both philosophers and people to a sort of nervous exaltation, which is especially recognizable in the exaggerated statements that soon filled the reports of the experiments of the German scientists. They seemed to be possessed with a feverish desire to intensify the strength of the discharge, and all their energies were directed to devising, for this purpose, improvements in the electrical machine. "Such a prodigious power of electricity," says Priestley, "could they excite from their globes, whirled by a large wheel and rubbed with woolen cloth or a dry hand . . . that if we may credit their own accounts the blood could be drawn from the finger by an electric spark, the skin would burst, and a wound appear as if made by caustic."

One result is that the records now become mere descrip-

[1] Phil. Trans., No. 476, p. 473, 1745.

tions of this and that apparatus—mostly experimental and of no consequence. Andrew Gordon, a Scotch Benedictine monk and a teacher in Erfurt, substituted a glass cylinder for the glass globe commonly employed. Johann Heinrich Winkler, professor of Greek and Latin in the University of Leipsic, replaced the dry palm of the hand with a leather cushion rubber adjusted to the glass by springs, and rotated the cylinder by a cord passing around its axle and connected at its ends respectively to a foot treadle and an elastic rod. By this means he managed to revolve his cylinder six hundred and eighty times per minute, and thus not only had the extreme satisfaction of producing brighter sparks and more severe shocks than any of his rivals, but of being able to do so, as he says, in any weather, no matter how damp.

He vied with Bose in wonder working, by lighting spirits in the presence of a large assembly, by sparks emitted from his fingers, and devised a machine which he termed a "Pirorganon," which is a tangle of little cylinders between which sparks are supposed to pass, and to form fanciful figures such as a winged wheel, etc. It was the first attempt to outline designs in electrical glow.[1] He was of an inquisitive mind, and I am inclined to think that he was the first who undertook to discover the speed with which electricity travels. At all events he suspended a cord 120 feet in length, which he considered amply long for his purpose, so that, returning on itself, the ends came within a few feet of one another, got some one to hold a tray of gold-leaf under one extremity, while he, fixing his eyes upon the gold-leaf and standing electrified upon his pitch cake, suddenly grasped the other extremity. "It is absolutely impossible," he says with great emphasis, "to distinguish any interval of time between the touching of the cord and the instant when the gold-leaf begins its movement."

Later, he devoted himself entirely to evolving new elec-

[1] Phil. Trans., No. 493, p. 497.

trical theories. He imagined a subtle electric matter forcing its way through bodies to which it is "proper," and in which it is inherent, and evaporating to form an atmosphere around them. This matter moving in right lines is not subject to central forces, runs like a fluid, and contains particles of fire. He speculated also concerning the elasticity of electricity, but settled nothing; and in fact the more he theorizes the less profitable becomes the task of summarizing his numerous treatises—so that it need not be further pursued.[1]

The popular demand for practical utilizations of electricity was growing more peremptory. The first to respond to it was the monk Gordon. It was Gordon who invented the electric bell—not the contrivance now known by that name, but two gongs and a metal ball suspended by silk lines in proximity to one another. The ball, on being electrified, moved to one gong, struck it, was repelled to strike the other, which again repelled it, and so on. Likewise it was Gordon who invented the first electric motor—curiously enough on exactly the same principle as the first steam motor—the ælopile of Hero of Alexandria. It was a metal star pivoted at its center, and having the ends of its rays slightly turned to one side, all in the same direction. The reaction of the electric discharge at the points whirled the star around on its pivot, just as the steam turns the ælopile of Hero, or the escaping water rotates a modern outflow turbine wheel. And Gordon also first used electricity for deadly purposes—for he killed many a chaffinch to show the power of the sparks from his machine. Nor did he disdain to compete in wonders with the wizard Bose, for when the latter conveyed electricity from one man to another over a distance of six feet by means of a jet of water, Gordon ignited spirits by a similar electrified stream, and left people lost in astonishment over the paradox of water setting things on fire.

[1] Winkler: Gedanken von den Eigenschaften, Wirkungen und Ursachen der Elektricität, Leipsic, 1745; Die Eigenschaften der Elektrischen Materie, Leipsic, 1745.

At about the same time, 1745, appeared the proposal to utilize electric light, made by Gottfried Heinrich Grummert, of Biala, Poland, who claimed to have found that a vacuum tube, after being set in glow through proximity to a powerfully electrified conductor, could, after a period of rest, be made to glow again without being re-electrified—probably by rubbing, as Hauksbee had done the same thing. This, however, he proposed to use "in mines and places where common fires and lights cannot be had," so that, in his notion, there is the germ both of electric illumination and the safety lamp. One other discovery closes the list of all that are worth especially noting among the many which fill the German treatises of the day, and that marks the first step in electro-chemistry. Krüger learned from Hausen of the sulphurous odor of the electric glow, due, as is now known, to the conversion of the oxygen of the air into ozone, and recalling the bleaching power of sulphur, determined to try whether electricity could cause discharge of color. He exposed red poppy leaves, and they quickly turned white; blue and yellow flowers blanched after some hours.

While the activity of the English philosophers had not been equal to that of the Germans, it had continued, and Dr. William Watson, apothecary and member of the Royal Society, made the first of his remarkable communications to that body in the spring of 1745. Watson's researches of that year, while mainly devoted to repetitions of the German experiments, were by no means barren of interesting results. He demonstrated the importance of the metallic conductor in the electrical machine in collecting the discharge, and concentrating it at a point. He ignited hydrogen by the electric spark (the beginning of electrical gas-lighting) and fired a musket by the same means; but perhaps his most important revelation for the time was that spirits electrified in a metallic spoon could be fired by the touch of a non-electrified person just as well as an electrified person could in like manner ignite non-electrified spirits. Watson called the last an effect of the attractive power of electricity, and the first a result of its repulsive power, an arbitrary hypothesis which served temporarily to satisfy curiosity.

But in the end the deductions tended to throw existing theories into greater confusion than ever. Here were inflammable substances ignited, not by sparks emanating from persons, and presumably due to the incorporeal fire set loose, but by sparks apparently engendered in the substances themselves and proceeding to persons, so that, by electrical means, not only could fire be driven out of one's body, but be equally well driven into it. The natural deduction was that if fire could in this way be poured into the human system, the less spirituous liquor contained in that system the better, if people did not want to be converted into involuntary bonfires. This was before the era of temperance agitation, otherwise the promoters of the cause might thus have found, ready at hand, a powerfully deterrent argument; for if one sort of fire would ignite a toper another kind might be equally efficacious, and a spark from one's pipe might do as much mischief as the flash from the electrical machine. In fact, however, it may be doubted whether any one drank a drop the less, despite the alarming possibilities suggested.

The philosophers of both England and Germany had now materially improved their electrical machines, which were yielding discharges hitherto unrivalled in strength. The similarity of the electrical flashes to lightning was commonly remarked, and when Gordon killed birds by them, another resemblance to Jove's bolt was recognized. As for explanatory theories, sentiment and opinion concerning not only electrical principles, but regarding the fundamental doctrines of matter and force, had undergone a great change. "At Paris," says Voltaire, referring to his visit to England in 1727, "you see the universe composed of vortices of subtle matter; at London we see nothing of the kind. With you it is the pressure of the moon which causes the tides of the sea; in England it is the sea which gravitates towards the moon . . . Among you Cartesians, all is done by impulsion; with the Newtonians, it is done by attraction of which we know the cause no better." In 1728, according to Voltaire, there were not twenty Newtonians outside of England. But now, sixteen years later, the mathematical prize questions proposed by the French Academy naturally brought the Cartesians and Newtonians into conflict, and not infrequently the Academy impartially divided its rewards between them. Its last act of homage to the Cartesian system was performed in 1740, when the prize on the question of the tides was distributed between Daniel Bernouilli, Euler, Maclaurin and Cavallieri—the last of whom endeavored to amend the Cartesian hypothesis on the subject. In 1744, Daniel Bernouilli declared himself even more Newtonian than Newton, for he expressed belief that matter may have been created simply through the law of universal attraction, without the aid of any gravific medium or mechanism.[1]

With the acceptance of the Newtonian doctrine came a tendency to imitate the mental attitude which had led to its conception. The logic of physical experiment was now more than ever the final arbiter. Newton's declaration "hypotheses non fingo" tended to check the inclination of speculative minds to evolve new electrical theories. Some like Krüger and Bose declined to formulate any, others sought to explain only specific happenings, and others offered hypotheses tentatively and in the interrogative form.

It might well be imagined that conceptions as to the nature and cause of electricity in such conditions would soon become involved in contradictions and confusion. Winkler's theory that a solid electrical matter inherent to bodies driven always in right lines from their pores by rubbing

[1] Whewell: History of the Inductive Sciences, ii., 198, et seq.

them, and forming a more or less dense atmosphere about them, has little resemblance to Nollet's hypothesis of the effluence and affluence of a subtle universal matter capable of self-inflammation by the "shock of its own beams"; while differing radically from both is Watson's provisional notion that electricity is a force analogous to magnetism, moving in right lines and not subject to refraction, and yet "in common with light, when its forces are collected and a proper direction given thereto upon a proper object, producing fire and flame." In all of them, however, can be traced something of the Newtonian ether—of that most subtle matter which Newton described as pervading and lying hid in all gross bodies; "by the force and action of which spirit, the particles of bodies mutually attract one another at small distances and cohere when in contact, and electric bodies operate at greater distances as well as by repelling and attracting the neighboring corpuscle, and by which light is emitted,"[1] and all sensation excited.

Such was the state of affairs when a discovery of the highest moment, made by different observers in different places, so nearly in point of time that the later observation happened to gain the earliest publicity, startled all civilized Europe.

In the fall of 1745, the German artisans, and especially those of Leipsic, probably recognized that the electric machine had come into good market demand. So simple was the apparatus, and so astonishing its effects, that people who made no pretence to being scientific bought it out of curiosity, and amused themselves by repeating at home the experiments which the philosophers publicly exhibited in the lecture rooms and laboratories. When a device is thus taken to the popular bosom, so to speak, the prediction may safely be hazarded that before long some one in an unexpected quarter will discover or invent something

[1] Principia, B. iii.

concerning it which the philosophers have never thought of or completely missed. And the more complex the intellectual gymnastics of a certain class of these erudite persons around it, the more certain it seems to be that the discoverer will be found to have solved the problem either by his simple wits or by accident and his wits combined.

It is not unlikely that among the more thoughtful students of electricity were some who did not look with favor upon the universal effort directed to the production of more and more powerful discharges. A maximum sooner or later must be reached—possible improvements in machines must terminate some time—and then what? There was nothing to show that the shocks which shook every joint in a man's body were capable of any effects, different in kind, from those which he could easily bear. Moreover, the electrical action came and went like the lightning—quicker than in the twinkling of an eye. Nothing could be more fugitive, nothing less utilizable, than force exerted under such conditions as this. Could it be imprisoned? Who would dare suggest the possibility? Who would risk the ridicule sure to follow the conception that the subtle electrical matter which, whether identified with the Newtonian ether or not, the philosophers agreed to be capable of penetrating all substances, could in some bonds be "cabined, cribbed, confined?" Even if one could imprison it, how was an explosive emanation, shooting in right lines in all directions and never moving continuously in a definite path, to be caught? The attempt would be as idle as trying to box a sunbeam in a soap bubble.

It being now, perhaps, sufficiently clear that not only did the knowledge of the time offer no way of practically confining or accumulating electricity, but that, on the contrary, the idea thereof would have been scouted on all sides as contrary to every respectable hypothesis and hence necessarily absurd, the conditions for the doing of the thing were manifestly ripe, and accordingly it was done.

On the 11th of October, 1745, Dean Von Kleist of the

Cathedral of Camin in Pomerania, completed certain experiments, concerning which on the following 4th of November he felt sufficiently sure to send an account of them to Dr. Lieberkuhn in Berlin. And in December he forwarded other descriptions to Dr. Krüger in Halle and to Archdeacon Swietlicki of the Church of St. John in Dantzic, and later to Winkler and others. Lieberkuhn[1] reported the facts to the Berlin Academy, Krüger printed the letter as an appendix to his book,[2] and Swietlicki, having communicated the intelligence to his dozen or so co-members of the little Physical Society of Dantzic, some of the latter tested the matter experimentally and sent back word to Von Kleist that his apparatus would not work.[3] All of the others kept silent, for they appear to have reached the same conclusion.[4]

Now what Von Kleist did, according to his own story, is this: Up to the present time, he says,[2] it has not been recognized that sparks and streams flow of themselves out of electrified wood, but that in order to make a light appear, something unelectrified must be approached. But all that is needed now to show the sparks is to insert a spool on which wire is wound in a glass tube. Wood and tube, however, must be warm and dry. If further an iron nail be placed in the spool, then the flames will stream sometimes from the metal and sometimes from the wood. That was the first step. The next was to place a nail or a wire in a narrow-necked medicine vial—shaped like a Florentine flask—and to electrify the nail. Strong action, he says, follows, especially if mercury or alcohol be in the vial; and when he takes the vial from the machine a flaming pencil of light breaks forth, which continues burn-

---

[1] Priestley: History of Elec'y. London, 1767.

[2] Krüger: Geschichte der Erde. Halle, 1746.

[3] Gralath: Nachricht von Einigen Electrischen Versuchen. Versuche, etc., der Naturforschenden Gesellschaft in Dantzig, vol. 1. Dantzic, 1747.

[4] Winkler: Die Stärke der Elektrischen Kraft des Wassers. Leipsic. 1746.

ing while he walks sixty steps. He can even electrify the apparatus, take it into another room and ignite alcohol with it. If, while electrified, the nail be touched with the finger, the resulting shock shakes the arm and shoulder. No one can imagine the strength of the shock: that yielded by a vial four inches in diameter containing liquid is so great that no man would care to endure it a second time. It scatters spirits without igniting them, and hurls the spoon from one's hand. The electrification lasts for twenty-four hours. Bose would never dare brave the kiss of a Venus so armed.

This is Von Kleist's own recital, merely condensed; and so far it is Hamlet without the prince. I have preferred, however, to present it in this way so as to show how Von Kleist himself regarded the matter. He first sees only electrified wood which gives sparks of itself; then a nail which is very powerfully electrified; then that something that is electrified can be carried about from one place to another; that it gives a flame, ignites alcohol, and delivers tremendously strong shocks, and that it holds its electrification for twenty-four hours.

But the strangest thing of all he keeps for the end. Electrify the bottle as strongly as you can, put it on the table, and touch the nail or wire entering it with the finger, and it only hisses. It cannot then be got to kindle spirits. In fact, none of these terrible shocks or bright sparks can be got from it *unless it is held in the hand.*

Small wonder that Von Kleist at once began to question what new capability of the human body had thus come to light, and that this aspect of the discovery should have seemed more important to him than the astounding revelation that electricity could apparently be bottled for a day at a time. How it affected the Dantzic philosophers can easily be imagined from the results. They undoubtedly regarded Von Kleist's warning that the bottle must be held in the hand as involving some delusion, for how could hands control the strength of any electrical discharge? So they put alcohol and a wire in a bottle and electrified it, and put it down and contemplated it, and saw nothing, and wrote to Von Kleist that his apparatus, whatever it was, must be of peculiar strength, as theirs would not work. And Von Kleist answers naively that he has never seen any apparatus but his own, and hence cannot draw comparisons, but that he has not found the least difficulty in his performances, and in fact has made an excellent little contrivance out of a thermometer tube four inches long, containing water and a wire tipped with a lead ball, which lights spirits satisfactorily and sometimes gives two discharges. Hitherto he has spoken of his device only generally as a machine; now he names it the "Electrical Thermometer," a designation which it has never borne.[1]

The title which it has received, and how it came so to be known, is now to be told. Meanwhile the Dantzic philosophers, with such new light as Von Kleist afforded, returned to the charge, and at their task for the present I leave them.

The two most eminent physicists of Holland, during the

[1] The weight of evidence from all sources examined is in favor of the foregoing account of the discovery of the Leyden jar; but a passage in one of Winkler's treatises (Die Eigenschaften der Electrischen Materie und des Electrischen Feuers, etc., Leipsic, 1745), which bears date the 20th of August, 1745, and hence some months prior to Von Kleist's formal communication of his experiment to Lieberkuhn and others, indicates that Von Kleist not only made the experiment considerably before this time, but essayed to describe it to Winkler. Winkler's understanding of it was evidently not clear, for in discussing the strengthening of electric sparks, he says that he placed iron and brass tubes of different lengths one upon another, and hung a large hollow copper ball from them, electrifying all together, and getting stronger sparks than when a single tube four ells long was employed. He then notes that Von Kleist has bound together two iron rods and got similar results, and adds: "The electrical sparks from metal were especially strengthened if the metal object were placed on silk cords in such a way that either the object itself or an iron rod hanging therefrom reached the surface of water, which in a thin glass vessel was electrified while resting upon a silk net."

period under review, were Wilhelm Jacob s'Gravesande and Peter Van Musschenbroeck. To them is due the introduction of experimental philosophy and the Newtonian doctrines into the country, and the establishment of systematic study of these subjects in the University of Leyden.

s'Gravesande was rather a mathematician than a physicist, and Van Musschenbroeck,[1] who was originally his pupil and protegé, became, under his guidance, a remarkably able teacher and experimentalist rather than an investigator. As an instructor, it may be said without exaggeration, that kings vied with one another for the possession of him. He held the chair of philosophy in the University of Duisberg; then in that of Utrecht. From the latter Denmark sought to entice him to Copenhagen, the English king to Göttingen, and the king of Spain vainly offered the tempting salary of 20,000 florins per year. The simple request of his native town proved more potent than all these allurements, and he left Utrecht to succeed Wittich as professor of philosophy in the Leyden University, where he remained for the rest of his life, adding to the number of his multifarious physical treatises, and attracting crowds of students from all over Europe, despite the dazzling inducements to abandon his chosen field held out by the king of Prussia and the empress of Russia. One recognizes something characteristically Dutch in the solidity of attainments and persistent fixity of purpose which Van Musschenbroeck above all else possessed, just as something typically French is apparent in the dazzling abilities and captivating style of the Abbé Nollet, whose celebrity at that time, in France at least, even exceeded that of the Leyden professor.

Jean Antoine Nollet was an abbé of the *ancien régime*, not even ordained a priest, but assuming a minor order, and with it the ecclesiastical garb and name of abbé, as many another brilliant man had done, not for the sake of the vocation, but for social distinction and security of posi-

[1] Nouv. Biographie Generale, 37.

tion about the court, which otherwise might prove unattainable to the simple student of science, letters or art. Dufay had been his preceptor, guide and friend, and left him stamped with his own charming qualities, to which Nollet added an individual genius for simplifying and expounding physical science, which made his lecture-rooms at Versailles the resort of the gay French court; and this, not because he had become tutor to Monsieur the Dauphin, nor even because his experiments were astonishing, but because his talk was delightful and witty. There is many an old print representing the Abbé in his curled wig and skull cap, with his black gown barely concealing the richly-laced coat and rapier beneath, daintily conducting Madame la Marquise to the electrical machine, where, to the edification of the other assembled grandes dames, she will receive, with a little grimace, a little shock which will not disarrange a patch on her face, nor disturb a fold of her furbelows; or, perhaps, inviting Monsieur le Comte to witness the spirits burst into flame beneath his sword point, or to laugh at the overthrow, by the fierce discharge, of some stolid serf wearing the king's uniform. Indeed there was no startling experiment of Hauksbee, Gray, Dufay or Bose which Nollet did not repeat, and in many instances on a scale greater than the originator had ever attempted.

There was a great contrast between this French philosopher of the salon and the Dutch philosopher pedagogue: as different from one another as both were from that German "wizard" Bose; and yet alike in each being a philosopher, which Von Kleist, whose discovery has contributed so much to the immortality of the memories of both of them, certainly was not.

But, at the time when Van Musschenbroeck wrote his famous letter to Réaumur, which Nollet made public in France, neither writer nor promulgator had ever heard of the Pomeranian Dean and his medicine vial. The British Magazine, the Universal Magazine, the London Maga-

ABBÉ NOLLET EXHIBITING GRAY'S EXPERIMENT OF THE
ELECTRIFIED BOY.[1]

[1] Reproduced in fac simile from the frontispiece of Nollet's Essai sur l'électricité des corps. Paris, 1746. The boy is suspended on silk lines and electrified by the excited glass tube held by the lecturer, so that his hand attracts bits of loose foil on the table below.

zine, the Gentleman's Magazine, even the Newcastle Journal and the Caledonian Mercury, and perhaps dozens of other public prints in England, were giving the new electrical discoveries as part of the regular news of the day, as fast as they were told by those who made them; but journalistic enterprise of that sort had not yet reached the Continent, and for quick intelligence the private letter was still the best and safest medium.

In January, 1746 (the Dantzic philosophers still puzzling over Von Kleist's instructions), Musschenbroeck wrote to Réaumur as follows:[1]

"I wish to inform you of a new, but terrible experiment, which I advise you on no account personally to attempt. I am engaged in a research to determine the strength of electricity. With this object I had suspended by two blue silk threads,[2] a gun barrel, which received electricity by communication from a glass globe which was turned rapidly on its axis by one operator, while another pressed his hands against it. From the opposite end of the gun barrel hung a brass wire, the end of which entered a glass jar, which was partly full of water. This jar I held in my right hand, while with my left I attempted to draw sparks from the gun barrel. Suddenly I received in my right hand a shock of such violence that my whole body was shaken as by a lightning stroke. The vessel, although of glass, was not broken, nor was the hand displaced by the commotion: but the arm and body were affected in a manner more terrible than I can express. In a word, I believed that I was done for."

He then proceeds to say that the shape of the vessel is unimportant, but that he believes that a thin white glass five inches in diameter would possibly give a shock strong enough to kill. The person receiving the discharge may

---

[1] Mémoire de l'Acad. Roy. des Sciences, 1746, Paris.

[2] Gordon imagined that he discovered that blue silk threads insulated better than any others, and for this reason every one about this time was using them.

stand on the floor, and must either hold the jar in one hand and excite sparks with the other, or he may place the jar on a piece of metal on a table, and touch the metal with his hand, bringing a finger of the other hand to the wire.

Of course this experiment is the same as that of Von Kleist, and goes further, for it eliminates the necessity of supporting the vessel in the hand, while making it clear that the seat of the effect is not in the body, as Von Kleist

THE LEYDEN EXPERIMENT.[1]

supposed, and as the Dantzic philosophers evidently refused to suppose, but in the apparatus, and that when one hand touches the wire which enters the jar (and extends down into the water therein), and the other hand touches the metal plate on which the jar rests, a path for the discharge is made through the body of the operator.

[1] Reproduced in reduced facsimile from Winkler's Die Stärke der Electrischen Kraft des Wassers in gläsernen Gefässen. Leipsic, 1746.

Other letters from Leyden, especially from Allamand,[1] Van Musschenbroeck's colleague and assistant, soon brought further details—for the Professor's first communication was evidently written while he was still suffering from the nervous prostration following the shock; and he was doubtless entirely in earnest in his remark that he would not undergo the experience again for the Crown of France; although he did do so, and with even worse results than before. The observation was made by accident, Van Musschenbroeck's object, some say, being to ascertain whether the charge on electrified bodies could be prevented from dissipation by contact with water—others that he was examining the capacity of water for receiving and propagating electricity. Allamand avers that the shock deprived him of breath for some minutes; but the most important part of Allamand's communication to Nollet is his ascription of the credit of the actual discovery to one Cunæus, a scientific amateur, who, he says, observed the effects while repeating at home certain experiments which Van Musschenbroeck and Allamand had shown him. The evidence, however, in support of Cunæus, is not only weak, but in details contradictory, and it seems safer to conclude with the Abbé de Mangin, who, in his history written contemporaneously with the event, declares that the claim for Cunæus is "a mere stratagem devised by people envious of Musschenbroeck for the purpose of depriving him of a part of the glory which was justly due him"—*pace*, of course, Von Kleist. At all events, whether originating with Van Musschenbroeck or Cunæus, it is certain that the attention of the world was first attracted to the discovery by the letter which Musschenbroeck wrote and Nollet published; and, as that information came from Leyden, the discovery became known sometimes as the Musschenbroeckian, oftener as the Leyden experiment, while the contrivance itself

---

[1] Mem. de l'Acad. Roy. des Sciences, 1746.

was called, and to this day bears the name of the Leyden jar.

To return now to the Dantzic Society, or rather to Daniel Gralath, who was at work in its behalf. In February of 1746, Von Kleist sent a final epistle, which seems to have clarified matters; so that ten days later, Gralath definitely finds that the jar must be held in one hand and its wire touched with the other, and ascribes the long delay to Von Kleist's failure to make this plain in the beginning. Gralath soon after hears of the Leyden experiment, and at once advocates Von Kleist's claim to the discovery, naming the proceeding the "Kleistian strengthening experiment," and the jar, the "strengthening machine." But it was too late—the infant had already been christened, and the world refused, justly or unjustly, to sanction the change of cognomen.

Gralath's experiments were, however, fraught with new discovery. In common with Musschenbroeck he records the great power of the discharge, which he says gave some people the nose-bleed, and acted like a lightning stroke; but announces that the thinner the glass of which the bottle is made, the stronger are its effects: that he has succeeded in retaining the charge in it for three days (but here Von Kleist excels him, for in his hands the bottle worked well even after eight days' inaction); and that, although the bottle might seem to be completely discharged so that no trace of electricity is manifest, nevertheless, after a short period of rest it once more yields vivid sparks.

The difficulty which minds moving in a rut always find in getting out of it, is well exemplified in the manner in which the philosophers dealt with the new apparatus. Despite its singular capacities, they saw in it only a contrivance for producing shocks stronger than their machines would yield, and bent their energies to testing the effects. Nollet killed birds with the discharge, noting that on dissection they exhibited the same condition of ecchymosis shown by people struck by lightning. Gralath destroyed life in beetles and worms ; but not succeding in so doing in birds, sought still further to intensify the discharge, and thus reached the idea of combining the effects of several jars, which he placed in metal pans, with their lead balls in contact with the prime conductor of his machine, while from each pan a wire proceeded to a copper globe placed within sparking distance of the conductor. This was the first grouping of electric generators in battery, in which they were obviously disposed in parallel, or multiple arc— an arrangement which for some time was the only one known.

Gralath now killed birds easily, and reports minutely on the physiological changes produced ; but, as he saw that whatever the effects of these strong discharges might be, no certain knowledge as to them could be obtained unless their strength could be measured, he turned his efforts to contriving a measuring instrument. But he soon found the difficulties insuperable. What should be the standard? What the unit? What was really to be measured—the attractive power of the charge, or the striking energy of the discharge? He arranged near a scale-pan, which he maintained in a non-electrified state, an iron rod which communicated with his machine—the rod being adjustable nearer to or further from the scale-pan, and attracting the latter when electrified. The attractive force of the rod was counterbalanced by weights in the opposite pan. His factors were the distance of the electric machine from the apparatus, the distance of the rod from the scale-pan and the balancing weights ; and he tabulates his results, arriving at the conclusion that with the rod distant half an inch from the scale-pan, the ratio of attractive force, when the electric machine was at maximum distance from the apparatus, to that existing when the machine was at minimum distance therefrom, was as 74 to 44 ; and that this inverted represented the relative strengths of the corresponding discharges. He had no faith in his deduction, which, he says, requires proof by long trials and experiment, and confines himself to remarking that, if a natural law connecting attraction and spark energy should be established, his contrivance would be an "Electrometer," adapted to measure both the attractive force and the sparks, and that it would be of great use, and free this branch of philosophy from many uncertainties. Such was the first attempt to measure electricity, to-day the most modern of all electrical arts.

The records of the Leyden jar experiments which now appear are devoted more to graphic descriptions of the physical sufferings of over-zealous philosophers than to the announcement of new discoveries. Winkler modified the apparatus by winding an iron chain around the bottle, and connecting it to a metal plate near the prime conductor of his machine; the wire from within the bottle also being connected to the same conductor. His letter to the Royal Society recounts his convulsions, the agitation of his blood, the supervention of an ardent fever, and the evil result attending the curiosity of his better-half, who, taking the shock a second time, was afflicted with nosebleed. Nollet entertained the French king by transmitting the discharge through 180 of his guards, "who were all so sensible of it at the same instant that the surprise caused them all to spring up at once." This however, was outdone by the performance of the Carthusian monks in Paris, who formed a line 900 feet long "by means of iron wires of proportionable length between every two, and consequently far exceeding the line of the one hundred and eighty guards. The effect was that when the two extremities of this long line met in contact with the electrified vial, the whole company at the same instant of time gave a sudden spring, and all equally felt the shock."

It is not difficult to understand why the electrical histories of de Mangin, Secondat, Priestley, d'Alibard, Gralath and others, written near to the time of the discovery of the Leyden jar, hail it as a great advance, because of its capabilities in the production of discharges of unprecedented strength. The writers of fifty years ago, however, find in the supposed storing or accumulating properties of the contrivance its chief value, and for that cause assign to it a high place among the great electrical inventions. From the modern point of view the historical importance accorded to the Leyden jar or condenser seems disproportionate when the relatively minor part which it plays in existing applications of electricity is recalled; but, on the other hand, the immediate reason for the great promoting influence which it exerted upon electrical progress at the time of its advent is not found wholly in the magnifying power and the accumulating property of the contrivance. As ensuing events soon showed this influence rests, and perhaps chiefly rests, upon the fact that by means of the Leyden jar came the first recognition of an electrical circuit.

The discharge of the electric machine, like that of the rubbed glass tube, had hitherto been delivered from the globe either directly to the object to be electrified or to a metal prime conductor (usually a suspended gun barrel), and thence to the object—the latter being insulated on a pitch cake or by suspension on silk strings. Because the Dantzic philosophers had supposed that the Leyden jar would act in the same way, they regarded it as a failure when, on being merely laid on the table, it refused to deliver its spark to an object brought near to it. As soon, however, as Gralath and others understood that the charged jar must rest in one hand, while the other touched the ball upon the end of its inserted wire, the recognition of a circuitous path, to which the electricity of the jar was confined, was complete. That path included both the jar and the human body. When, for the holding hand, a metal pan in which the jar rested, or a chain enwrapping the jar, was substituted, communicating by wire with the ball, then the path became a metallic circuit, and the supposed influence of the human body *per se* was eliminated. In that path the electricity seemed to be present, and not elsewhere. Here then was the electric matter not only seemingly shut up in a jar, but retained in a definite continuous circuit.

It was not long before experiments on the Leyden jar, due to mere curiosity to witness the strength of the discharge, began to give place to more rational investigation. Then it became manifest that a great change had taken place, and that the progress of thought was to be different from ever before. The field of research had suddenly broadened, and now new paths, opening in all directions, lost themselves in the mists of the new horizon. The old way continued onward in far perspective, but nowhere on the shadowy circle was there a sign to show that the goal whither it was believed to tend—the discovery of the ultimate nature of electricity—was a whit the nearer. Of these paths one led apparently to the solution of the mysteries of the wonder-working jar; another to the revelation of the capabilities of the force confined in the circuit; to the hiding-place of a potent weapon wherewith to combat all ills and diseases, a third seemed directed; while a solitary finger-post pointed, not to the low-lying fog, but straight upward to the clouds—upward to the very home of the lightning. Many such ways stood open, but of them all, to the foregoing were apparently the most inviting. Into the first two flocked the electricians, and into the third the physicians. The fourth remained for awhile untraversed, awaiting the advent of a philosopher who just then was regulating municipal affairs for the staid citizens of far-off Philadelphia.

The European electricians worked assiduously. Their experiments are legion. The records especially of those devoted to correcting their own mistakes would fill huge volumes, but do not, thanks to the mercilessly winnowing rules of the Royal Society and the French Academy— ordinances which were enforced by public opinion outside of the circles of those learned bodies as effectively as within them. There are no big folios and massive quartos of the eighteenth century devoted to electrical treatises. People who wrote on electricity were forbidden to write sermons; and people who wrote sermons could not write on electricity. Dr. Priestley was the only exception; but even his ponderous history is a mere brochure for the author of 346 books, mainly theological.

It is not necessary here to epitomize, however briefly, the contents of the many little books which appeared in Europe during the two years immediately following the advent of the Leyden jar. Some, like the anonymous Venetian work,[1] which begins by recounting the gallant adventures of a pair of young soldiers in the charming city of the Adriatic, and ends by making them listeners to an elaborate dissertation on electricity as a part of the polite conversation in somebody's palace, are quaintly curious. Nollet is developing more and more startling experiments with what seems to be constantly augmenting ingenuity. He turns to the electric light, and places glass flasks exhausted of air directly upon the metal conductor, which he connects by a chain with the globe of his electric machine. Sometimes the flask bursts into glow, and luminous aigrettes shoot from the metal cap and stop-cock. If he brings his fingers to the exterior the flames divide as if to meet them. Sometimes a single powerful stream flows from the end of the rod, and when he touches the latter with his finger a spark leaps forth, and "at the same instant the vessel is filled with so brilliant a light that all objects near it are made distinctly visible," so that he adds, enthusiastically, "A more natural representation of the lightning flashes which precede or accompany thunder could not be found."

The effect of electricity on vegetables and animals he essays to test by direct experiment. He plants mustard seeds in two receptacles, and maintains one in an electrified state for eight days. "The electrified seeds," he says, "had all sprouted at the end of that time, and had stalks

[1] Dell' elettricismo ossia delle forze elletriche. Venice, 1746.

fifteen or sixteen lines in height, while but two or three of the non-electrified plants had appeared above ground, and even these had stems not more than three or four

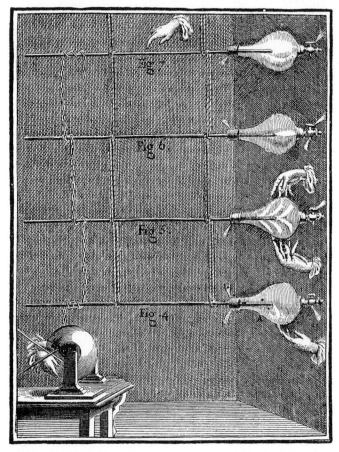

ABBÉ NOLLET'S EXPERIMENTS ON ELECTRIC LIGHT IN VACUUM FLASKS.[1]

[1] Reproduced in reduced fac simile from Nollet's Recherches sur les Causes Particulieres des Phenoménes Électriques. Paris, 1749. The glass globe being set in rapid rotation is electrically excited by the pressure of the hands against it, and the electrification passes from globe to chain and thence to the conducting rods, and exhausted flasks.

lines high." He has no doubt that electricity accelerates vegetation, although it seems to him that the plants thus forced are not as hardy as those which grow under natural conditions.

ABBÉ NOLLET'S EXPERIMENTS IN ELECTRIFYING ANIMALS AND VEGETABLES.[1]

[1] Reproduced in reduced fac simile from Nollet's Recherches sur les Causes Particulieres des Phenoménes Électriques. Paris, 1749. The hand on the right holds a vessel of water, which trickles slowly from the spouts turned away from the electrified chain, but is projected in jets from those in proximity thereto.

Then he turns to animals. Two cats, "each four months old, of nearly the same size, and fed alike," are placed in cages, one of them being near the conductor of the electric machine, which is excited for some hours. Both the electrified cat and the non-electrified cat lose weight, but the electrified cat loses the most, about 54 grains. Nollet thinks this may be due to "difference in temperament," although he admits that the cats went placidly to sleep, except when he gave them shocks. Then he electrifies pigeons and small birds, and finally persons, and concludes that in all cases there is a loss in weight due to "transpiration;" but when he attempts to treat actual maladies he fails. "The paralytics, experiencing no relief which would sustain their patience (for some is necessary in order that they may undergo this sort of torture), complained bitterly," and the Abbé abandons for the time his high hopes of thus relieving suffering humanity.

The great majority of experiments now contemporaneously recorded, however, are of little interest. A better idea of the thought and achievement of this period can be had by following the work of a few men, whose superior intelligence, or better facilities, or both, led their thought, for a short time only, to bring forth all the fruit that is worth garnering.

Winkler discovered that when electricity had several paths to choose from, it appeared to traverse the one which was composed of the material which conducted best, and that is all that need be said now about him. The two philosophers who most attract and hold attention are rivals —Louis Guillaume Le Monnier, the younger, in France, and Dr. William Watson in England. At this time no one pretended to understand why the Leyden jar behaved as it did. First, it could be electrified by the ordinary globe machine or rubbed tube; second, it yielded an extraordinarily strong shock and bright spark; and third, it did this last only when its exterior was connected in circuit with its interior. In entering upon a new inquiry, it is often as efficacious, for the purpose of starting, to challenge an existing theory as to propound a new one. Thus did Le Monnier. Dufay had stated in substance that conductors cannot be electrified unless supported on non-conductors. The Leyden jar, says Le Monnier,[1] must be an exception to Dufay's principle, for it can be electrified, although it is supported on the hand, which is a conductor. That shows us, at once, that Le Monnier—and he probably reflected the idea of the French philosophers generally—simply considered the jar as an electrified mass, regardless of its diverse materials. For him the objective point is less the jar, than the circuit. He states the principle of it: "All bodies are electrified by means of a vial of water fitted to a wire," if "placed in any curved line connecting the exterior wire and that part of the bottle which is below the surface of the water"—but passes at once to something remarkable. Hitherto everything to be electrified was insulated on pitch cakes or silk supports. What astonishes Le Monnier now is, if 200 men be placed hand in hand—the end individuals touching the inserted wire and the bottom of the bottle respectively—a violent concussion is felt by all at once; and this equally well, whether they are all mounted on cakes of resin or stand on the floor; equally well when they are connected by iron chains; equally well whether the chains dip in the water or lie on the ground, and the electricity runs equally well—now he abolishes the men—through a wire a league long, "though a part of it dragged on the wet grass, went over charnel hedge or palisades and over ground newly ploughed up." He even bends a bar of iron to touch the two points of the jar, and observes that it does not acquire more electricity when held by silk lines than when supported in the hand. Strange, he thinks, how "the electricity will stay in the path thus made for it, without either running off or becoming absorbed."

[1] Phil. Trans., No. 481, p. 247, 290. Memoirs de l'Acad. Roy. des Sci., 1746.

There was an ornamental octagonal pond in the Tuileries gardens of those days, which measured about an acre in extent. Around the semi-circumference of this Le Monnier disposed an iron chain so that its ends came diametrically opposite one another. These ends were held respectively by two observers, one of whom dipped his disengaged hand in the water. The other, across the basin, held in his free hand an electrified Leyden jar, the inserted wire of which he thus presented to an iron rod which entered the water, and was supported on a cork float. Thus early in 1746, a circuit was made including both water and a metallic conductor, over which passed the discharge of the jar, so that the two observers were simultaneously shocked.

Like other experimenters who had dealt with long conductors, Le Monnier sought to measure the velocity with which the electric matter ran over them, but without avail; nor was he any more successful in finding out what impelled it at a speed which he estimates to be at least "thirty times that of the velocity of sound in air." He made up his mind, however, that the electric matter is communicated to bodies in proportion to their surfaces, and not in proportion to their masses.

The conclusions of Le Monnier appear to have been regarded by the English electricians as a challenge. Watson was now their leader, and his response was ready. Le Monnier had dealt with the jar as a mere electrified mass, operating to increase the shock or spark, for some reason, unknown. Watson,[1] in reply, declares that it owes its capabilities to the accumulation of electrical matter within it—this happening because the glass acts as a barrier and prevents the electrical matter escaping from the water as it is supplied thereto by the inserted wire. We shall see changes in these notions soon. Meanwhile, as for the rest of Le Monnier's observations, they merely prove, says Wat-

[1] Phil. Trans., No 482, p. 388. Watson: Exp'ts. and Obs'ns. on Elec'y. 3d ed. London, 1746.

son, "what I have myself found out," that the electricity will always describe the shortest circuit between the electrified water and the wire of the vial which contains it, "and this operation respects neither fluids nor solids, as such, but only as they are non-electric (conducting) matter. Thus this circuit," he adds, tracing it and using the word to name the path provided by Le Monnier, "consists of the two observers, the iron chain, the line of water and the iron rod in the floating cork."

Watson[1] is now well in the van. The German and the French electricians, preferring to follow the leadership of Winkler and Nollet, are devoting themselves chiefly to contriving variations of experiments already decisive, and so to heaping up a great mountain of cumulative proof. Watson shows that if the amount of water in the jar is increased even to four gallons, the stroke is not augmented in strength ; that iron filings therein answer as well as water, and mercury as well as iron filings. The specific gravity of the material in the jar he thus discovers has no influence. He states that the Leyden vial "seems capable of a greater degree of accumulation of electricity than anything we are at present acquainted with . . . by holding its wire to the globe in motion, the accumulation being complete, the discharge runs off from the point of the wire as a brush of blue flame."

Watson now, as the result of all his observations, propounds a theory which was generally accepted by the English philosophers. Historically, and in the light of immediately ensuing events, it is of especial importance.

The hypothesis affirms the existence of an electrical

[1] Watson's papers of this period in Phil. Trans. are : No. 478, p. 41, read February 6, 1746 ; No, 482, p. 388, read January 29, 1747 ; No. 484, p. 695, where there is added to his paper of Februry 6, 1746, "A Sequel to the Experiments and Observations," etc., read October 30, 1746. The principal papers were separately published. Experiments and Observations, 3d Ed., London, 1746. Sequel to Experiments and Observations, 2d Ed., London, 1746 : An acccount of the experiments made by some gentlemen of the Royal Society, etc. London, 1748.

ether, much more subtle than common air, and passing to a certain depth through all known bodies. It has the property of air, of moving light objects, and is likewise elastic, this last fact being shown by its extending itself around excited electrics, by its increasing the motion of fluids, by the apparent influx of electrical fire to all bodies and by its giving violent shocks to the human frame.

With this ether all bodies are normally charged. If, however, a body be excited, then the normal conditions are disturbed, so that the ether in the nearest unexcited non-electric tends, by its elasticity, to move to the excited body where it accumulates. In so doing it carries light bodies with it, which accounts for electrical attraction.

Applying this idea to an electrified Leyden jar held in one hand of an observer, who touches with the other the metal gun-barrel on which it is suspended by its inserted wire, Watson maintains that, on the explosion which follows, the man (nearest conductor) instantly parts with as much fire from his body as is accumulated in the water of the jar and in the gun barrel; the fire rushing violently through one arm to the water, through the other to the barrel. Then as much fire as the man has lost is immediately and with equal violence replaced from the floor of the room. Hence, and for both reasons, the shock. This flux, he further says, may be prevented, and its effects are not seen, when the glass containing the water is too thick, or if the man stand on an insulator, or if the points of contact between his (conducting) hand and the jar which it holds are fewer. The last limitation, it may be observed in passing, proved suggestive; for Dr. Bevis, a member of the Royal Society, promptly showed that the greatest number of contact points would be obtained by coating the exterior of the jar with sheet lead or so-called tin foil. This suggestion was adopted, as it was found that a person who merely touched this coating with a small wire obtained as strong a shock as if the whole hand rested against the exterior of the uncoated bottle.

Watson's "sequel" is dated October 20th, 1746, and was read before the Royal Society ten days later. Reduced to a few words, his theory is simply that the exciting of an electric causes the advent thereto of fire from the nearest adjacent conductor, and that the latter regains an amount equal to that lost. "By asserting," he adds, "that that we have hitherto called the effluvia does not proceed *from* the glass or other electrics per se, I differ from Cabæus, Digby, Gassendus, Brown, Descartes and the very great names of the last as well as the present age."

It may be conceded that Watson supposed that what he calls the elastic electric ether became more dense in one body and less dense in another; but it will be observed that there is no principle of equilibrium here involved. He imagined that the man touching the charged Leyden jar parted, immediately with his fire, and immediately regained it from the floor. But no matter how highly charged the jar, if, according to Watson's notion, he stood on a pitch cake, or even had dry soles to his shoes, the flux to him from the floor would thereby be prevented, and the jar would give him no shock—which is of course erroneous; for the man's body, no matter on what it is supported, obviously closes the circuit between the inside and outside of the jar.

Enough has now been stated to show what Watson's theory actually was in the fall of 1746. I shall recur to it hereafter.

The physical advance accomplished may now be noted. Van Musschenbroeck had found, and Watson had likewise recently re-verified the fact, that the thinner the glass of the jar the stronger the shock. Watson alone had found that the greater the area of the conductors in contact with the glass, again the stronger the shock. Two of the three conditions upon which depend the capacity of a condenser had thus been discovered: namely, the thinness of the dielectric stratum between the coatings, and the size of the coatings themselves. The third (specific inductive capacity of the dielectric) was still far in the future.

So much for the jar. Now as to the circuit. By midsummer of 1747, Watson had gained a comprehensive idea of the law of resistance, and states it thus:

"This circuit, where the non-electrics (conducting substances), which happen to be between the outside of the vial and its hook, conduct electricity equally well, is always described in the shortest route possible; but if they conduct differently, this circuit is always formed through the best conductor, how great soever its length is, rather than through one which conducts not so well, though of much less extent," in other words, he had established and now announced that the resistance of a conductor to the passage of electricity is proportional to its length, and, other things being equal, depends upon the material of which it is composed.

DR. FRANKLIN.

From an old Engraving by Fisher
from the Mason Chamberlin portrait.

CHAPTER XVI.

THE printer's boy, who had landed hungry, footsore and all but penniless at the Market Street wharf in Philadelphia, after a hard journey by both sea and land from Boston, was now, twenty-three years later, the chief citizen of the growing town. To no one did that community then owe so much as it did to Benjamin Franklin.[1] The once runaway apprentice had organized its police, founded its school (destined afterwards to become one of the great universities of the world) devised for it a system of fire protection, established its Philosophical Society and its public library (the first in the colonies), printed its books and its newspapers, supplied it with concentrated worldly wisdom in the maxims of Poor Richard, served it in various official capacities, and invented for it the stoves to which it still clings. Of the magnificent services which he was later to render, not to his town, but to his country, Franklin, at forty years of age, had doubtless no anticipation. The time seemed to him near at hand when he might relinquish some of the many tasks imposed upon him—when the grind of money-getting might cease, and when with the modest fortune which tireless endeavor and patient frugality had brought to him, he might turn, not to idleness, but to work which, through the pleasure it afforded, bore no resemblance to toil. As his inclinations were to philosophic study, this it was now his ambition uninterruptedly to pursue.

[1] I have followed the autobiography of Franklin as edited by the Hon. John Bigelow, in his fine edition of Franklin's Works, N Y., 1889. Parton's Life and Times of Franklin, New York, 1864, has a chapter (vol. I, c. ix.) devoted to "Franklin and Electricity," but the errors in it are many. Weems' biography is chiefly a work of pure imagination.

In the year 1746, while revisiting Boston, Franklin met there a Doctor Spence, lately arrived from Scotland, who exhibited to him some crude electrical experiments. Spence's apparatus was meagre, and his skill small; but the subject was entirely a new one to Franklin, and it surprised and delighted him.

Meanwhile the circulating library which he had established several years before had attained the dignity of a corporation under a charter granted, in 1742, by John Penn, Thomas Penn and Richard Penn, "absolute Proprietaries of the Province of Pennsylvania and the counties of Newcastle, Kent and Sussex upon the Delaware," and was known as the Library Company of Philadelphia.[1] As a matter of course, this institution drew its supply of books from England—for colonial publications were few and far between; and it was especially fortunate in possessing in London, rather as its benefactor and friend than as its agent, Peter Collinson, a merchant having extensive business relations with the American colonies, and a member of the Royal Society. Collinson was in the habit of gathering, not only books, but news and transmitting the same to the Library Company; and occasionally the members of the latter, in return, would send to Collinson accounts of remarkable natural events occurring in their vicinity. It was a common custom in those days for foreigners and non-members of the Royal Society to report such happenings or the results of their own new experiments to members, so that the latter might offer them to the Society, which, if it approved, caused the accounts to be published in the official transactions. In this way for instance, Joseph Breintnall, a member of the Library Company, communicated through Collinson to the Royal Society, under date of Feb. 10, 1746, his experiences following a rattlesnake bite. Collinson himself was a botanist of high reputation. Through him a system of exchange

---

[1] A catalogue of the books belonging to the Library Company of Phila. 1769.

of horticultural products was maintained between England and the colonies, and into the latter[1] on his recommendation, the culture of flax, hemp, the silk-worm and the wine grape was introduced.

The electrical experiments of Dr. Watson and his new theory accounting for them, created no small stir among the British philosophers, as may readily be imagined, and in fact stood unrivalled as a topic of scientific interest. In the fall of 1746, Watson republished his "experiments and observations" and also his sequel thereto in book form, and to the former added a preface in which he urges the prosecution of similar investigations by others, while replying to the still-prevalent cry of "what is the use of it?"[2]

"It must be answered," he says, "that we are not as yet so far advanced in these discoveries as to render them conducive to the service of mankind. Perfection in any branch of philosophy is to be attained but by slow graduations. It is our duty to be still going forward; the rest we must leave to the direction of that providence which we know assuredly has created nothing in vain. But I make no scruple to assert that notwithstanding the great advances which have been made in this part of natural philosophy within these few years, many and great properties remain undiscovered. Future philosophers (some perhaps even of the present age) may deduce from electrical experiments uses entirely beneficial to society in general." Furthermore, in order to show with what facility such research can be conducted, he states that his experiments "were all made with glass tubes about two foot long, the bore about an inch in diameter," and gives some simple directions as to warming and drying the tube before rubbing it.

Watson's books were sent over to the Library Company by Collinson, together with such a tube as Watson de-

---

[1] Stephen: A Dict'y of Nat'l Biography. London, 1887, vol. xi.
[2] Watson: Experiments and Observations. London, 1746.
Watson: Sequel to Experiments and Observations. London, 1746.

scribes, probably very soon after the reading of Watson's sequel to the Royal Society in October, 1746. He added directions for using the glass. Franklin, being already interested, eagerly seized the opportunity of repeating the experiments which he had seen in Boston; and then, as he gained skill, performed those described in Watson's pamphlets. The attention of his friends to whom he exhibited these wonders became enlisted to such a degree, and the news of them spread so widely, that before long, his house was continually filled with curiosity seekers. As he had no fancy for indefinitely repeating these performances merely as a show, and a very decided one for pressing ahead to discover new marvels, he presented several tubes which he had caused to be blown at the glass house to his friends, and invited them to "divide a little this incumbrance with him." The advice given by Watson doubtless acted as a spur to others as well as to Franklin; but instead of each pursuing his own researches, those most interested came together, and before long, a quartette composed of Franklin, Ebenezer Kinnersley, Thomas Hopkinson and Philip Sing united their efforts. If Kinnersley was not Franklin's equal, in point of scientific knowledge and experimental ability, he ranked but little below him. He had been educated in England, and had emigrated to Philadelphia, where he was eking out a rather precarious existence teaching school, at the time he became Franklin's coadjutor. The letters of Franklin to Collinson bear frequent testimony to his ingenuity, and as will hereafter be seen, he played no inconsiderable part in spreading knowledge of the new science throughout the colonies. Hopkinson was the first president of the American Philosophical Society, and Sing was one of its members. These four men were the "we" to whom Franklin alludes in his early letters as directly participating in the "Philadelphian experiments."

By the latter part of March, 1747,[1] Franklin, having be-

[1] Experiments and Observations on Electricity made at Philadelphia, in

come satisfied that his colleagues and himself had made some really new discoveries, wrote, under date of the 28th instant, to Collinson, so advising him and expressing the intention of soon sending him an account of them; adding that he "never was before engaged in any study that so totally engrossed my attention and my time as this has lately done," and that, during some months past, he has had "little leisure for anything else."

On July 11, 1747, Franklin fulfills his promise, and the story is told to Collinson of the first electrical discoveries made in America. Immediately—at the very threshold— is foreshadowed the great achievement which left Franklin's name immortal. The initial announcement refers to "the wonderful effect of pointed bodies both in drawing off and throwing off the electrical fire."

It was no novelty to electrify pointed conductors. Von Guericke had done so, and Gray and Dufay and the Germans. Hauksbee had seen the glow at his finger-tips. The fire hissing from the ends of iron rods is abundantly pictured in the old engravings of the apparatus of Winkler and Nollet and Watson. But that was not the achievement which Franklin relates.

He electrified a small cannon ball, and suspended a bit of cork near to it by a silk string. The cork, after touching the ball, was repelled to a few inches' distance and maintained in that position. When he brought the point of a steel bodkin, held in his hand, in the vicinity of the ball, however, the cork fell back against the ball and was no longer repelled by it. The little metal rod thus seemed to conduct the electric atmosphere away from the iron: to draw it off, as Franklin says.

There is no doubt in Franklin's mind as to the part taken by the sharpened end of the rod. In the dark, a light gathers around it like that of a fire-fly or glow-worm;

America, by Benjamin Franklin, LL. D., and F. R. S. The fifth ed., London, 1774. A list of the various editions of Franklin's electrical papers will be found in Mr. P. L. Ford's Bibliography of Franklin.

but if the extremity be blunt, then the light is not seen unless it be brought very near to the globe.

The circumstances surrounding this attack upon the problems of electricity were novel. In Europe men had become skillful electricians, apparatus had been brought to a condition of refinement, and the keenest philosophical minds had seemingly exhausted their powers in proposing explanatory theories. Every investigator of electricity worked under the potent influence of this highly-developed environment.

On the other hand, in the colonies there was virtually no environment at all in any wise corresponding in character. The knowledge of past achievement possessed by Franklin and his colleagues was probably all drawn from Watson's pamphlets, Collinson's brief letters and Spence's crude experiments. And this was perhaps fortunate, for had they been better posted, they would probably have deemed impracticable, in the beginning, efforts which ultimately ended in success. They seem to have copied nothing from European sources—not even the electrical machine, which they re-invented. The tubes which Franklin caused to be made at the glass house were of common green glass, thirty inches long and as large as could be grasped in the hand. Rubbing them with buckskin, as he says, was fatiguing exercise; and it was for greater convenience that Philip Sing made the glass into a globe, and taking the hint from his grindstone, gave it an axle and crank.

Not being aware of the multitude of earlier theories, and unable to reconcile Watson's hypothesis with the showing of experiment, it was inevitable that the Philadelphia experimenters should seek for themselves some other explanation of the strange and novel effects before them. Thus came into existence, at the very outset of their research, the Franklinian theory, and it is first announced in the same letter to Collinson in which is described the "drawing off" action of the pointed rod. It gained a wider ac-

FRANKLIN'S THEORY OF ELECTRICITY.

ceptance than any electrical hypothesis hitherto proposed. It may almost be said to have become the world's theory, and to have retained a certain ascendency even to the present time; for it is the most easily thinkable of all to the non-mathematical mind. There is probably no electrical fluid running along conductors and accumulating like water in a tank; but that idea of it is imbedded in the language and in every-day thought, and the hydraulic analogies maintain the vitality of the conception. Indeed, whether the time will ever come when the world will cease to imagine electricity as an actual fluid, may well be doubted.

The theory which Franklin announced assumed the electrical fire to exist in all bodies as a common stock. If a body acquired more than its normal amount, he termed it "plus" or positively electrified. If it lost some of its normal amount, he regarded its condition as "minus" or negatively electrified. The common stock of electrical fire in all bodies he held to be in a state of equilibrium, and into this common stock the fire from a positively or over-electrified body will flow, while from the common stock the fire will flow to a negatively or under-electrified body. Thus, imagine, says Franklin, three persons, each having his normal equal share of electrical fire. "A, who stands on wax and rubs the tube, collects the electrical fire from himself into the glass; and his communication with the common stock being cut off by the wax, his body is not again immediately supplied. B (who stands on wax likewise), passing his knuckle along near the tube, receives the fire which was collected by the glass from A; and his communication with the common stock being likewise cut off, he retains the additional quantity received. To C standing on the floor, both appear to be electrified; for he having only the middle quantity of electrical fire, receives a spark upon approaching B, who has an over quantity; but gives one to A, who has an under quantity. If A and B approach to touch each other the spark is stronger, because the difference between them is greater. After such touch there is no spark between either of them and C, because the electrical fire is reduced to the original quantity. If they touch while electrizing, the equality is never destroyed, the fire only circulating." B therefore is positively or plus electrified, and A negatively or minus.

The fire may be circulated, says Franklin, and "you may also accumulate or subtract it upon or from any body as you connect that body with the rubber or with the receiver (tube), the communication with the common stock being cut off."

Franklin's chief concepts, therefore, are first, the normal state of equilibrium of the common stock of electrical fire in all bodies; second, that this equilibrium may be disturbed, so that a body, by reason of the disturbing action, may have fire given to it or taken away from it; and third, that, after the disturbing action ceases, the reaction is a transference of the fire back to the original state of equilibrium. The fluid analogy readily suggests itself. The common stock of electrical fire may be represented by the atmosphere. If air be accumulated above atmospheric pressure in a vessel, it will escape therefrom into the aerial ocean until the pressure without and within the vessel is equalized. If air be exhausted from a vessel, the atmosphere from without will rush into that vessel again until the pressure outside and the pressure inside are the same.

The staid people of Philadelphia, however, do not flock to Franklin's house to listen to his theories, but to witness his experiments; and, indeed, he and his colleagues are as alive to the marvelous aspect of it all as Bose himself. The electrical fire leaps "like lightning," writes Franklin, around the gilt ornaments on china plates, or on the sides of books, or around the mirror and picture frames. Philip Sing contrives little pasteboard wheels which are driven like wind-mills when brought near the rubbed tube. Franklin lights candles just blown out, by drawing a spark amidst the smoke between the wire of the Leyden vial and the snuffers. By the aid of the vial he "increases the force of the electrical kiss vastly," and even the cork ball, vibrating like the end of Guericke's thread, between the Leyden jar and a conductor near by, is blackened and given legs of linen thread, to make it into a counterfeit spider which appears, to quote the genial philosopher once more, "perfectly alive to persons unacquainted." The words are almost identical with those which Porta, nearly two centuries earlier, had used to describe the strange behavior of the iron filings in the magnet field, and the astonishment which their movements created among the bystanders.

Two months later, Franklin sends Collinson a second letter, in which he describes the Leyden jar as electrified positively within and negatively without, and marvels that these two states of electricity—the plus and minus—should be "combined and balanced in this miraculous bottle! situated and related to each other in a manner that I can by no means comprehend!" He also connects his plus and minus theory with the phenomena of attraction and repulsion, by stating that "when a body is electrified plus it will repel a positively electrified feather or small cork ball. When minus (or when in the common state) it will attract them, but stronger when minus than when in the common state, the difference being greater."

This is all hypothetical; yet it leads, through Franklin's conviction that the equilibrium of the bottle is restored by exterior communication between its inside and outside, to a discovery of great moment, though he himself never lived to realize its importance.

Here is the experiment. A wire is fastened to the lead coating of the jar and extends upwards so as to stand parallel to the wire which enters the jar. A cork, suspended on a silk thread, is placed between these wires, and the jar is electrified and placed on wax. Then, says Franklin, the cork "will play incessantly from one (wire) to the other 'till the bottle is no longer electrified; that is, it fetches fire from the top to the bottom (inside to outside) of the bottle 'till the equilibrium is restored."[1]

Never before has the electrical fire shown itself in the circuit other than as a spark, a shock, an explosion, instantaneous "with a violence and quickness inexpressible." Now Franklin is effecting this restoration of equilibrium slowly. He is breaking up the explosion, so to speak, into a great many little successive explosions. A very small amount of the fire passing from wire to ball is enough to electrify the latter, so that the wire will repel it. It swings over to the opposite wire, to which it delivers its charge, and swings back again. And thus it may go on vibrating to and fro, until it has ferried over all the fire which disturbs the electrical equilibrium between the outside of the jar and the inside. With this experiment (commonly cited as illustrating "electrical convection") begins the evolution of the electric current; the forging of the link between the Leyden jar and the voltaic cell.

The state of political affairs in Philadelphia when this second letter was written (September, 1747) had become critical. Trouble had arisen, several years before, between England and Spain, as to the right to gather salt at Tortugas and cut logwood at Campeachy. Volunteers had been raised in Pennsylvania for an invasion of Cuba, but the colony would not take any measures to put itself in a state of defense, even when war had broken out, not only with Spain, but with France also. The Quakers of Philadelphia, in pursuance of their peculiar tenets, would neither fight themselves, nor openly provide means for others to fight.[2] On the day following that on which Franklin's first letter to Collinson is dated, a French privateer, anchored off Cape May, and her crew plundered houses within twenty miles of Philadelphia. Still the Quakers refused to provide any means of defense. Shortly after-

---

[1] See Fig. II. of Franklin's illustration on page 561.
[2] McMaster: Benjamin Franklin as a man of Letters, N. Y., 1887.

wards another French privateer sailed up the Delaware, and within a fortnight Spanish privateers followed. The city was terror-stricken, but the assembly remained obdurate and would provide neither men nor money, arms nor forts. The result was that Franklin stopped making electrical experiments, wrote "Plain Truth," a pamphlet which depicted the horrors of war in a way that mightily stirred up the people; raised money; built a battery and organized a regiment. Fortunately no occasion rose for testing the efficiency of the safeguards, for the war was ended by the treaty of Aix-la-Chapelle in October, 1748.

Franklin now definitely determined to retire from business and devote himself to the study of electricity. He sold his newspaper, almanac and printing-house to David Hall. The sum thus realized, added to the fortune which he had amassed, and the revenue derived from places which he held under the crown and the colony, gave him abundant resources to enable him to live the life he most desired, and which he described as "leisure to read, study, make experiments and converse at large with such ingenious and worthy men as are pleased to honor me with their friendship or acquaintance." Meanwhile he had purchased the apparatus which Dr. Spence had imported, and had added some better instruments which the Proprietaries had sent over from London. Thus well equipped and relieved from all pressing cares, Franklin renewed his researches; and at this task I leave him, in order to note the progress which in the interim the European philosophers had made, and the reception which his plus and minus theory encountered in England.

Watson's hypothesis prevailed in Great Britain. On the continent, where international animosities had full play, there was confusion. "As the French say," writes a member of the Royal Society in the fall of 1746,[1] "there

[1] Phil. Trans., No. 481, p. 247.

are so many of what they term "bizarreries," or unaccountable phenomena in the course of electrical experiments, that a man can scarce assert anything in consequence of any experiment which is not contradicted by some unexpected occurrence in another;" and the same correspondent quotes the famous naturalist De Buffon as saying, that the whole subject of electricity is "not yet sufficiently ripe for the establishment of a course of laws, or indeed of any certain one, fixed and determinate in all its circumstances," which is significant in view of the persistence with which Abbé Nollet was advocating his favorite effluence doctrine, to which allusion has already been made.

Lemonnier had (probably in Watson's eyes) committed the indiscretion of announcing discoveries which Watson insisted he himself had *in petto*. The natural philosopher while sometimes, like other humanity, apt to indulge in the wish, "Pereant male qui ante nos nostra dixissent,"[1] has a better method of self assertion, at hand which has the merit of being useful. It consists simply in making additional experiments, which, even if they go to support the discovery of one's rival, completely eclipse, by their magnitude or striking character, those on which the latter has rested his conclusions. There is much sagacity in this, because human nature is very apt to link the great results to the great object lessons, and not to the little ones, especially after time has befogged the chronology. Thus did Watson, with respect to Lemonnier; in dealing with Franklin, as we shall see later, he adopted a different course, equally favorable to himself, and equally tinctured with worldly wisdom.

[1] Less sententiously, but perhaps as well said in Chevalier D'Aceilly's version :

"Dis-je quelque chose assez belle
L'antiquité tout en cervelle
Prétend l'avoir dite avant moi,
Cést une plaisante donzelle!
Que ne venait-elle après moi?
J'aurais dit la chose avant elle!"

Lemonnier had apparently caused electricity to traverse the pond in the Tuileries gardens. This Watson determined to outdo; and, not without some misgivings, prepared to make the "commotion," as he calls it, felt across the River Thames. With the aid of several members of the Royal Society he laid a wire along Westminster Bridge—a distance of some twelve hundred feet—and carried its ends to the water edge. On the Westminster side of the river one of the company held the wife in his left hand and touched the water with an iron rod held in his right. On the Surrey side, a second person held the extremity of the wire in his right hand and a charged Leyden jar in his left —the ball of the jar being touched by a third observer, who also grasped an iron rod dipping into the river. All three individuals felt a smart shock the instant the circuit was closed, and alcohol on one bank of the stream was fired by electricity discharged on the other.

This experiment, which was repeated with various changes in detail, was made in July, 1747. Martin Folkes, then president of the Royal Society, the Earl of Stanhope, and other distinguished persons, took part in it; and this alone would have attracted public attention even if the results had not been of such great philosophical interest. Watson, however, cared nothing for the sensational or popular side of the achievement. The observation which seemed to him of most importance was the great advantage which wire, as a conductor, possesses over chain —for "the junctures of the chain not being sufficiently close . . . caused the electricity in its passage to snap and flash at the junctures where there was the least separation, and these lesser snappings in the whole length of the chain lessened the great one at the gun barrel," which formed a terminus of the line. This suggested to him the possibility of sending the discharge over circuits of wire and water even greater than 2400 feet in length; so he changed the scene of his operations to Stoke-Newington, where the windings of the New River gave him (although the two extreme points were distant, in a straight line, but 2800 feet) a water course nearly 8000 feet in length. Here a wire, from the outer coating of a Leyden jar, disposed in the window of a house overlooking the river, was led over the meadows to the distant point, where, as before, an observer held its end in one hand, and established communication with the water with the other. A second wire from the window went directly to the river, so that it was necessary merely to bring the house end of this wire to the ball of the jar to discharge the latter. The experiment was successful—but a new question arose from it, because it had been noticed that the "commotion" traveled over the circuit even when the distant end of the wire did not communicate with the water but with the land, touching the earth at a distance of even twenty feet from the stream. Was the electrical circuit formed throughout the windings of the river, or by way of the much shorter path through the meadows? Tests showed that the meadow-earth would conduct, and this was supposed to be due to its damp condition. At all events, thought Watson, the matter must be tested. So observers, at the ends of a wire about 500 feet long, were insulated on pitch cakes and told to touch the ground with their iron rods. The shocks from a jar in the circuit were felt smartly by both. That, and similar trials, settled the matter of the feasibility of making the earth a part of the circuit, and made further experiments on long water-courses needless.

Watson had noticed that when the wire running across Westminster Bridge touched wet stones the shock transmitted seemed to lose strength, and that the same result happened when it lay on wet grass. He surmised at once that a leakage of the charge thus took place from the wire. He now provided a circuit nearly four miles in length, being two miles of wire supported on dry sticks and two miles of earth. The observers at the distant stations fired muskets to notify the man at the jar when they wanted the discharge to take place. The shock was so severe that some of them demurred to receiving it through their bodies, although they found amusement in the antics of the astonished countrymen whom they persuaded to join hands with them.

Successful transmission over a four-mile circuit—"'a distance without trial too great to be credited"—left Watson wondering how far the commotion would actually manifest itself, and what experiments he should try in order to find out. If he could determine the velocity of electricity, then perhaps he could form some idea of the length of circuit which would serve to test the matter. He attacked that problem very much as Lemonnier had done, by endeavoring to make a comparison between the speed of the commotion and the velocity of sound; but the effort was as unavailing as that of his French rival, and his conclusion the same; that the transmission of electricity "over any of the distances yet experienced is nearly instantaneous."

None the less, however, had Watson invented and used the circuit of wire and earth which, in later years, proved of such great value in long telegraph lines. But no intelligence was sent electrically over Watson's wire. The shock of the jar made the observers jump—and that was all. No one thought of transmitting shocks at varying intervals so as to signal intelligence by them. There was not the slightest notion of telegraphic communication present in Watson's mind. He was merely seeking to discover how far the "commotion" would travel, and in that way to obtain some knowledge of its strength and speed.

Next to having one's discoveries prematurely made by another, nothing is more disconcerting than to have somebody else bring home the conviction that the fundamental hypothesis upon which one has based a whole series of creditable deductions and experiments is probably wrong. However excellent the last may be in themselves, they are left in the air, so to speak, and something must be done without delay to replace the shattered underpinning, a task often requiring much ingenuity and some subtlety. Watson had already suffered the first annoyance at the hands of Lemonnier. When Collinson gave him Franklin's letters, he found that the second was also to be encountered. He could not dispute Franklin's conclusions, because he was himself convinced that they explained matters very much more reasonably than did his own. He felt instinctively that if he had only thought of them he would have promulgated them without hesitation. Unfortunately he had not done so. In brief, he was willing to admit the validity of Franklin's theory, but unwilling to concede the invalidity of his own.

The communication which Watson sent to the Royal Society in January, 1748, would have been more in harmony with the reputation of its brilliant and ingenious author had he shown in it greater candor. As it was, his chosen course precipitated a controversy which has retained vitality to the present time, and which has engendered dissensions exhibiting British insularity in some of its least agreeable phases. Without seeking to revive it here, it will suffice to say that Watson found, in his own mind, arguments which justified him in affirming that his theory, as a whole and radically, had always been the same as that propounded by Franklin, although a suspicion of salving his conscience is unavoidable when it is found that afterwards he really reverses his hypothesis in detail to make it accord. His partisans saw in the first proceeding reason for ascribing to him, rather than to Franklin, the full credit for originating the plus and minus doctrine; and in the second, only proof of ingenuous willingness on the part of the most eminent philosopher in the kingdom to defer to any one, however humble, rather than permit conclusions presented by him to retain the semblance of inaccuracy. But even an advocacy which included that of the all-knowing Whewell, and left its mark in the abridgments of the Philosophical Transactions, cannot overcome the plain meaning of Watson's own words, written before he had ever heard of Franklin, which have remained in the records of the Royal Society. The unprejudiced student of to-day will perhaps find in the idea of electrical equilibrium in all bodies a sufficient distinction between the Franklinian and Watsonian theories, even if, in view of other differences, he does not finally regard the two hypotheses as diametrical opposites.

New theories now began to crop up on every hand. Benjamin Wilson supposed an electric matter, composed of Newtonian ether, light and other material particles "that are of a sulphurous nature," existing more or less in all bodies, and moving with such exceeding velocity that, when that motion is checked by the near approach of another body, a sudden rarefaction of the air causes an explosion attended with the dissipation of the electric matter in flame. John Elicott asserted that electric phenomena are due to effluvia which are attracted by all other bodies, but the particles of which are mutually repellent. Boulanger conceived an electric fluid, consisting of the finer parts of the atmosphere, which crowded upon the surfaces of electric bodies when the grosser parts had been driven away by the friction of the rubber. Nollet further amplified his doctrine of the affluence of electric matter driving all light bodies before it by impulse, and its effluence carrying them back again, and supposed in every body to which electricity is communicated the existence of two sets of pores, one for the emission of the effluvia, and the other for the reception of them; for the spirit of Descartes was still lingering in France. Du Tour improved upon Nollet's theory by assuming a difference between the affluences and effluences, and considered that the particles are thrown into "vibrations of different qualities."

It would be easy, but useless, to add to this list. Priestley well describes the condition of affairs in saying that many hypotheses were no more than the beings of a day, and were no sooner started than their authors found themselves compelled, upon the appearance of a new fact, to remodel or reject them. They were, as a rule, the offspring of limited knowledge, promulgated because they happened to fit specific phenomena with which their proposers were acquainted, and not reached by any rigorous system of induction from accumulated and well-chosen facts.

Through all of them, however, one clearly defined idea now begins to show itself—that of an electrical fluid; first regarded as identical with fire, afterwards distinguished therefrom. And because Franklin dealt with this fluid in the simplest possible way, considering merely the quantity of it, recognizing no varieties in it, and, in brief, treating it from a purely material, almost mechanical, standpoint, his conception replaced all others and survived them.

To return, however, to Watson, who had resumed experimenting upon the Leyden jar, and who was now endeavoring to increase its strength, which, as he says, he succeeds in doing to an astonishing degree by using three vials coated with sheet lead and containing each some fifty pounds of shot. The wires which entered these were connected by an iron rod which in turn communicated with the gun-barrel prime conductor of an electric machine. The coatings were also connected by small wires, all of which were united to a tail wire. Watson amused himself—in fact, from this time on nearly all electrical experiments assume rather an entertaining character—by concealing the charged jars in his room and running the tail wire from them through the carpet, so that it would be invisible to any one standing on it, and then completing the circuit by touching the gun barrel with his finger. In this way he astonished his visitors with unexpected shocks coming from no visible source. It will be observed that his three jars were still connected in multiple arc, or parallel, relation, a fact of especial significance in view of the steps taken by Franklin immediately after receiving Watson's pamphlet, in which this experiment is described.

Meanwhile Dr. Bevis, who had advised Watson to coat the outside of the jar with sheet lead instead of holding it in his hand, again tells him of another and capital improvement. Bevis had coated both sides of a thin pane of glass about a foot square with leaf silver, and had found that, after charging the glass in the usual way, a person touching both silver coatings received a shock as strong as from a half-pint vial of water. Watson had hitherto supposed that the strength of the discharge of the jar was due solely to the "great quantity of non-electric (conducting) matter" contained in it; but here only about six grains of silver had been used to cover the glass, so that the quantity was exceedingly small—and thus that hypothesis fell. But the Leyden jar, in the shape in which it is still commonly known, resulted. Watson coated a cylindrical jar of thin glass with leaf silver inside and out, and obtained an explosion equal in strength to that of his three lead-covered vials in parallel; and evolved a new theory, which ascribed the effect "not so much to the quantity of non-electrical matter contained in the glass, as to the number of points of non-electrical contact within the glass and the density of the matter constituting those points, provided this matter be in its own nature a ready conductor of electricity."

The more powerful discharges which still larger jars gave him and the ease with which they traversed non-insulated conductors, encouraged Watson to make another attempt to find out the velocity of electricity by bringing both ends of a long circuit wire to a single observer; but, although the circuit measured 12,276 feet in length, he was again obliged to record the fact that the passage of the "commotion" cannot be regarded as other than instantaneous.

Watson's account of these latest experiments was published in book form in the fall of 1748, and the diligent Collinson duly dispatched it to Franklin. The avidity with which it was read by that philosopher and his colleagues is plainly shown by the contents of the next letter which Collinson received. The Americans now saw that they were fully as far advanced as the British electricians, and that each party was as likely to make important discoveries as the other. Franklin had recognized this fact in his preceding letter, wherein he stated that the rapidity of the progress in England half discouraged him from writing further on the subject, lest his communications should contain nothing new or worth reading. The news in Watson's paper that Dr. Bevis had already devised the pane of glass coated with sheet metal as a substitute for the jar, is, therefore, something of a disappointment; and Franklin even excuses himself for mentioning it, although he thought to have communicated it as a novelty since "we tried the experiment differently, drew different consequences from it, and as far as we know, have carried it farther."

There seems to be no reason for this diffidence on Franklin's part. It does not appear certain that Bevis' invention antedated his own—on the contrary, the multiplicity of Franklin's experiments go to show that he may well have used the coated glass before Bevis. But the American colonist of those days had a respect for the mother country that was controlling, and which made it almost instinctive for him to assume that knowledge moved westward, and not in the reverse direction.

Franklin's letter of 1748, in point of historical interest, is of the highest importance. Kinnersley's discovery that the Leyden jar can be electrified as strongly by sparks delivered to the outside as to the inside, begins it; so that it opens with an assertion than which nothing could be more disconcerting to the European electricians who still persisted in the belief that the electrical fire entered the water within the jar, and became somehow entangled there. Franklin, following, shows how, if the inside of one insulated jar be connected to the outside of another, an explosion and shock follows, and both jars are discharged; how half the charge in an electrified jar will go to a non-electrified jar; how jars are oppositely electrified according as the charge is imparted to the inside or the outside, and how the suspended cork ball will continue vibrating between the hooks on the ends of the inserted wires of two oppositely-charged jars, "fetching the electric fluid from the one and delivering it to the other, till both vials are nearly discharged."

It will be remembered that when Winkler or Watson desired to combine the strengths of two or more Leyden jars, they arranged the latter in the parallel or multiple arc relation. Franklin, studying the charging process, now suspends "two or more vials on the prime conductor, one hanging on the tail of the other, and a wire from the last to the floor." "An equal number of turns of the wheel," he says, "shall charge them all equally, and every one as much as one alone would have been; what is driven out at the tail of the first serving to charge the second; what is driven out of the second charging the third, and so on. By this means a great number of bottles might be charged with the same labor and equally high with one alone, were it not that every bottle receives new fire and loses its old with some reluctance, or rather gives some small resistance to charging."

That is the first announcement of the arrangement of electrical sources in the series or tandem relation—an invention which, as will now be seen, Franklin immediately turned to practical account.

Meanwhile he made a little series of experiments which, for neatness and exquisite ingenuity, remind one of Dufay, which show incidentally how the coated pane came to the inventor under circumstances entirely different from those which led to its suggestion by Bevis, and which ended in the famous Franklinian battery.

The question which especially puzzled Franklin was where the charge went in the jar, and how there could be produced therein, at the same time, a plenum which "presses violently to expand, and the hungry vacuum (which) seems to attract as violently in order to be filled." He had theorized about it: so had Watson and everybody else since the jar had been discovered: just as the world had theorized about the amber before Gilbert; just as the world always finds it so much easier to explain Nature's workings by the vibration of its own brain molecules than to let the workings explain themselves. Where is the charge in the Leyden jar? In the man who holds it, said Von Kleist. In the water, said Musschenbroeck. In the inner conducting coating, said Watson—and so on. Franklin proceeded to pull the jar to pieces.

First.—He put it on glass, so that the charge could not run away during the dissection. Then he pulled out the cork and the inserted wire, and taking the bottle in one hand, put a finger of the other near the water within. A spark passed. Therefore the cork and wire had nothing to do with the matter.

Second.—He recharged the bottle, put it again on glass, drew out cork and wire and poured out the water into another jar, also standing on glass. Now, if the charge was in the water, that second jar should give a shock. It did not. There was no electricity in the water at all. It must either have been lost by decanting, or must still remain in the first jar. If it was in the latter, the jar should give its shock when fresh water was poured into it. He poured some in "out of a tea-pot." The jar worked perfectly. So the water had nothing whatever to do with the matter, and the charge must be either in the glass or in the outer coating of the jar (either the hand or lead foil), for the simple reason that there were no other parts of the apparatus left.

Third.—He laid a pane of glass flat on his hand, and put a lead plate on it: the glass, like the wall of the jar, now stood between two conducting layers—hand and lead. He electrified it, and got a shock on touching the lead plate. The form of the jar was therefore immaterial.

Fourth.—He placed the glass between two plates of lead less in area than the pane, and electrified the glass between them by electrifying the uppermost lead. Then he took the glass from between the lead plates and found that, on touching it here and there with the finger, he obtained "very small pricking sparks," but a great number of them might be taken from different places. There was no sign of electricity in the lead. The moment he put the glass back between the plates and connected the latter through his body, a violent shock ensued.

And so he concludes that "the whole force of the bottle, and power of giving a shock, is in the glass itself: the non-electrics in contact with the two surfaces serving only to give and receive to and from the several parts of the glass: that is, to give on one side and take away from the other," and he compares the metal coatings to the "armature of a lodestone to unite the force of the several parts."

The road was now clear to the construction of the battery. It was made of eleven large plates of sash glass armed with thin leaden plates, with the giving side of one pane connected to the receiving side of the other, but provided with a contrivance "to bring the giving sides after charging into contact with one long wire and the receivers with another, which two long wires would give the force of all the plates of glass at once through the body of any animal forming the circle between them." As Franklin supposed that the greatest effects would be gained with the plates in parallel, he placed them in series for charging, and so encountered a resistance which he says "repels the fire back again on the globe;" and thus, in the beginning, the battery did not prove as efficient as he expected. Afterwards, however, he wrote "there are no bounds (but what expense and labor give) to the force man may raise and use in the electrical way: for bottle may be added to bottle, and all united and discharged together as one, the force and effect proportioned to their number and size. The greatest known effects of common lightning may, I think, without much difficulty be exceeded in this way."

Kinnersley, the ingenious, now appears with a variety of amazing toys. He has made a magic picture of King George with a golden crown on his head, and arranged Leyden jar fashion, so that he who touched the gilded frame and at the same time irreverently sought to grasp the crown received a violent shock. "God preserve him," says the loyal Franklin, in mentioning the King's name; but a few years later, men were indicted in Philadelphia for sedition for saying just the opposite. There is also the electrical jack—a horizontal, wooden, pivoted disk, having insulated brass thimbles around its edge which successively touch the wire of a charged jar and are repelled, thus turning the disk; "and if a large fowl," adds Franklin, "were spitted on the upright shaft, it would be carried round before the fire with a motion fit for roasting."

A much more elaborate electric motor was made from a circular sheet of glass, coated on both sides and pivoted to turn horizontally. The coatings alternately communicated with bullets fixed at equal distances on the circumference of the glass. Fixed near the disk were glass supports carrying brass thimbles, and near these last the bullets passed as they were carried around by the disk. The wheel being charged, the bullets were alternately repelled and attracted by the thimbles. Franklin says that it ran for half an hour at a time at a speed of twenty turns a minute, and Kinnersley applied it to ringing chimes and actuating orreries. This was the first application of electricity to performing useful mechanical work. Father Gordon's little reaction wheel had merely spun around and driven nothing.

The summer of 1749 was now at hand, and the Philadelphia experimenters determined to suspend work until after the hot weather. Franklin, who does not conceal his regret that "we have been hitherto able to produce nothing in this way of use to mankind," ends his letter to Collin-

FRANKLIN'S EXPERIMENTS.

son with his oft-quoted forecast of an electrical pleasure party when "turkey is to be killed for our dinner by the electrical shock, and roasted by the electrical jack before a

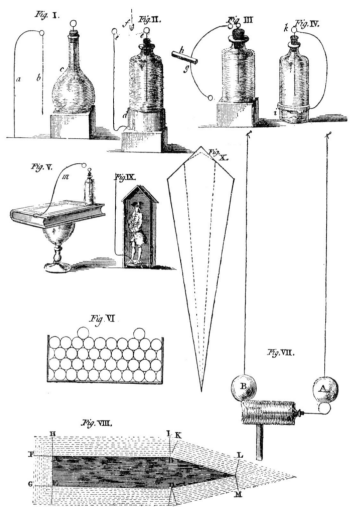

[1] FRANKLIN'S ILLUSTRATIONS OF HIS EXPERIMENTS.

[1] Reproduced in reduced fac simile from the folding plate in Franklin's New Experiments and Observations on Electricity made at Philadelphia

fire kindled by the electrical bottle; when the healths of all the famous electricians in England, Holland, France and Germany are to be drank in electrified bumpers, under the discharge of guns from the electrical battery."

He had then no prescience of the great discovery which he was so soon to make. It may be said that the fulness of time was at hand, the environing conditions were all favorable, and that if the identity of the lightning and the electrical spark had not been shown by Franklin, others,

in America—Part I., 2nd ed. London, 1754. This picture became the frontispiece in the later editions of the work. Figs. I, II, III, IV, and V, are described in Franklin's letter to Collinson, dated July 28, 1747; Fig. VI in letter IV to Collinson, and Figs. VII, VIII, IX and X, in the "Opinions and Conjectures" sent Collinson in 1750. Fig. I represents a Leyden bottle (c) which whenever touched by the finger attracts the thread (b) suspended from the wire (a). Fig. II shows a suspended cork (c) vibrating between the wires (e) (e), one of which enters the bottle and the other is connected to a ring of lead upon which the bottle stands. In Fig. III, the bottle rests on wax and is discharged by electrically connecting the interior and exterior by means of the wire (h) held in a sealing wax handle (g). Fig. IV represents a bottle surrounded by a ring of lead (i) connected by a conductor with the knob (k) on the inserted wire—such a bottle says Franklin, "cannot be electrified; the equilibrium is never destroyed." In Fig. V, the jar rests on a book having a gilded design on its cover; a wire (m) touches the gilding and may be brought into contact with the knob of the bottle. "Instantly," says Franklin, "there is a strong spark and stroke and the whole line of gold which completes the communication between the top and bottom of the bottle will appear a vivid flame, like the sharpest lightning."

Fig. VI is intended to show that particles at the surface of water are less strongly held by cohesion than others in the body of the fluid, and hence when the water is electrified are more easily repelled and thrown off. Fig. VII illustrates Franklin's description of the partition of a charge or "electrical atmosphere" from a Leyden jar to two suspended "apples or two balls of wood" and between the objects themselves. Fig. VIII is in illustration of Franklin's supposition that "electrified bodies discharge their atmospheres upon electrified bodies, more easily and at a greater distance from their angles and points than from their smooth sides." Fig. IX is the first representation of the lightning rod. Fig. X represents Franklin's "electrical fish"—a piece of Dutch metal, cut in the shape shown, which flies to the prime conductor of the electric machine and keeps "a continued shaking of its tail like a fish so that it seems animated."

ANCIENT SUPERSTITIONS ABOUT LIGHTNING.

by sheer force of circumstances, would have proved the fact at about the same period. When an invention is claimed to have leaped full-armed from the mind, when it has no discernible ancestry or evolution, this argument may find some application. But can any one read the preliminary experiments which have now been detailed, or note Franklin's reflections on the drawing power of points or the tremendous force of the battery discharge, without recognizing that, guided as by some controlling power, he was unerringly moving toward the goal, even if he knew it not himself?

Jets of blinding flame leaping across black and angry skies, deafening peals of thunder reverberating from the mountain-sides and echoing amid the clouds, the swift obliteration of life, buildings bursting into flame, great rocks and trees shattered—all this to the old world meant the warfare of gods upon gods, or the fearful retribution visited by offended Deity upon rebellious man. The thunderstorm became the symbol of divine wrath. Its tremendous effects offered the only realization of the majesty of the divine presence. The law is given amid the thunders and lightnings of Sinai. The voice of God is the thunder and he "directeth his lightnings to the ends of the earth."

As the ages passed the superstitions clustered thicker and thicker about the thunderbolt, and the dread of it deepened. "From lightning and tempest, from plague, pestilence and famine"—so the prayer for delivery comes down to us, with the lightning first on the lips. It was for naught that the philosophers sought to explain it by natural causes. Zeus launched his fiery missiles long after Aristotle described them as moist exhalations bursting out of moist clouds. What protection could there be against so fearful a visitation save invocation of the divine mercy—for what shield could the puny arm of man interpose against the might of the Almighty?

We shall go far astray if the great force of such a belief be not kept steadily in view throughout any effort we may make to discover possible knowledge of the ancients[1] regarding the nature of lightning or the provision of safeguards against it. It is useless to search the annals of an intensely religious or superstitious people to discover either physical explanations or material defenses. Equally unavailing is it to assume that a wrong interpretation of "the cause of thunder" could result in an adequate means of protection. The only rational basis for entertaining the notion that some knowledge may have existed in remote times must be found in the supposition that chance circumstances under which the lightning-stroke was apparently warded off were recognized and reproduced empirically. If, for example, it were perceived that the lofty trees of a forest were struck oftener than those of less growth, it would not be difficult to conclude that the erection of high towers, spires, minarets, or obelisks in the vicinity of low buildings might result in providing scape-goats or sacrifices upon which the celestial fire might wreak its vengeance, leaving the more important human habitations unharmed.[2]

Nothing is so difficult as prophecy before, nothing so easy as prophecy after the event; and the history of the lightning-rod has many an instance of the latter. The Temple of Jerusalem, says one archæologist,[3] was fully protected, because Josephus[4] records that on the roof there

[1] See Salverte: Philosophy of Magic. N. Y., 1847, vol. ii., p. 150, in which there is an extended discussion on the electrical phenomena employed by the magicians, with many references to ancient writings.

[2] Under such a theory as this it may perhaps be possible, for example, to account for the two thousand ancient pagodas, which are now falling into ruin in China, and which, although apparently useless, act as the Chinese geomancers claim, "to draw down every felicitous omen from above, so that fire, water, wood, earth and metal will be at the service of the people, the soil productive, trade prosperous, and everybody submissive and happy." Williams: The Middle Kingdom, ii, 747.

[3] Michaelis: Mag. Scient. de Göttingen, 3d yr., 5th No., 1783.

[4] Josephus: Bel. Jud. adv. Rom., Lib. v, c. xiv.

were many points, similar to those which appear on the Roman temples of Juno, and that pipes ran from the roof to caverns in the hill on which the building was situated. The Jewish historian assigns for the points the somewhat prosaic function of perches for the birds, and it requires no especial effort to conceive that the pipes served to lead off rainwater. But, says the acute inventor of the new hypothesis, the temple was never struck by lightning during a thousand years; it cannot be conceded that those points were put there for the benefit of the birds; the ignorance of Josephus in this respect is merely "proof of the facility with which the knowledge of important facts is forgotten;" and indeed, it is inconceivable "that the advantage to be derived from them (the points) had not been calculated upon."

Such prophets always take unnecessary pains. It would have been far simpler to have said that King Solomon, out of his exceeding wisdom, knew all about lightning-rods, just as earlier writers asserted his familiarity with the mariner's compass: although any supposition in the premises has the fatal defect of ignoring the sacrilege which the profoundly-devout Jew would surely have seen in such an attempt to make the roof of the temple into a sort of sieve to keep out the troublesome manifestations of the Deity who dwelt in its sanctuary.

The folk-lore of almost every nation has its legends recounting the drawing of fire from heaven. The skill of Prometheus in bringing down the lightning (a fable which sets Rabelais wondering what has become of the art), the death of Zoroaster by lightning in response to his own prayer, the descent of the vestal fire from the clouds, have furnished many a poet with a fertile theme. Occasionally the old writers are curiously suggestive: Lucan,[1] for instance, when he says that "Aruns collected the fires of lightning dispersed in the air and in the midst of noise buried them in the earth;" or Ctesias, in his description

[1] Pharsalia, i., 606.

of the Indian iron which dissipates clouds, hail and whirlwinds. But they are all vague and shadowy. If one attempts to follow the roads down which such stories have come, he will find that they all lead, not to Rome, but to her great rival, Etruria.

As I have already pointed out, in tracing the history of the compass, the Etruscans were indefatigable students of meteorology. Their augurs were weather-wise, and exceedingly skillful in recognizing impending changes, and in predicting them. They distinguished three kinds of lightning-flashes, according to the gravity of their effects; and eleven different species of lightning itself. Certain lightnings, they held, came out of the earth and rose perpendicularly—others shot from the sky and struck obliquely. Under the guise of worship of Jupiter Elicius, they claimed actually to bring down the lightning, and taught the secret of it to King Numa, whose successor, Tullus Hostilius, seeking to repeat the ceremony from the instructions left by Numa, made some error and paid the penalty with his life.[1] Tarchon, the founder of the ancient Etruscan theurgism, is said to have protected his dwelling by surrounding it with white bryony—a belief in the efficacy of which plant, after the lapse of ages, still prevails in modern India.[2]

Gradually there grew up a sort of pseudo-fulgural science. Constantine the Great, several years after his conversion to Christianity, made a law authorizing the Romans to consult the aruspices when an edifice had been struck by lightning. Later still (A. D. 408), when Rome was besieged by Alaric the Goth, certain Etruscan magicians offered to extract the lightning from the clouds and direct it against the camp of the Barbarians. Innocent, the bishop, was willing that the experiment should be tried; but the Senate—here literally "more pious than the pope"

---

[1] Pliny, ii., 55.
[2] Columella: lib. x., 346, 7. Salverte. Phil'y of Magic. N. Y., 1847, ii., 152. "Fulmineo periit imitator fulminis ictu." Ovid: Met., xiv., 617, 618.

—refused to sanction an act which appeared to it almost equivalent to the public restoration of Paganism.[1] Down through the Middle Ages the control of thunder and lightning was a part of the recognized equipment of the sorcerer or witch—so that Prospero's

> To the dread rattling thunder
> Have I given fire, and rifted Jove's stout oak
> With his own bolt,

bespeaks no more than ordinary thaumaturgic skill.

Literature, ancient and modern, abounds in allusions to atmospheric electrical phenomena long before their true nature was known. Pliny describes St. Elmo's Fire as well as Shakespeare does.[2] Seneca,[3] and Cæsar,[4] and Livy[5] all record spears with flames at their points in the Roman camp, and Fynes Morison sees the same fires on the staves of Montjoy's horsemen at the siege of Kinsale, in 1601;[6] the ancient Romans and the modern Scot being as ignorant, one as the other, that electricity had anything to do with the strange appearance. It is hardly credible that the force of inverted prophecy could carry any one to the extreme of finding in this mention by Cæsar and Seneca of the fiery spears pre-knowledge of the Franklinian discovery; but such is the fact, and that the argument was sufficient to convince so profound a philosopher as Arago,[7] will always furnish an excuse for others whose sense of similitude proves capable of overbalancing their judgment.

As the world divested itself of the influence of supersti-

---

[1] Gibbon: Decline and Fall, ii., 122.

[2] "I boarded the King's ship, now on the beak,
Now in the waist, the deck, in every cabin
I flamed amazement: sometimes I'd divide
And burn in many places; on the top mast,
The yards and bowsprit, would I flame distinctly;
Then meet and join. The Tempest. Act i., Sc. 2.

[3] Quest. Nat., i.
[4] De Bello Af., 6.
[5] Hist., c. ii.
[6] Phil. Trans., Vol. 48, 754.
[7] Arago: Eloge Hist. de Alex. Volta, Acad. des Sci., 26 July, 1831.

tion, the material theory of the lightning which Aristotle had propounded gradually replaced the imaginative one. The flame came to be regarded as fire—not ordinary fire, but what Jerome Cardan[1] calls the "fire of fires." Fire, he says, which is hotter than any other—differing from any other because the mere touch of it kills—the "fire of thunder." It will melt the very money in your purse, and yet so subtly as not to harm the purse itself. It enters the metal and tears it asunder. Then he has this curious passage:

"And this kind of fire must necessarily have great velocity in solid matter. Indeed, why does the thunder never touch columns or sink ships? It seldom touches them, although once I saw in Florence, at the great church, a column broken and shattered by the thunder; but it does not strike them often, nor throw them down, because the blow glances because of the rotundity. Similarly, it seldom strikes the bottoms of ships, because it cannot penetrate more than five cubits below the surface of the earth; and the bottom of the ship is low and the mast is high, and this last is often struck. A certain remedy against thunder is to hide in deep caverns, and this is more sure than to crown oneself with sealskin or the skin of an eagle, or to carry a hyacinth stone; for it is said that these things are not touched by thunder. But I have known a laurel to be injured by thunder in Rome."

Observe that in Cardan's time the idea of possible protection against lightning had become thinkable—thanks, perhaps, to the Reformation—and the power is supposed to lie in the hyacinth stone, "which protects men from the thunder;" and this "is no small power, seeing the many noble personages who have thus suddenly perished —Zoroaster, King of the Bactrians; Capaneus, in the Theban War; Ajax, after the destruction of Troy; Anastasius, the Emperor, in the 27th year of his reign; Carus, also, and other emperors. Let us consider how this can

[1] De Subtilitate, Lib. ii.

happen. Either the hyacinth prevents the thunder from coming, or it directs the judgment of whoever carries it, or it simply prevents him who has it from being injured, even if struck. These are the only ways. To be struck by thunder and not hurt is incredible and, besides, the authors have not said this, but that the thunder does not touch the possessor. To hinder the thunder coming is a still greater miracle;" and finally the wise Cardan arrives at the conclusion that the stone can act only by making the heart strong and wise and joyful, so that the owner thereof keeps out of peril. Such was lightning protection and the philosophical notion of the nature of lightning in 1556.

A few years, and it is Shakespeare's time. Note the question which he gives to the crazed Lear—

> "First let me talk with this philosopher,
> What is the cause of thunder?"[1]

and answers through Brutus, and Ariel, and Volumnia—

> "Exhalations whizzing in the air."[2]
> "The fire and cracks
> Of sulphurous roaring."[3]
> "To charge thy sulphur with a bolt
> Which should'st but rive an oak."[4]

The lightning was then believed to be a burning sulphurous vapor; sulphurous because it caused the air to smell like sulphur—a circumstance which Boyle noticed in the rubbed amber, and made no more mental connection between it and the lightning odor than Krüger did after him. A little later and we shall find that the idea of guarding against the lightning crosses Ben Jonson's erratic orbit—

> "Sir—shall I say to you for that hat . . it is *proof Against thunder* and enchantment "—[5]

[1] King Lear, Act iii., Sc. 4.
[2] Julius Cæsar, Act i., Sc. 2.
[3] Tempest, Act i., Sc. 2.
[4] Coriolanus, Act v., Sc. 3.
[5] Cynthia's Revels.

and afterwards the seeker for such allusions will find a harvest of them which space does not permit me to gather here.

By the end of the seventeenth century the explanation of lightning which prevailed up to Franklin's time was fully formulated. Dr. Wallis believed it to be due to the detonation of a mixture of nitrous and sulphurous vapors in the air—the conditions being similar to those occurring during the explosion of gunpowder, in which substantially the same elements are present. Later opinions differed as to the nature of the exploding gases and their mode of generation in the atmosphere; but the general consensus regarded the lightning and the thunder as the celestial artillery—the explosion and the report occurring in the same way as in earthly fire-arms.[1]

Early in the eighteenth century Hauksbee compared the flickering lights in his globe to the lightning flash, and Dr. Wall saw in the cracklings and sparks of rubbed amber a resemblance, in some degree, to thunder and lightning; and Gray, following the same thought, conceived the electric fire and lightning to be "of the same nature." But these were the merest conjectures. That of Wall is equally true of the discharge of a fire-arm. Gray's conception that the electric fire is "of the same nature" as the lightning is consonant with the common belief that fire is an element, and therefore the same everywhere; so that his assertion amounts to nothing

[1] To show how a precisely similar idea is often reached by entirely different paths, it may here be noted that Dr. Dionysius Lardner, writing in 1844 (see Manual of Electricity, ii., 165), after noting the many instances collected by Arago of the sulphurous odor following a lightning stroke, and the detection by Liebig of nitric acid in rain water, says "it would be a curious and interesting result of scientific investigation to demonstrate that the thunder of heaven elaborates in the clouds the chief ingredients of the counterfeit thunders which man has invented for the destruction of his fellows."

more than the same indication of resemblance made by his predecessors.

What these men really did was to make so happy a suggestion that other men were led to seek reasons in support of it. And as the truth was in it, it lived.

Early in 1746, John Freke,[1] of the Royal Society and surgeon to St. Bartholomew's Hospital, announced the first hypothesis asserting, and attempting physically to explain, the actual identity of lightning and electricity. Observing that there was no change produced in the instruments for electric generation due to their production of electricity, he maintained that they had no more to do with the development of the electrical matter than a pump has with the development of water. The electrical matter he regarded as fire composed of similar particles, tending to adhere at certain distances apart, and impregnating the air. If the particles, however, be forced together, reducing these intervals, then the fire becomes more or less violent according to the degree of compression.

"Now," he says, "as by human contrivance here is more of the fire crowded together than in its natural state, it is no wonder in this confinement, if that which, as water unconfined, should be gentle and beneficent, should, with all the power that belongs to it, break out at the first door which is opened for its passage from this tortured state. . . . Lightning, which is produced by a great quantity of the elementary fire driven together, is of the same nature with electricity (which is no other than factitious lightning), for it will kill without a wound and pass through everything, as this seems to do." The celestial fire, he says, amassed by any cause and enveloped, perhaps, and retained in this disturbed state, discharges itself finally with the explosion which we call thunder.

[1] Freke: An Essay to show the cause of Elec'y, etc. Lond., 1746. See also British Magazine, Oct., 1746, 300; London Magazine, Nov., 1746, 573. Essai sur la Cause de l'Elec. (Trans. of 2d Ed., with supplement), Paris, 1748.

It is needless to discuss Freke's supposition, because, almost immediately following him, came Winkler,[1] with a theory still more highly elaborated, in which the likenesses between electricity and lightning were pointed out with remarkable detail. He demands "whether the shock and spark of strengthened electricity is a kind of thunder and lightning," and proceeds to answer the query at length.

The lightning-stroke, he explains, is enormously more powerful than the electric spark; but that is no proof that they are of different natures. Even if a man had never seen fire and explosion except from a cannon, would he fail to recognize, in the discharge of a boy's pistol, the same effects, but in weaker degree? The lightning-stroke and the electric spark are alike in rapidity. Add together the detonations of many electric sparks, and the noise may be augmented. So in the lightning, which may consist of an immense number of such sparks, the combined explosion of all causes the sound of thunder. Lightning moves through the air in a zigzag or serpentine path; so does an electric spark in passing over moist insulators. The lightning-flash is sometimes multiple— like a collection of rockets; so is the discharge between iron cylinders. The lightning-flash will lay hold of solid bodies and melt them even when enclosed and without injuring the envelope (Cardan's coins melted in their purse); the spark will reach non-electrics, through insulators. It will pass through one's clothes, or electrify metal enclosed in paper. True we cannot burn houses and trees, or kill men and animals, by the spark; but not all kinds of lightning do this, hence the spark may resemble some particular variety.

Winkler's conclusion is that the atmosphere contains matter in great quantities, derived from exhalation and evaporation, going on at the earth's surface. It abounds

[1] Winkler: Die Stärke der Electrischen Kraft des Wassers in Gläsernen Gefässen, etc. Leipsic, 1746, c. x. (Preface dated Sept. 6, 1746.)

in sulphurous, mercurial and nitrous vapors, the particles of which, rising and falling, are continually rubbing against one another. The rubbing of the sulphurous particles generates electric matter, which may lie quiet until some chance shock develops the conditions for discharge and explosion, and then the sulphurous and nitrous vapors are burned, and there is besides a conduction of electricity along the particles which are non-electric. This is the substance of a long and somewhat obscure dissertation which ends with the statement :

"It appears, therefore, that the electric sparks which through art may be excited are the same in material, nature and mode of production as the lightning flashes and strokes, and that the only difference exists in the relative strengths and weaknesses of their operation."

Such was the development of the idea in England and Germany. In France, nearly two years later (August 9, 1748), Abbé Nollet published a treatise on physics[1] in which, in due course, he deals with the nature of lightning. He describes the "matter of thunder" as a "mixture of exhalations capable of self-ignition on fermentation or by shock, and the pressure of clouds which the winds violently agitate and drive together. When a considerable portion of this mixture takes fire, it causes an explosion stronger or weaker according to the quantity or the nature of the ignited materials, or according to the obstacles which present themselves to its sudden expansion." He regards also the lightning-stroke as due to an ignited gas which always rends the cloud like an explosive bomb. With this theory, however, he is not satisfied, and merely gives it as the one which is generally accepted. Then he adds the following oft-quoted passage:

"If any one should take upon him to prove from a well-connected comparison of phenomena, that thunder is in the hands of nature what electricity is in ours, that the wonders which we now exhibit at our pleasure are little

[1] Nollet: Leçons de Physique. Paris, 1746, vol. iv., 315.

imitations of those great effects which frighten us, and that the whole depends upon the same mechanism; if it is to be demonstrated that a cloud, prepared by the action of the winds, by heat, by a mixture of exhalations, etc., is opposite to a terrestrial object; that this is the electrized body, and at a certain proximity from that which is not: I avow that this idea, if it was well supported, would give me a great deal of pleasure; and in support of it, how many specious reasons present themselves to a man who is well acquainted with electricity. The universality of the electric matter, the readiness of its action, its inflammability and its activity in giving fire to other bodies, the property of striking bodies externally and internally, even to their smallest parts, the remarkable example we have of this effect in the experiment of Leyden, the idea which we might truly adopt in supposing a greater degree of electric power, etc.; all these points of analogy, which I have been some time meditating, begin to make me believe that one might, by taking electricity for the model, form to oneself in relation to thunder and lightning more perfect and more probable ideas than such as have been offered hitherto.''

This was written in the year 1748. It adds nothing new toward the solution of the question of whether the lightning and electricity are the same or even similar. It is simply to the effect that the keen and skilful electrician who wrote it has concluded that the arguments before him are sufficient, in his opinion, to warrant some one in beginning experiments to determine whether the idea has any foundation in truth or not. It certainly does not aver that he himself has done anything in the premises beyond meditating, or has made a single experiment in pursuit of the object, or even knows what experiments to make or how to attack the matter. The sum and substance of it all is that the problem is not on its face absurd and is worth investigating.

Such, in brief, were the conditions which existed when

Franklin began his immortal work. Yet it has been contended, over and over again, that there was really nothing left for Franklin to do; that if Gray and Wall had not substantially discovered the identity of electricity and lightning, Nollet (for Freke and Winkler seem to have been generally overlooked) certainly did so. So easy is it thus to argue, when stimulated by pride of opinion and international rivalries.

Of the rise and progress of the idea in Europe Franklin probably had no knowledge. In 1737[1] he quotes with approval the theory of Dr. Lister, that "the material cause of thunder, lightning and earthquakes is one and the same, namely, 'the inflammable breath of the pyrites, which is a subtle sulphur and takes fire of itself.'" Lister regarded thunder and lightning as due to sulphur fired in the air, and earthquakes to the same substance ignited underground. "Why there may not be thunder and lightning underground in some vast repositories," comments Franklin, "I see no reason, especially if we reflect that the matter which composes the noisy vapor above us exists in much larger quantities underground." The conception of the electrical nature of lightning seems to have come to him at the very outset of his electrical studies, and then to have been formulated in writing; but he refrained from communicating it to Collinson until experiment brought him assurance. Then, in the early summer of 1749, he despatched the first of the two famous papers in which his discovery is made known to the world.

The theory developed in this essay is interesting, not because of its inherent truth—for Franklin himself abandoned it not long afterwards—but as showing the path over which his mind moved. It furnishes, moreover, a striking instance of the deductive, as distinguished from the inductive, method of reasoning. The ocean is assumed

[1] Pennsylvania Gazette.

to be a compound of water particles (non-electric) and salt particles (electric per se), which, rubbing together, produce electrical fire, which collects on the surface and is luminous at night. It is also assumed that because the surface is electrified the particles of water are repelled, and these, rising, carry with them the electrical fire and form clouds which retain their electrified state until an opportunity arises for them to communicate their fire to other bodies.

This is obviously pure supposition, although a remarkably bold and dexterous one, as we shall see by following it a little further.

The electrified cloud is swept over the land by the wind. It encounters a mountain which, being less electrified, attracts it. The electrical fire at once leaves the mass of vapor, with a sudden flash and report, while the particles of water instantly coalesce and fall in rain. If a succession of such clouds become dammed by a mountain ridge, then the first cloud, after yielding its own fire to the peaks, takes the fire of the second cloud, and thus the fire passes from cloud to cloud as far back as they may extend. Like effects occur when an electrified cloud, rising from the sea, meets a non-electrified cloud rising from the land—the particles of the first cloud open on losing their fire, the particles of the second close on receiving it, and then follows a concussion, a flash and downpour.

Such is the bridge which Franklin built. It may be asked wherein it differed from the equally ideal structures which Winkler and Freke had reared. In this; that it led somewhere. All three suppositions were far wide of the truth—all three frail and imperfect; but two of them were mere piers jutting out into an unknown gulf, while the third spanned it, at least long enough for some knowledge to be gained of the new land beyond.

Now Franklin moves forward beyond all the world, and the result is best given in his own quaint words:

"As electrified clouds pass over a country, high hills

and high trees, lofty towers, spires, masts of ships, chimneys, etc., as so many prominences and points, draw the electrical fire and the whole cloud discharges there.

"Dangerous therefore, is it to take shelter under a tree during a thunder-gust. It has been fatal to many, both man and beasts.

"It is safer to be in the open field for another reason. When the clothes are wet, if a flash in its way to the ground should strike your head, it may run in the water over the surface of your body; whereas if your clothes are dry, it would go through the body, because the blood and other humors containing so much water are more ready conductors. Hence a wet rat cannot be killed by the exploding electrical bottle, when a dry rat may."

The great generalization is here—yet encumbered with a tremendous "if." The course of orderly evolution from his very first experiment, which proved the capacity of points, when placed in the vicinity of electrified bodies, to draw off the electric fire noiselessly and quietly, had now led him to the belief that the same result would happen if the electrified body were a cloud and the point a tree or spire; *if* lightning and electricity were the same, *if* both were under control of the same laws.

But where was the physical proof? Where was the evidence that clouds are ever electrified, or that the fire in the sky is an electric flash, or that there is in fact any electricity in the atmosphere at all? He had merely supposed all this. No one knew better than he that the sea-born clouds, bursting with electrical fire, floated about only in his imagination. No one could better anticipate the derision which would be provoked by the unsupported assertion that the fierce blazes of the thunder-gust lay latent in the soft depths of the snowy couriers of the air, ready to obey the same control as the little sparks and crackles yielded by his globes and jars. There was the crux. *If* the lightning be electricity, then—but is it?

How was that to be found out?

From the bold conjectures of Wall or Gray or Hauksbee? From the fruitless hypotheses of Winkler or Freke—no better, if as good as his own? From Nollet's wish that somebody would attempt the test? Do not all the prophecies after the event which Franklin's detractors have solemnly made on the strength of these prior speculations seem destitute of substantial basis?

Whether Franklin at the outset fully realized the magnitude of the consequences depending upon the resolution of the question, may be doubted. A more emotional nature than his might have done so and been overwhelmed when success became apparent. As it was, he probably never looked upon himself as one set apart to dispel a terror of the ages, to destroy the power of a scourge which had lashed all humanity since the world was peopled: he was prosaically trying to utilize the new knowledge. Furthermore, if there is any one characteristic of the man which shines forth in all his writings, at least until great age weakened his faculties, it is that of thinking straight. He sees the problem before him clean and clear—he never loses sight of it—he never grows misty or confused—he makes for the goal steadily, persistently, and by what seems to him to be the most direct path. And, at the very threshold, he differs from European philosophers.

Look back at their experiments and their theories. Every one of them has his eyes fixed on his globes and his jars. Every one of them is making the electricity which he produces the model. Every one of them is hunting for resemblances to the lightning in exhausted receivers or on the edges of wet tea cups, or in his little circuits. And the most they got out of it all was that the little sparks and the little crackles and the little glow and the lightning seemed to be of the same "nature"—the meaningless answer of both the ignorant and the philosopher. Their question was—is electricity lightning in miniature?

Such Franklin did not do. He turned from the bottles and the wires and looked straight into the face of Nature.

Of a summer's evening he watched the soft radiance glimmer fitfully from one pink vapor wreath to another, or silently bathe the distant horizon again and again in golden glow. He saw the cold fires of the Aurora waving like fingers beckoning men to find them in the frozen North-land. He saw the great, heavy clouds sweeping in from the sea and forming their serried columns upon the mountain-flanks. He saw them crowd into the gorges, and break against the peaks, rolling back in scattered fragments to form new cohorts to hurl themselves once more upon the eternal rock. He saw the lightnings shivering and seething in their fleecy masses, or leaping out, hissing and snake-like, to rend the stone battlements and send the avalanches rattling down the precipices. He heard the crash of the warring forces of earth and sky, the fury and turmoil of the never-ending battle of the clouds and the mountains, roaring and resounding from steep to steep until its deep diapason died away amid a thousand echoes and left the earth shuddering.

The great poet of his race had already idealized the spirit of the air that did these things. Franklin's invocation lay not to the imprisoned imps in the bottles, but to "Ariel and all his quality."

All through the summer of 1749 he kept at work, resolutely holding himself aloof from political allurements. Kinnersley helped him. His procedure is methodic; his trials and conclusions are noted without a shadow of emotion or a sign that their author knew himself to be speedily nearing his goal. His question was not, What is electricity? but, What is lightning? His object, a physical mode of making nature herself answer; not a collection of analogies and resemblances from which a reply might, with more or less certainty, be inferred. These, however, it was necessary to gather in order to perceive what the crucial experiment ought to be. Therefore, he seeks out every feature in which the electrical effects produced by his machines agree with the lightning, and sets them all down categorically. Here is the list noted in his diary under date of November 7th, 1749:

"Electrical fluid agrees with lightning in these particulars: 1. Giving light. 2. Color of the light. 3. Crooked direction. 4. Swift motion. 5. Being conducted by metals. 6. Crack or noise in exploding. 7. Subsisting in water or ice. 8. Rending bodies it passes through. 9. Destroying animals. 10. Melting metals. 11. Firing inflammable substances. 12. Sulphurous smell."

In all these things the agreement of lightning and electricity is perceptible by the senses. Yet it does not necessarily follow that they are identical. That can only be resolved by determining whether they obey the same laws under the same conditions.

There is one fact which he has recognized to his complete satisfaction, and that is, that the so-called electrical fluid of his jars and globes is attracted and drawn off by points. Here there is a direct apparent control of the fluid. He has no evidence that lightning possesses what he calls a similar property of being attracted. But if it has, if it will come out of a cloud to go to a point, as the electric fluid seemingly comes out of his glass globe, that shows that the same fluid is in the cloud and in the glass. The necessary test for the identity of lightning and electricity is now plain. He ends his minute thus: "Since they agree in all the particulars wherein we can already compare them, is it not probable that they agree likewise in this? Let the experiment be made!"

In July 1750, Franklin sends to Collinson the most elaborate and longest of all his communications. It is one which he regarded as of especial importance, and for that reason asks Collinson to convey it to "our honorable Proprietary," to show to him that his "generous present of a compleat electrical apparatus" had been put to good use. In it, he describes the making of the proposed experiment, though only in miniature. But the results so completely confirm his anticipations, that he is willing to base upon it definite instructions how to make the actual trial itself, and to leave the performance to others, whose facilities for carrying it out might be better than his own. This self-abnegation shows itself everywhere throughout Franklin's scientific career. No one could have been more destitute of pride of opinion than he, no one more totally free from the desire of profit in any form to himself, no one more purely single-hearted in the devotion of his genius to the good of all men.

The little experiment, as usual, was made with homely apparatus. He hung up an old-fashioned pair of brass scales by a twisted cord attached to the middle of the beam so that the pans, as the cord untwisted, would move round in a horizontal circle. He suspended the pans from the beam by silk threads instead of the usual chains, so as to insulate them. On the floor, and in such position that the scale-pans would move over it in their path, he set up an old metal punch, on end. Then he electrified one scale-pan.

Now, as this pan came over the punch, it was attracted and moved down to the iron, and when near enough the charge would pass from pan to punch with a snap and crack. But if a sewing-needle were "stuck on the end of the punch, its point upwards, the scale, instead of drawing nigh to the punch and snapping, discharges its fire silently through the point and rises higher than the punch. Nay, even if the needle be placed upon the floor near the punch, its point upward, the end of the punch, though so much higher than the needle, will not attract the scale and receive its fire, for the needle will get it and convey it away before it comes nigh enough for the punch to act." Of course, the scale pan here represented the electrified cloud, and the punch the building or mountain which might be struck by the spark, did not the needle draw it harmlessly off.

This description is the preface to the two famous paragraphs which were destined to place Franklin first among living philosophers.

"I say," he declares, "if these things are so, may not the knowledge of this power of points be of use to mankind in preserving houses, churches, ships, &c., from the stroke of lightning, by directing us to fix on the highest parts of those edifices upright rods of iron made sharp as a needle, and gilt to prevent rusting, and from the foot of those rods a wire down the outside of the building into the ground, or down round one of the shrouds of a ship and down her side till it reaches the water? Would not these pointed rods probably draw the electrical fire silently out of a cloud before it came nigh enough to strike, and thereby secure us from that most sudden and terrible mischief?

"To determine the question, whether the clouds that contain the lightning are electrified or not, I would propose an experiment to be tried where it may be done conveniently. On the top of some high tower or steeple, place a kind of centry-box big enough to contain a man and an electrical stand. From the middle of the stand let an iron rod rise and pass bending out of the door, and then upright twenty or thirty feet, pointed very sharp at the end. If the electrical stand be kept clean and dry, a man standing on it when such clouds are passing low, might be electrified and afford sparks, the rod drawing fire to him from a cloud. If any danger to the man should be apprehended (though I think there would be none) let him stand on the floor of his box and now and then bring near to the rod the loop of a wire that has one end fastened to the leads, he holding it by a wax handle; so the sparks, if the rod is electrified, will strike from the rod to the wire, and not affect him."

When Collinson received that paper, he recognized at once that here was no ordinary discovery, and that however ingenious or interesting Franklin's ideas may hitherto have been concerning the nature of the electric fire, the behavior of jars and such matters, this announcement reduced every past item of electrical knowledge to comparative insignificance. For not only was this the first great utilization of everything that had been learned from the rubbed amber and its posterity, but the importance of it as a safeguard to life and property was inestimable. Hitherto the Royal Society had not been unfavorably disposed to Franklin, and even Watson, in appropriating his honors, did so in a considerate and even laudatory way. But when Collinson came with this story and wanted the Society to consider it, he met with prompt rebuff and even derision. The whole matter was regarded as too visionary for serious discussion by the Society, whatever individual members might think about it.[1]

The calm indifference with which Franklin accepted this turn of affairs found no reflection in the breast of Collinson, who, on the contrary, developed a most unquakerlike spirit of antagonism. He was now determined that not only should these last papers of the American philosopher be published, but that the earlier letters already received should go to the world, whether the Royal Society put their imprint on them or not. And to this he was urgently incited by Dr. Fothergill, who cordially undertook to assist him.

So he offered the letters to Cave—Cave, the lordly owner of the Gentleman's Magazine; Cave, the typical Grub Street publisher, who regarded £50 as an adequate bait for the highest literary genius—the Cave of Dr. Samuel Johnson, who looked upon his very abode at St. John's Gate with respectful awe;—and Cave refused them place in those sacred pages, although he was filling the latter with long diatribes from nobodies about the latest humbugs in "medical (!) electricity." But Cave had an eye to profit, and while unwilling to imperil the fortunes of his magazine by admitting such heterodox matter,

[1] Nevertheless a brief notice of Franklin's electrified cloud theory found place in the transactions very shortly afterwards, through a report on it by Dr. William Stukely, who had heard the first letter to Collinson publicly read at some gathering. Phil. Trans., 496, 601.

he saw no reason why he should not issue it as a separate publication—price two shillings and six pence; especially as no outlay on his part was required, and all the revenue was to come solely to him. Thus the collection came to be published in 1751.

Meanwhile Franklin was pursuing the even tenor of his way, and not only all the Philadelphians, but the people of far-distant Boston and New York and Charles Town were manifesting increased interest in his astonishing proceedings. Cadwallader Colden, in New York, had opened correspondence with him and had become practically his disciple; so had James Bowdoin, in Boston, afterwards Governor of the colony. If his house had hitherto been a rendezvous for all the sight-seers in Philadelphia, it was now more attractive than ever. He killed turkeys with the discharge from large Leyden jars, and once, by accident, in the same way, nearly killed himself. To Colden he writes that he has "melted brass pins and needles, inverted the poles of the magnetic needle, given a magnetism and polarity to needles that had none, and fired dry gunpowder by the electric spark." He dwells upon the powerful effects of the Leyden jar battery, and adds: "So we are got beyond the skill of Rabelais' devils of two years old, who . . . had only learned to thunder and lighten a little round the head of a cabbage." Then people got the notion, probably from news of some curious discoveries said to have been made in Italy, that electricity was the universal panacea; and Franklin found himself besieged by invalids. Governor Belcher, of New Jersey (aged 70, drinks small beer and half a bottle of Madeira daily, and is "tremulous"), begs Franklin to send him the electrical apparatus in order that he may treat himself, and bewails its breakage on the road.[1] Paralytics come to him in large numbers, and he gave them all the same remedy—the united shock of two six-gallon glass jars through the affected limb, three times a day; but he never

[1] N. Y. Col. Records, viii., 7.

saw any advantage after the fifth day, when the patients "became discouraged, went home, and in a short time relapsed." In fact, Franklin is not disposed to accord to his shocks even the first small improvement which appeared; which he thinks rather due to the "exercise in the patients' journey and coming daily to my house, or from the spirits given by the hope of success enabling them to exert more strength in their limbs."[1]

By this time he determines that something must be done to assuage popular curiosity in a more wholesale manner. Kinnersley, who had been assisting him in his experiments, needed remunerative employment. He was well familiar with all Franklin had accomplished. The plan developed is told in the following advertisement which soon appeared in the Pennsylvania Gazette:

"Notice is hereby given to the Curious that on Wednesday next, Mr. Kinnersley proposes to begin a course of Experiments on the Newly Discovered Electrical Fire, containing not only the most curious of those that have been made and published in Europe, but a considerable number of new ones lately made in this City, to be accompanied with methodical lectures on the nature and properties of that wonderful element."

There were two of these discourses which Franklin had written. Kinnersley himself fitted up the apparatus with characteristic ingenuity, and thus equipped, the first lecturer on science in the New World began his tour. From Philadelphia he went to Boston, where the venerable walls of Faneuil Hall resounded with the cracks and snaps of his jars and globes, long before they echoed the impassioned eloquence of the orators of the Revolution. His experiments, writes Governor Bowdoin to Franklin, "have been greatly pleasing to all sorts of people that have seen them." In New York and in Newport the exhibitions created a genuine sensation—the citizens especially mar-

[1] Letter to Pringle, Dec. 21, 1757.

veling at his showing of how houses and barns could be (perhaps) protected from the lightning.

In the intervals of his lectures, Kinnersley used his apparatus for further experimenting, and in the spring of 1752, he re-discovered the different electricities of sulphur and glass—the resinous and vitreous electricities of Dufay—concerning which neither he nor Franklin appears to have had any earlier knowledge. He communicated this at once to Franklin, who repeated the experiments, and at first concluded that the different attractions and repulsions observed proceeded rather from the greater or smaller quantities of the fire obtained from different bodies, than from the fire being of different kinds or having different directions; but subsequently he conceded that a glass globe charges positively and a sulphur one negatively. He did not probe into this, however, with his accustomed energy. Another and weightier matter was on his mind, and he had no relish for new research until the question which it raised could be settled. His letters had been published in Europe, but as yet no one had made the experiment. Could he not do it himself?

He had already canvassed the possibilities of doing so, but had given up the idea because there were no hills or other natural elevations about Philadelphia, and no edifices higher than ordinary dwelling-houses. He believed it necessary to place his pointed rod on some lofty peak or high tower; but in all Philadelphia there was not even a church-spire; indeed, he might have traversed the whole province of Pennsylvania without finding one. True, the vestry of Christ Church by slow degrees had made up its collective mind some time to build a steeple, but that notion had faded into the dim distance when the war broke out. Franklin was now seeking to revive the project. A lottery had been established to procure the needful funds, both for the structure and the bells, and he watched with impatience the incoming subscriptions and the taking of chances, in the hope that enough money would soon be raised to erect his long-desired pinnacle. But the receipts were small and their advent slow. Waiting was tedious— all other experiments seemed so tame beside this one. Kinnersley was drawing off the popular excitement; the attractions of political life began to look once again very fascinating.

The summer came—a bad season for electrical experimenting, as he was well aware. He would put it all aside until the old interest should revive with different conditions, when—the news came from across the sea that the experiment *had been tried!* Tried by the first philosophers in France under the auspices of the French King himself. Tried with magnificent and unquestionable success, and that all Europe was ringing with it.

He needed all his philosophy now.

How had they done it?

Cave, with characteristic prudence, had issued but a few copies of the pamphlet containing Franklin's letters; and now, as nearly a year had elapsed, and no notice of it had been taken, was doubtless applauding his own foresight. Some one, however, had sent a copy to De Buffon in Paris, and he perceived instantly that here was something both extraordinarily novel and extraordinarily strange. He persuaded D'Alibard to translate the work into French—a task very imperfectly done, but not so obscurely as to prevent the quick-witted Frenchmen from seeing the important nature of the discoveries, and the logical skill which had been exhibited in their announcement. D'Alibard's book sold tremendously—doubtless to the agony of Cave, who got no profit out of it. The probability of success of the Philadelphian experiments was the staple of conversation everywhere; from the meetings of the philosophers it spread to the gatherings of the *beau monde*— from the salons to the Court—to the King—and the result was his Majesty's command that Franklin's experiments should be repeated before him.

The Duke D'Ayen placed at the King's disposal his chateau at St. Germain. De Lor, master of experimental philosophy, was selected to make the exhibition, and Louis watched with the keenest interest the great sparks from the cascade (series) battery and the performances of the various ingenious contrivances which Franklin describes in his early letters to Collinson.

Meanwhile, De Lor, De Buffon and D'Alibard, having got together, found themselves agreeing that the experiment of all others was that of the pointed rod. They did not show that to the King, doubtful perhaps of its success; but De Lor and D'Alibard each separately undertook to test the matter.

D'Alibard, in a garden at Marly la Ville, about eighteen miles from Paris, had erected a sharply-pointed iron rod an inch in diameter and forty feet high. This rod was insulated at its base, which rested upon a table, arranged within a small cabin, to the posts of which last the rod was also secured by silk ropes. A thunder storm not being immediately at hand, D'Alibard employed an old dragoon, one Corffier, to watch the apparatus, and provided conveniently at hand a brass wire mounted in a glass bottle for a handle, with which to draw off the sparks from the rod, if it should become electrified, as he hoped would be the case. Some days elapsed, and when the thunder-gust did come, Corffier was on guard alone. Instead of waiting for D'Alibard's arrival, he concluded to try the experiment himself, and so, grasping the wire, he presented it to the rod. The sparks flew, with loud reports. Corffier dropped the wire in terror, and shouted to his neighbors to send at once for the village priest, for the fierce flame and the sulphurous odor were clearly infernal.

The ecclesiastic came in full run, with the villagers in throngs at his heels. The hail-storm was terrific, but, as all believed Corffier had been killed, no one minded it. Corffier, however, was found uninjured, and, as the good Prior of Marly had no fear of the machinations of the fiend, he began to experiment for himself by drawing sparks with the brass wire.

"I repeated the experiment at least six times in about four minutes in the presence of many persons," he writes to the absent D'Alibard, "and every time the experiment lasted the space of a *pater* and an *ave*." He managed to touch the rod himself and got a rather severe shock; but he wrote the letter to D'Alibard and sent it off by Corffier before he left the scene.

"Franklin's idea ceases to be a conjecture," says D'Alibard, in concluding his report to the French Academy— "here it has become a reality."

De Lor, in Paris, followed, on May 18th, with an iron rod 99 feet high, from which he drew off sparks freely during a thunderstorm.

Such was the intelligence which reached Franklin. It is not difficult to imagine the amazement with which he received it. True, these French philosophers had ostensibly made the experiment—but how?

With rods, one of which would not overtop buildings in Philadelphia, and the other, though twice as high, still, in his belief, far from being sufficiently lofty. That sparks had been drawn from rods which ended in the air close to the earth's surface, and not within hundreds of feet of the clouds was not conclusive. This was the experiment in one sense, and yet, in another, it was not. It showed that the rods had become electrified—but not necessarily that the lightning had electrified them or had passed over them.

Again the question pressed upon him—could he not make the test himself? This time a way flashed across his mind—one of the boldest conceptions ever imagined by man. Why not cause the fierce fires of the heavens to descend so that he may place them side by side with the puny sparks and flashes of his globes and jars—and so see the identity? Why not send up a kite into the very heart of the thunder-cloud, and bring the lightning down on its cord?

Two light strips of cedar placed crosswise, and a large thin silk handkerchief secured to them at its corners—such was the kite. To the top of the upright stick of the cross was fastened a sharp wire about a foot long. The twine was of the usual kind; but he provided a piece of silk ribbon and a key—the first to attach to the twine and to hold in his hand after he had raised the kite, as some possible protection against the lightning running through his body—the other to be secured at the junction of ribbon and twine, to serve as a convenient conductor from which to draw sparks—if they came.

He had not long to wait for a thunderstorm in that hot summer weather. As he saw it gathering he betook himself—accompanied only by his son, then a young man twenty-two years of age—to the open commons. He desired no other assistant—he had confided his intentions to no one else. The experiments of the Frenchmen had convinced them, perhaps, but not him. He proposed to be satisfied now.

It has been said that he kept his own counsel concerning this, because he feared ridicule should he fail. There is no basis for that. Why should he, who had borne already with perfect equanimity the derision of the Royal Society, fear more of it? Many a time before he had done things, many a time afterwards he did others, which made him the very butt of sneers and scoffs; but his serenity remained unbroken. Why should he fear ridicule now? Nor was there anything else that he feared—not even death. And with death he now believed he was to stand face to face.

All his past work had taught him this. He had seen the furious shock blot out life from animals, he had felt it in his own body rack him almost into insensibility. He had said, over and over again that if potent enough it would be instantly fatal. He was now going to lead into his hand the fearful fire of the thunderbolt.

He knew nothing of the laws of conduction. If the

lightning could descend that cord, how much of it would so come there was nothing to tell. Every presumption pointed to an out-pour of living flame which would infallibly kill. And yet, if his theory was right, the electrical fluid should be drawn from the cloud and flow down with harmless vigor.

No man ever confronted what he must have believed to be terrible danger with more superb heroism. No man ever led a forlorn hope, or faced a hail of bullets, with more unflinching bravery. No man ever so calmly, so philosophically, staked his life upon his faith.

The great clouds roll up from the horizon, and the gusts grow fitful and strong. As Franklin and his boy disentangle the kite from its cords and tail, and get it in position for ascent, the thunder mutters nearer, and the rain begins to patter upon the grass. A swishing blast comes over the meadows, the kite feels it and rises swiftly, swooping this way and that as the air-currents catch it. The rain now falls heavily, and the mist begins to close in. There is a friendly shed at hand, and Franklin, drenched, takes refuge under it. The kite, heavy with water, is sailing sluggishly, except when the gusts set it moving in spirals.

A huge low-lying black cloud traveling over him suddenly shoots forth forked flame, and a crash of thunder shakes the very earth. The pour is now in sheets; again the blaze, again the rattling explosion. The kite is moving upward, for Franklin is quickly unwinding the cord. It is soaring straight into the black mass, from which the flashes are now rapidly coming, and in which it soon becomes invisible.

Quietly Franklin is arranging the silk ribbon and the key. This done, he watches the cord close to him. There is no sign yet to guide him. Has he failed? Suddenly he sees the little loose fibres of the twine erect themselves. He has not failed, but the moment has come. Without a tremor he advances his knuckle to the key. And then a little crack, a little spark—the same little crack and the same little spark which he had taken a hundred times from his glass tube—and the great discovery is complete, his name immortal.

As the kite dashes through the masses of vapor hurrying over him, he touches the key, and again and again the conquered lightning returns, as it were, a caress—even submitting to be caged in the Leyden jar like the common electricity from his rubbed globe.

And, as the storm abates, the thunder dies away on the horizon, the clouds sweep off toward their ancient enemies, the mountains, and the kite moves lazily in the blue; while on the thankful, upturned face of the man gleams the glad sunshine which he had thought never to behold again.

"It is a dogma of faith that the demons can produce wind, storms, and rain of fire from heaven. The atmosphere is a battlefield between angels and devils . . . The aspiring steeples around which cluster the low dwellings of men are to be likened, when the bells in them are ringing, to the hen spreading its protecting wings over its chickens: for the tones of the consecrated metal repel the demons and arrest storms and lightning."

So wrote the Angelic Doctor centuries before the days of Franklin. Those whose minds were still filled with the superstitions of a bygone age, clung to their belief in the efficacy of the church bells, and denounced the lightning rod as a sacrilege; nay, even as an awful defiance of Heaven, if it were placed upon a house of worship. Abbé Nollet, forgetting the philosopher in the ecclesiastic, declared it to be "as impious to ward off God's lightnings as for a child to resist the chastening rod of the father." In vain it was urged that as the rain fell alike upon the just and unjust, so the thunderbolt shattered with equal impartiality the steeple of the Christian church or the minaret of the mosque of Mahomet. Good people, in their zeal for their convictions, retorted that the presence of the infernal contrivance *de*consecrated the lofty spire and invited its destruction; and this not only because of the affront offered the celestial powers, but for the purely physical reason that the lightning sought the conductor, and so became directly attracted to the place which otherwise it might harmlessly pass by. The failure of some of the first-erected rods to protect, through defective construction or imperfect earth connections, gave color to these arguments. Nevertheless, the spikes bristled on the pinnacles, and man learned "to sleep fearless of the thunder."

The epoch of the intellectual rise which I seek to chronicle, here reaches its end. The establishment of the identity of electricity and lightning marks its conclusion, and at the same time brings to culmination the long series of events whereby the single incomprehensible effect observed in the lodestone and the amber gradually grew into recognition as a world force, subject to universal law and pervading all nature. It had lived and persisted and grown mighty, steadily rising over all antagonisms, even as the points of Franklin's rods reared themselves toward the clouds, far above the jarring clangor of the bells.

"Vivos voco,
Mortuos plango,
Fulgura frango,"

sang the resounding throats in the steeples as of old, while the lightning, laughing at their vociferations, silently and safely followed the slender iron to the ground.

"The truth of science has ever had not merely the task of evolving herself from the dull and uniform mist of ignorance, but also that of repressing and dissolving the phantoms of the imagination, which ever rise up in new and tempting shapes, and which, not being of her, crowd before and around her and embarrass her in every way.'"[1] So said the Master of Electricity. Nor does he picture all her task; for there is mental inertia to be overcome and conditions to be created, whereby minds are rendered willing to conceive as possible, things contradicting experiences or habits of thought, long established and familiar.

How fraught with these difficulties, how impeded by these obstacles, has been the intellectual rise in electricity, and yet how persistently, how inevitably it has moved onward—of this, some imperfect idea may perhaps be gleaned from these pages. After all, they recount but one of the many struggles of the human mind clearly to perceive, and so perceiving to understand, something which it intuitively recognizes as written in the great book of Nature. Whether it be a woman of Syria, in a bygone age, curiously watching the chaff leap to her amber spindle, whether a degraded Indian of the Orinoco idly rubbing the dry stalks of the Negritia to see them attract lint; whether a Franklin striving to fathom the secret of the clouds, the perception is the same, the effort to understand the same; and the object of all is the deciphering of Nature's message told in the amber and the vine and the atmosphere. It is in intellectual force alone that the differences appear. It is before the steadily augmenting power of the intellect that Nature yields one by one the keys to her enigmas.

Before those still unopened we may wait and wonder; wondering as the savage Hurons wondered before the magnet which the Jesuits brought to them; wondering as the Greeks wondered before the mysteries of Samothrace; wondering as we wonder now, when beyond the little horizon of our knowledge we think we discern the great dim shadow of the universal all-pervading force.

Men wait for times, but times oftener wait for men. The intellectual advance is not marked by the almanac, but by change in mind. At its extremes stand the savage and the sage—not yesterday and to-day. So in the future,

[1] Faraday.

as in the past, as the intellect waxes greater and more potent, will it read ever new, ever greater teachings in the eternal handiwork.

Thus the lesson of this record, and of all kindred others in the broad fields of science, may well be taken to heart, for none is more reassuring. Man-made systems may fall. Apostles of degeneration may find, in the things which make up the environment of the hour, signs of impending decay. But he who turns to the history of intellectual endeavor in the study of Nature will learn that when mind thus faces the purity of the Infinite it does not and cannot degenerate. Rather will he see in the constant effort to reveal the truth, an influence always making for the good —always neutralizing the tendency to evil—always vast in uplifting power.

Nor will this be but a safe and complacent optimism; for his too will be the abiding faith, that while ignorance and error and superstition may hinder, while the light of science falsely so called may mislead, until progress may appear to cease and even the way seem lost, still the advance of the intellect is continuing—constantly, surely, steadily, and in God's own time it must show.

When electricity and lightning were known to be one, the end seemed to have come, and the tidings which the amber and the magnet had to tell were believed of all men to have been told to the last syllable. But the book had only been opened. We have read much—very much —from it since. As the rise in ourselves continues, so, with equal pace, shall we read on.

INDEX.

A.

Abarbanel, 29.
Academy, del Cimento, exp'ts of, 432; French, on Great Pyramid, 58; of Sciences, French Royal, formation of, 378, 450.
Acmon, 22.
Addison, refers to Strada's magnetic teleg'h, 384.
Adsiger, Peter, 192.
Advancement of Learning, criticisms on Gilbert in Bacon's, 328.
Ægean Sea, iron on coasts of, 22.
Aelfred, King, 114.
Affaitatus, Fortunius, 211.
Age of iron, 20.
Agricola, George, describes gnomes, 25 ; on amber, 241.
Air pump, Boyle's exp'ts on, 414; Guericke's, 389 ; Hauksbee's, 457.
Akenside, refers to Strada's teleg'h, 384.
Akkadians, connection of, with Chinese, 63.
Albertus Magnus, on lodestone, 158, 308.
Aldrovandus, collections of, 342.
Alexander of Aphrodiseus, 93, 303.
Alexander the Great, campaigns, 38.
Alexandria, Univ'y of, 44.
Alfonso X., laws of, 111.
Allamand, describes Leyden Jar, 521.
Altaic nations, 61.
Amalfi, 116.
Ambassadors, Chinese legend of the, 69.
Amber, ancient trade routes of, 15; ancient use of, for decoration, 16; and lyncurium identical, 43; Asclepiades on, 52; attraction, Agricola on, 247; attraction, Cardan on, 249; attraction, distinguished by St. Augustine, 89; attraction, Fracastorio on, 241; attraction, first Chinese knowledge of, 74; attraction, first obs'd by Syrian women, 17; attraction of water by, 310; attraction, Plutarch on, 50; Baltic, 15; black, 43; Boyle distils to caput mortuum, 419; called "harpaga," 17; deposits of, 15; disc'y of, in lake dwellings, 11; Galen on, 52; Greek legends of, 74; in ancient China, 74; in ancient Greek literature, 16; in Egyptian temples, 52; insects in, 17; lack of, in Egypt, 52; -soul, not conceived by Thales, 34; St. Augustine distinguishes attraction of, 89; Theophrastus on, 40.
Amplitude, sun's, 198.
Amulets, lodestone, 25.
Anatomy, in Gilbert's time, 262.
Anatomy of Melancholy, Burton's, 368.
Anchor, invention of, 59.
Animals, Nollet's exp'ts on electrifying, 527.
Annals of China, national, 65.
Anthony of Bologna, 188.
Antiphyson, 93.
Apellikon of Teos, 43.
Apollonius Pergaeus, 44.
Appulus, William, 117.
Apuleius, on Thales, 35.
Arab compass, 110.
Arabian Nights, story of magnetic rocks in the, 96.
Arabs, in Spain, 108; on magnetic rocks, 100; early navigation of, 103.
Archimedes, 44; sphere of, 166.
Aristotle, Arabic treatises of, 157; deductive method of, 39; establishment of school of, 38; on Thales, 33, 34; on water soul, 34; phil'y of, followed by Gilbert, 282; referred to by Bacon, 279; relation of Gilbert to, 270; says nothing about amber-soul, 35.
Armature, first vibrating, 49, Galileo on the, 345, 348; Gilbert's, 288.

(597)

INDEX.

Arsinoe, lodestone roof in temple of, 44.
Arundel, Lord, his magnet, 333.
Aryans, the, 61.
Ascham, on Italian learning, 334.
Asclepiades on use of amber, 52.
Astrolabe, 179, 215.
Astronomy of Chinese, 79.
Attraction, theories of, prevalent in 17th century, 433.
Attractive point, Norman's, 214.
Aurora, Bose suggests elec. origin of, 499.
Averrhoes, on amber attraction, 156.
Azieros, 22.
Aztec ignorance of iron, 21.
Azimuth compass, 181.

B.

Babylonia, connection of Chinese with, 63.
Bacon, Francis and Descartes, 356; and Gilbert, 317; criticisms on Gilbert, 321–322; definition of heat, 417; indebtedness to Gilbert, 319, 329; inductive method of, 330; invective against Gilbert, 327; observes shock of torpedo, 401; on learning of his time, 334; on medicine, 253; on truth, 311; physiological remains, 324; reference to Aristotle, 279; reference to electrics, 325: relation to old and new phil'y, 331; treatise on magnet, 326.
Bacon, Roger, 160; on Peregrinus, 165.
Bailak Kibdjaki, 110; on mag. rocks, 100.
Bak tribes in China, 63.
Bandi, Countess, spont. combustion of, 503.
Barlowe, Dr. W., 316, 336, 337; controversy with Ridley, 338; mag. discoveries and relations to Gilbert, 340.
Barometer, Picard's luminous, 453.
Battery, Franklin invents series, 556, 559; first electrical, 523.
Beal, telegraphic predictions, 387.
Beauvais, Vincent de, on magnet, 158.
Bede, on magnet, 115.
Bellerophon, suspended horse of, 46.
Bell, Gordon invents elec., 507; Schwenter's mag., 383.
Bells, alleged lightning protection by, 592.

Belus, magnets in temple of, 29.
Bercy, Hugo de, 157.
Bernouilli, John, 455, 469.
Betham, Sir W., 56.
Betulae, 56.
Bevis, Dr., imp'ts on Leyden jar, 534, 555.
Bible of Guyot de Provins, 152.
Birds as guides at sea, 106, 113.
Blanco, Andrea, map of, 197.
Bleaching, elec. discovered by Krüger, 508.
Bologna stone, 454.
Bond on compass variation, 446.
Bononian stone, 454.
Boodt, De, disputes mag. teleg'h, 383; on lychnites, 43.
Bose, George Matthias, 493; "beatification," 498; electrical poem, 498; electrifies water jets, 499; experiments, 495; ignites gunpowder by elec'y, 497; invents prime conductor and obtains powerful discharges, 496; suggests elec. origin of Aurora, 499.
Bowdoin, correspondence with Franklin, 585.
Boy, Gray's exp'ts with suspended, 476.
Boyle, Robert, 404; additions to electrics, 419; compared with Hooke, 427; correspondence with American scientists, 425; corpuscular theory, 416; disputes Gilbert, 417; elec. and mag. theories, 415; electrical doubts, 421; Evelyn's estimate of his character, 424; experiments, 420; first scientific chemist, 414; not original discoverer of elec. repulsion, 420; observes mutual att'n of electrics and rubber, 418; observes odor of rubbed electric, 415; on mech'l production of elec'y, 418; on Van Helmont, 375; primary concepts, 415; sermonizing, 424; theory of elec. repulsion, 419.
Brand, discovers phosphorus, 454.
Bremond on Mahomet's coffin, 47.
Britain, magnet known in ancient, 114.
Browne, Sir Thomas, 380; exp'ts on mag. telegraph, 386; on garlic myth, 143.
Bruno, Giordano, 267, 333.
Buffon, De, declares elec'y unripe for fixed laws, 548; on Franklin's exp'ts, 587.

INDEX.   599

Burrowes, on compass variation, 446.
Burton, Robert, 371, 377.

C.

Cabæus, Nicolaus, 349; criticises Gilbert, 350; discovers elec. repulsion, 351; elec. theory, 351; on Garzoni's discoveries, 229; on mag. spectrum, 352; on mag. telegraph, 385; theory compared with that of Boyle, 419.
Cabiri, 23, 25.
Calamitico, el, 204.
Cambridge Univ'y in time of Gilbert, 261.
Canal, Necho's failure to build, 58.
Cardan, Jerome, 243; differentiation of amber and magnet, 249; Gilbert's attitude toward, 280; on lightning, 568.
Cardinal points, Chinese and Chaldean names similar, 63; Etruscan inv'n of, 59; named by Charlemagne, 133; named by Flemish sailors, 133.
Cart, Chinese south-pointing, 67, 69, 71, 72, 73, 81.
Casciorolus discovers Bologna stone, 454.
Cassini, astronomical obs'ns, 452; on mercury light, 456.
Castor and Pollux, 23.
Catullus, des'n of spinning, 18.
Cave publishes Franklin's papers, 585.
Cecco d'Ascoli, 203.
Cedrinus, on mag. suspension, 45.
Celmis, 22.
Cephisis, Lake, 17.
Cesare, disc'y of magnetism induced by earth, 227.
Cesi, founds Lyncei Academy, 342.
Chadids, 382.
Chain of lodestone, 24.
Chaldeans, Chinese civilization from, 63.
Challoner, Sir T., 339.
Chamberlain's letters, Gilbert mentioned in, 264.
Chariot, south-pointing—see Cart, Chinese south-pointing.
Charge, Dufay on distribution of, 483.
Charlemagne, names cardinal points, 133.
Charles II., interest in physical science, 406, 407, 408.
Charleton, Dr. Walter, 373, 376.

Charter-house, the, 470.
Chaucer on compass points, 191.
China, amber in ancient, 74; burning of books in, 66; first ships built in, 78; first south-pointing chariots in, 67; iron in ancient, 73; magnetic rocks on coasts of, 98; nucleus of, 64; original settlers of, 63; pagodas in, 564; Phœnician voyages to, 77; south-pointing carts, lost art in, 71; Tchoü dynasty in, 68; voyages to, in 675 B. C., 56.
Chinese, ancient navigation of, 77; astronomy, 79; characteristics, 77, 81, 82; chronology, 65; connection with Akkadians and Babylonians, 63; discover compass variation, 76; first knowledge of amber, 74; same of lodestone, 72; geomancers, 76; junks, 77, 78; inventions, 80; legend of ambassadors, 67; mariner's compass, 75, 76, 85, 189; south-pointing chariots, 67, 69, 71; superstitions about compass, 105; voyages to Japan, etc., 78; worship of magnet, 80.
Cherif-Edrisi, 100.
Chow, King of, 70.
Chronology, Chinese, 65–67.
Circuit, first elec., 525; Lemonnier's water and metal, 532; Watson's, across the Thames, 549.
Claudian, poem on magnet, 93; parodied by Strada, 383.
Clayton, letter to Boyle, 425.
Clement of Alexandria, 45.
Clutcher, name for amber, 17.
Clycas, 45.
Coition, magnetic, 276.
Colchester, 260.
Colden, Cadwallader, 585.
College, The Invisible, 379.
Collegium Naturale Curiosorum, 490.
Collinson, Peter, 538, 583.
Colonne, poem of Guido, 156.
Columbus, Christopher, 195; magnetic discoveries of, 200, 202; theory of compass, 199.
Combustion, cases of spontaneous, 503.
Compass, Mariner's, alleged use in building Great Pyramid, 57; Ancient Finn, 141; Appulus on, 117; Arab, 110; attributed to Egyptians, 57; to ancient Greeks, 54; to King Solomon, 55; to Phœnicians, 54;

Compass (continued.)
to various ancient people, 53; Azimuth, 181; boxing the, 187; Chinese, first marine, 189; Chinese obs'n of variation of, 76; Columbus' alteration of, 196; Columbus' disc'y of variation of, 200; Columbus' theory of, 199; derivation of word, 133; design of card, Etruscan, 60; De Vitry on, 154; dip or inclination of, 210; disc'y of dip of, 209; early Spanish, 111; errors in Columbus', 201; evolution of, 131; Finn, 140; first des'n of, 128; garlic effect on, supposed, 143; Gilbert on storage in meridian, 313; Gilbert's electroscope resembles, 304; governed by earth's poles, 277; Guyot de Provins on, 153; in time of Peregrinus 179; Lully on, 191; Neckam's des'n of, 128; non-mag. metal in, 183; Norman's disc'y of dip, 2:5; Norse penalty for falsifying, 144; not Chinese inv'n, 85; not derived by Arabs from Chinese, 105; old mode of using, 130; Peregrinus', 180; Porta on protecting needle of, 238; punishment for tampering with, 144; secular variation of, 446; suggestion of telegraphy by, 239; telegraph, Schwenter's and others', 382; unknown to Saracens, 109; variation of, 196; William the Clerk's poem on, 150; Wisbuy origin of, 146.
Condenser, Franklin's plate, 556.
Conduction, electric discovered by Guericke, 399; magnetic, Gilbert on, 289.
Conductor, Desaguiliers proposes name, 488; magnetic, first suggestion of, 47; prime, inv'd by Bose, 496; or non-electric, 482.
Constantine, law of, concerning lightning, 566.
Convection, electrical, 545.
Copernican theory, 267.
Copernicus, Nicolas, 267.
Corffier shocked by lightning, 588.
Corrichterus, his mag. unguent, 37.
Corybantes, 23.
Cowley, poem on R. Society, 413.
Creagus, 159.
Creation, prehistoric account of, 164.
Crows, as guides at sea, 113.
Ctesias, suggestion of lightning protection, 565.

Cunaeus, inv'n of Leyden jar ascribed to, 521.
Curetes, 23.
Current, first suggestion of magnetic, 47.

D.

Dactyls, Idean, 22.
Dalance, treatise on magnet, 448.
D'Alibard translates Franklin's papers, 587; exp'ts on lightning-rod, 588.
Dantzic, philosophers, exp'ts of, 514; physical society of, 513.
De Augmentis, criticisms of Gilbert in Bacon's, 328.
De Beauvais.—See Beauvais.
De Bercy.—See Bercy.
De Boodt.—See Boodt.
De Buffon.—See Buffon.
De Fantis.—See Fantis.
De la Hire.—See La Hire.
De Lor.—See Lor.
De Magnete, Bacon's "remains" taken from Gilbert's, 325.
De Monmor.—See Monmor.
De Natura Rerum, Lucretius' poem, 47; Neckam's treatise, 123.
Denmark, Iron Age in, 21.
D'Epinois, Gautier, poem of, 156.
Desaguiliers, Dr. Joseph, 470; exp'ts of, 488; on atmospheric elec'y, 489.
Descartes, René, 356; copied by Digby, 378; magnetic theory, 359; mag. theory compared with that of Plutarch, 51; method compared with that of Bacon, 356; on electrics, 364; on mag. spectrum, 362; theory abandoned in France, 510; theory compared with that of Lucretius, 48; vortex theory of, 357.
De Subtilitate, Cardan's work, 246.
Diamond, alleged attraction of, by iron, 281; alleged magnetism of, 238; alleged screening effect, 88.
Diaz, Bartholomew, voyage of, 205.
Digby, Sir Kenelm, 376; elec. theories of, 378; replies to Browne, 380.
Digges, Madam, her sparkling frock, 425.
Digges, Sir Dudley, 339.
Digges, William, letter concerning Mrs. Sewall, 425.
Diocles, 41.
Diogenes Laertius, 34.
Dionysius, 59.

Dioscorides on magnet, 92; on ligurius, 42.
Dioscuri, the, 23.
Dip of magnetic needle, 209, 210; Affaitatus' supposed disc'y of, 211; Gilbert on, 213; Hartmann's disc'y of, 209; Norman's disc'y of, 215, 217.
Drebbel, Cornelius Van, 44, 192.
Dufay, Charles François. 478; broad view of elec'y, 487; discovers vitreous and resinous elec'y, 484; electrifies himself, 483; electrifies metals, 479; exp'ts on colored objects, 481; on distribution of charge, 483; no distinction between electrics and non-electrics, 479; sends letter to R. Society, 485; recognition of Gray's work, 485; tribute to Gray, 487; uses solid insulators, 482; verifies Gray's exp'ts on conduction, 479.
Du Tour on Nollet's theory, 554.

E.

Earth, field of force of, Gilbert on, 292; magnetism induced by, 227; return circuit disc'd by Watson, 550.
Eclipse at time of Thales, 34.
Effluvium, electric, Cabæus on, 351; Gilbert on, 308; magnetic, 292.
Egypt, absence of Science in ancient, 31; iron in, 28, 58; lack of amber in ancient, 52; mag. suspension in, 45; opening of, to commerce, 30; religion of ancient, 31; vending machines in ancient, 87.
Egyptians, ancient, ignorance of magnet of, 27; alleged knowledge of compass by, 57; voyages of, 58.
Electorius, 42.
Electrical, first use of word, 339.
Electric and magnetic motion compared, 311; attraction, theories of, 307; attraction, Gilbert on, 308; bell, inv'd by Gordon, 506; light, see Light, electric; Machine, Gordon's, 506; Hauksbee's, 461; Guericke's, 395; Winkler's, 506; Motor, Gordon's, 507.
Electricity and lightning, Franklin's exp'ts on identity of, 580, *et seq.;* Freke on identity of, 571; Nollet on identity of, 573; Winkler on identity of, 572; and magnetism linked by Newton with gravity, 439; atmospheric, Desaguiliers theory of, 489; Franklin's theory of, 576; beginning of modern, 299; Boyle on mech'l production of, 418; Digby on, 378; dual nature of, found by Dufay, 484; and by Kinnersley, 586.

First application to medicine, 501-2; first attempt to measure, 523; first book on, in English, 420; first distinguished from magnetism by St. Augustine, 89; first notice of, by R. Society, 402; first use of word, 373.

Gordon kills animals by, 507; Franklin's theory of, 643; Germans regard as fire, 492; Greene's poetical references to, 369; Hausen's theory of, 494; Jonson's reference to, 368; new theories of, in 1747, 553; Quelmalz, theory of, 503; resinous and vitreous, discovered by Dufay, 484; s'Gravesande's definition of, 488; speed of, Lemonnier's attempt to measure, 532; Watson's attempt, 551; Winkler's attempt, 506; Watson's theories of, 507, 534.
Electrics, and non-electrics, Dufay on, 479; Bacon on, 325; become non-conductors, 482; Boyle's additions to, list of, 419; Boyle observes mutual attraction of electric and rubber, 418; Browne's exp'ts on, 381; Cabaeus' additions to list of, 350; Descartes on, 364; Gassendi on attraction of, 418; Gilbert's list of, 299; Gilbert on nature of, 307; mutual attraction of, obs'd by Acad. del Cimento, 433; naming of, 302; *per se,* 488.
Electrida, 17.
Electrides, 16, 17.
Electro-magnetism, word coined by Kircher, 365.
Electrometer, first use of word, 524.
Electron, 16.
Electrum, in Egypt, 52; lake, 17.
Elicott, John, elec. theory of, 554.
Elizabeth, Queen, learning in time of, 332-334; legacy to Gilbert 265.
Emerson, R. W., on genius, 426.
Emperor First, Chinese, 66.
England in time of Elizabeth, 334; in time of Hauksbee, 463.
Ephesus, mag. suspension in temple of, 46.

Epicurus on attraction, 51.
Erasmus on ligurius, 42.
Eridanus, amber on shores of, 17.
Erigena founds scholasticism, 118.
Ether, Newtonian, 511.
Ethiopia, iron in, 22
Etruria, amber in, 15.
Etruscans, the, 59; design object like compass card, 60; genesis of, 62; on lightning, 566.
Euclid, 44.
Euripides, Oeneus of, 24.
Eustachius, 262.
Evax, 42.
Evelyn, John, 378, 407, 424.

F.

Fabri, Honoré, 420.
Fallopius, 262.
Fanshawe, Lady, on Digby, 379.
Fantis, Antonio de, 192.
Faraday, and Gilbert, 293; efforts to connect gravity and magnetism, 442; on Newton's lines of force, 442.
Fathers of Church, on magnet, 90.
Ferabosco, the, visits R. Society, 407.
Ficino, Marsilio, mag. theory of, 240.
Field of force, Descartes on, 359; development of, 218; Dufay on, 483; Gilbert on, 272, 291; iron filings in, 50; Lucretius on, 48; magnetic and electric, 434; Maxwell on, 440; Newton on, 440; Peregrinus on, 172, 207; Porta on, 235; Sarpi on, 227.
Figure-head, inv'n of the, 59.
Finland, conquest of, 137.
Finns, the, 59, 83; and Lapps, 139; magic of, 138; superstitions concerning, 139; use of compass by, 139.
Finno-Ugric family, 59.
Fire, ancient records of atmospheric, 568; electrical, 509; Germans regard elec'y as, 492.
Flesh magnet, the, 159.
Fludd, Dr. Robert, 375.
Fluid theory of elec'y, Franklin's, 544.
Form, Aristotelian, 276, 282, 419.
Fountain, Desaguiliers' electrified, 489.
Fracastorio, Jerome, amber theory of, 241; Gilbert's attitude toward, 280; on mag. rocks, 204.

France, condition of phys. science in, in 17th cent'y, 452; Franklin's exp'ts repeated in, 587; learned societies in, 378.
Franklin, Benjamin, 537; advises Collinson of discharging effect of points, 541; correspondence with Colden, Bowdoin and Belcher, 585; electrical exp'ts on Leyden jar, etc., 543, 544, 545, 556, 558; on identity of lightning and elec'y, 580; on points, 541; on suspended scale pan, 582; experiments repeated in France, 587; invents plate condenser and series battery, 556; kite experiment, 590; lightning rod, 582; papers rejected by R. Society and published by Cave, 583; regards lightning as fired sulphur, 575; retires to devote himself to elec'y, 547; theories of elec'y, 543, 576; theory claimed by Watson, 552; uses battery for curative purposes, 585.
Fra Paolo, 224.
Frederick I. of Prussia founds Berlin Society, 490.
Freke on identity of lightning and elec'y, 571.
Froude, on genius, 426.

G.

Gagates, the, 43, 126.
Galen on amber, 52; on lodestone, 92.
Galileo, abjuration of, 355; condemnation of, 356; correspondence with Duke of Tuscany, 345; experiments on magnet, 345; on Gilbert's discoveries, 344–345; on magnetic teleg'h, 385.
Garlic, alleged effect on compass, 143; Matthiolus on, 281.
Garzoni, alleged mag. discoveries of, 229.
Gassendi, elec. theory of, 418.
Geomancers, Chinese, 75.
Germany, physical science in, 490, 492.
Gibbon on Mahomet's coffin, 46.
Gilbert, William, 258; amber attraction of water, 310; amber questions, 295; and Aristotle, 270, 275; and Barlowe, 340; and Guericke, their mag. theories compared, 393; and Kouopho, 311; attitude to predecessors, 279; authorities quoted by, 280.

INDEX.

Gilbert (continued.)
Compared by Bacon to Xenophanes, 328; comparison of the poles, 277; conception of gravity, 437; condemns mag. fallacies, 281; continued as court physician by James I., 315; copied by Van Helmont, 373; correlation of elec. with other motions, 309; correlation of gravity and magnetism, 293; cosmical philosophy, 269; cosmical system, 294; cosmical theory accepted by Kepler, 354; cosmical theory compared with Newton's, 435, 438; criticised by Bacon, 321-322-327; death and burial place, 315; declares earth a magnet, 276; De Magnete, his treatise, 260; De Magnete rec'd in Italy, 343; De Mundo Novo, his treatise, 260, 316, 318; discoveries recapitulated, 312-313.

Education of, 259; elec. effect of atmospheric conditions noted, 305; elec. and mag. motions compared, 311; electroscope, 303; embellishments in De Magnete, 268; errors as to variation, 273; failures in observation, 312; field of force discussed, 272, 291; form and matter theory, 276; free philosophizing of, 310; generation of lodestone, 287; inductive method of, prior to Bacon, 330; influence of Aristotle on, 282; insulation, 308-310; list of electrics, 299; list of non-electrics, 305.

Magnetic discoveries of, 288; magnetic repulsion, 285; magnetic theory of, 276; disputed by Boyle, 417; Matter and Form, 284; Meteorologia of, 329; methods of thought of, 266; "nature" discussed, 285; nature of electric, 307; negative conclusions regarding elec'y, 306; nomenclature, 301; orb of virtue, 272; compared with obs'ns of Porta and Peregrinus, 351; Owen's epigram on, 341; predecessors referred to, 287; portraits and works, 260; posthumous volume, 316, 318; proposed addition to De Magnete, 316; referred to by Bacon, 318; residence and society, 263; relations to Queen Elizabeth, 262, 264; relations to Sarpi, 344.

Scaliger's criticism, 341; terrella of, 277; terrestrial attraction of moon, 292; theory of elec. attraction, 308; theories, cosmical, 269, 294; theories compared with those of Peregrinus, 278; condemned by Kircher, 366.

Gioja, Flavio, 187.
Glanvil, encyclopædia of, 160; telegraphic predictions, 387.
Gnomes of Middle Ages, 25.
Goddard, Jonathan, 404.
Goose, Kircher's genesis of solan, 365.
Gordon, Andrew, elec. inventions, 506 *et seq.*
Gottland, 134.
Gralath, Daniel, exp'ts on Leyden jar and elec. measuring inst's, 522 *et seq.*
Grandamicus, mag. theory of earth, 405.
Graunt, John, refused admission to R. Society, 409.
Gravity, and magnetism, Gilbert on, 293; Newton co-ordinates elec'y and mag'n with, 442.
Gray, Stephen, 470; and Dufay, 486; Dufay's tribute to, 487; his friends, Godfrey and Wheler, 473; exp'ts on brush discharge, 486; on charge resident on surface, 476; on conduction, 474; on elec. induction, 477; on glass tube, 472; on hair, etc., 471; on similarity of elec. discharge to thunder and lightning, 486; planetary theory and death, 487.
Greeks, amber in literature of, 16; amber trade of, 16; compass attributed to, 54; emigration to Egypt, 30; iron working of, 23; nature worship of, 31.
Greene, Robert, literary references to mag'n and elec'y, 369.
Grote, on philosophy of Aristotle, 39; of Thales, 37.
Grummert, utilization of elec. light, 508.
Guericke, Otto von, 388; and Gilbert compared, 393; believes earth to be animate, 393; discovers discharging effect of points, 398; elec. conduction, 399; elec. light, 402; sound due to electrification, 402; elec. repulsion obs'd by, 397; elec. terrella of, 395; forgotten in 18th cent'y, 491; hypothesis of virtues, 392; invents air-pump,

H.

Guericke (continued.)
  414; invents elec. machine, 395–6;
  treatise de Vacuo Spatio, 391.
Guilford, Lord, sells barometers, 406.
Guinicelli, poem of, 155.
Gunpowder ignited by elec'y, 497; lightning compared to explosion of, 570.
Guyot de Provins, 152.

Hair, elec. attraction of, obs'd by Boyle, 422.
Hakewill, 384.
Hale, Lord, on hydrostatics, 406.
Halley, Dr. Edmund, 447.
Hammering, magnetization by, 290.
Hartmann, disc'y of dip, 209.
Harpaga, 17.
Hauksbee, Francis, 457; electric machine, 461; exp'ts on elec. induction, 467; on elec. light, 460; on lines of force, 467; on luminous fountain, 459.
Hausen, Christian August, 493.
Healing by first intention, Browne on, 381.
Heat defined as mode of motion, by Bacon, Boyle, Locke and Hooke, 417; destruction of magnetization by, 227, 237.
Hebrews, iron working by, 29.
Heliades, legend of, 16.
Helmont, John Baptist Van, 372.
Henry, Prince, the Navigator, 194.
Heraclea, 27.
Heraclean stone, 24, 27.
Herculean stone, 24, 27.
Hero, 44.
Herodotus on Thales, 34.
Hesiod, amber mentioned by, 16; brass mentioned by, 21.
History of stones, Theophrastus', 39.
Hippalus, 104.
Hipparchus, 44.
Hippias, 35.
Hoang-ti, legend of, 67; reign of, 65.
Hoar, Leonard, corresponds with Boyle, 425.
Hobbes, attacks R. Society, 378.
Homer, amber mentioned by, 16; iron seldom mentioned by, 21; knowledge of compass attributed to, 54.
Hooke, Robert, 427; on elec. light, 430; on heat as vibration, 417; inventions of, 428-9; phonograph suggested, 429; spiral spring of, 429.
Hopkinson, Thomas, 540.
Horus, magnet termed bone of, 28.
Huet, Bishop, on Solomon's voyages, 55.
Humboldt on Columbus, 200.
Humor, elec. attraction ascribed to, 308.

I.

Iceland, discovery of, 113.
Ida, Mt., mag. legend of, 19.
Idean Dactyls, 22.
Ignition, electric, 496, 507, 508.
Inclination of compass, 210.
Induction, electric, Gray's exp'ts on, 477; Hauksbee's exp'ts on, 467; magnetic, Descartes on, 361; Norman on, 219; of earth, 227; Peregrinus on, 176; St. Augustine on, 87.
Inductive method, beginning of, 38; compared with deductive, 356; Gilbert's use of before Bacon, 330; Nichol's definition of, 330.
Innocent, Bishop, 566.
Insulation, first use of term, 482; Gilbert on, 308.
Insulators, solid, used by Dufay, 482.
Inunction, 130.
Invisible College, 404.
Ion of Plato, 24.
Iolinus on lychnites, 42.
Iron, acquired magnetism of, 289; age, 12, 20, 21; Aztec and Peruvian ignorance of, 21; -clad ships, Norse, 98, 112; decay of, in Egyptian soil, 28; -filings in mag. field, 50, 352, 412; food for magnet, 238; Hebrew knowledge of, 29; in ancient China, 73; in ancient Egypt, 28, 58; mag. screening effect of, 238; mag. separation of, 159; mag. suspension of, 45; mentioned by Homer, 21; by Hesiod, 21; miners in Samothrace, 23; mines, ancient, 22, 27; natural state of, 20; -ore, Gilbert on, 287; Greek deposits of, 25; workers, the first, 22; working by Finns, 138.
Irving, W., on Columbus, 201.
Israelites as iron workers, 28.
Italy, learning in in 17th cent'y, 342.

INDEX.   605

J.

Jade traffic in China, 67.
Japan, first Chinese voyages to, 78; south-pointing carts first used in, 81.
Jefferson, Thomas, exp'ts on heat, 389.
Jeroboam, golden calves of, 29.
Jesuits and Sarpi, 228-9.
Jet, 43, 126, 369.
Jews, mag. knowledge of, 29.
John of London, 162.
Jonson, Ben, refers to magnetism, 368; to protection against thunder, 509.
Josephus, on protection of temple from lightning, 565.
Junks, Chinese, 77-78.

K.

Kalevala, the, 138.
Kepler, John, 354.
Kinnersley, Ebenezer, 540; elec. lectures of, 585; exp'ts on Leyden jar, 556; invents magic. picture, elec. jack and elec. motor, 560; re-discovers vitreous and resinous elec'y, 586.
Kircher, Athanasius, condemns Gilbert's theories, 366; criticizes Gilbert and Kepler, 355; genesis of Solan goose, 365; Hebrew use of magnet, 29; invents words "magnetism" and "electro magnetism," 365; works, 365.
Kite experiment, Franklin's, 590.
Kleist, Dean von, discovery of Leyden jar, 512.
Kouopho, 89; and Gilbert, 311; on amber attraction, 74.
Kratzenstein, Christian Gottlieb, 502.
Krüger, Johann Gottlob, 501; observes bleaching effect of ozone, 508; Von Kleist describes Leyden jar to him, 513.

L.

Laertius, Diogenes, 34, 35.
La Hire, De, on mercurial light, 456.
Lake dwellings, amber in, 11.
Lange, elec'y for curative use, 503.
Lapis lyncurius, 41.
Lapis solaris, 454.
Lapps and Finns, 137.
Latini, Brunetto, 162.
Leakage, magnetic, Descartes on, 361.
Learning, in England and Italy contrasted, 341.
Legends of magnet, 219.
Leibnitz on loss of Galileo's magnet, 349.
Lemonnier, Louis G., 530.
Leonardus on black amber and lyncurium, 42, 43.
Leopoldine Society, 490.
Leyden Jar, Bevis' improvements in, 534, 555; described by Musschenbroeck to Réaumur, 517; discovered by Von Kleist, 512; first elec. circuit recognized in, 525; Franklin's exp'ts on, 545, 558; Gralath's exp'ts on, 522; improved by Watson, 554, 555; in battery, 523; invention of ascribed to Cunæus, 521; origin of name, 522; Watson's theory of, 534.
Library Company of Philadelphia, 538.
Light, electric, Bernouilli on, 455; compared to lightning, 459, 469; Dufay finds identical with fire, 485; Grummert utilizes, 508; Guericke discovers, 402; Hauksbee's luminous fountain, 459; Hausen differentiates spark, brush and glow, 494; in vacuum, 460; Ludolff on, 497; Nollet's exp'ts on, 527; Picard observes in barometer, 453.
Light, Milton on magnetic nature of, 438; theories of Hooke, Descartes, Newton and Young, 431.
Light magnet, the, 455.
Lightning, and electricity, identity of, Franklin on, 580; Freke, 571; Gray, 486; Hauksbee, 459; Nollet, 573; Wall, 469; Winkler, 572; Cardan on, 568; compared to powder explosion, 570; deaths by, 568; Franklin's early views on, 580; Lester on, 575; Shakespeare on, 569; Wallis on, 570.
Lightning protection, accidental, 564; ancient suggestions of, 565; Ben Jonson's reference to, 569; Franklin on, 576, 582; St. Thomas Aquinas on, 592.
Lightning rod, denounced by Nollet, 593; erected by D'Alibard and De Lor, 588, 589; Etruscan knowledge of, 566; Franklin's des'n of, 582.

Ligure or ligurian stone, 42.
Lilly of compass, 60.
Lincurius, 42.
Lines of force (see Field of Force), Descartes on, 359; Hauksbee on, 467.
Lines of magnetic direction ex'd by R. Society, 412.
Lines of no variation, world divided on, 204.
Lister, Dr., on lightning, 575.
Livio Sanuto on mag. rocks, 204.
Locke, defines heat as motion, 417.
Lodestone, (see Magnet), and Greek phil'y, 33; Albertus Magnus on, 308; Chinese worship of, 80; Dioscorides and Galen on, 92; disc'y of, after iron, 20; Egyptian knowledge of, 58; Egyptian name for, 28; field of force about, 48; first Chinese knowledge of, 72; first mention of attractive property, 24; Gilbert's armed, 288; Gilbert's terrella, 277; Gilbert's theory of generation of, 287; Greek mystery of, 22; Israelite use of, 29; Lady, in Jonson's play, 368; legend of disc'y of, 19; nature of, 19; Patristic writings refer to, 90; polarity of, 127; prehistoric knowledge of, 20, 63, 83; repulsion by, 49; rings as amulets, 25; St. Augustine on, 87.
Lor De, exp'ts on lightning, 588–589.
Louis XIV., endows Royal Academy, 450; physical science at court of, 451.
Louvois, dealings with Royal Academy, 451.
Lucan, suggestion of lightning protection, 565.
Lucera, siege of, 165.
Lucian, amber mentioned by, 16.
Lucretius, on Bronze age, 20; mag. theory of, 48; on derivation of word "magnet," 25; on filings in mag. field, 50; on jumping rings, 49; on mag. field, 48; on "Nature of Things," 47; on Samothracian rings, 24; on vibrating armature, 49.
Ludolff, ignites spirits by elec'y, 496; shows mercury light to be electric, 497.
Lully, Raymond, 190.
Lychinus, 42.
Lychnites, 42.

Lykeum, Aristotle's, 38.
Lyncei, Academy of, 342.
Lyncis, 42.
Lyncurium, 41.
Lyngurius, 42.
Lynx stone, 41.

M.

Magdeburg experiment, 385.
Magellan, voyage of, 206.
Magic, Finn, 138; rise of, 95.
Magnesia, foundation of, 26.
Magnesians, 25.
Magnet, (see Lodestone), Albertus Magnus on, 158; artificial, made by Sellers, 446; Bacon's treatise on, 324; compound, 290; de Beauvais on, 158; Dioscorides and Galen on, 92; Cardan on, 249; field of force of, shown by Peregrinus, 208; flesh, 159; Galileo on, 345; Gilbert regards earth as, 276; known in early Britain, 114; Latini on, 162; light, 455; Lucretius on derivation of name, 25; medical uses of the, 255; myths of the, 219; Paracelsus' curative use of, 222; Patristic writings on, 90; Peregrinus on selection of, 169; on testing, 170; on finding poles, 170; Porta on measuring strength of, 238; prehistoric knowledge of, 83; Roger Bacon on, 161; so called by Euripides, 24; St. Augustine on, 87; worshipped by Chinese, 80.
Magnetes, tribe of, 26.
Magnetic, cure for wounds, 372; healing, Browne on, 381; intercommunication, 382; Lady, Jonson's play of, 368; Nuntii, 373; supposed place of pole, 204; Rocks, 313; legends of, 367; Fracastorio on, 204; Livio Sanuto on, 204; Maurolycus on, 204; Oviedo on, 204; Ptolemy on, 203; saturation, 290; spectrum, Cabæus on, 353; Descartes on, 362; Wren on, 412; statesman, Digges so-called, 339; synonymous with Herculean, 27.
Magnetical Animadversions, Ridley's, 339.
Magnetism, and electricity linked by Newton with gravity, 439; animal, 372; at end of 17th cent'y, 448; by earth induction, 289; Descartes on, 361; destruction of, by

fire, 227; Digby on, 377; Greene's references to, 369; induced by earth, 227; Jewish knowledge of, 29; mineral, 372; Michell on law of, 236; Peregrinus on law of, 173; Porta on law of, 235; Sarpi on, 226; Shakespeare on, 370; term first used by Kircher, 365.
Magnetite, in ancient China, 73; nature of, 19.
Magnetization, Gilbert on, 289; of compass needle, 130; of iron in air, 289.
Magnetizers, the, 372.
Magnetometer, Gilbert's, 312.
Magnetotherapy, 24.
Mahomet's coffin, myth of, 46.
Maimonides, 29.
Manetho, on magnet in Egypt, 28.
Marbodeus, 42; on jet, 43.
Marcellus Empiricus on lodestone, 93.
Marco Polo, 189.
Mariner's compass, see Compass.
Mather, Rev. Cotton, 463.
Matter, Aristotelian, 276, 282.
Matthiolus, garlic theory of, 281.
Maundeville on magnetic rocks, 99.
Maurolycus on magnetic rocks, 204.
Mausoleus, 46.
Measuring instrument, first electrical, 523.
Medicine, first application of elec'y to, 501–502.
Melancthon on magnet, 143.
Mercurial phosphorus, 455, 490.
Mercury light, obs'd by Picard, 453.
Mersenne, 378.
Meteorites as talisman, 57.
Meteorologia, Gilbert s, 329.
Michell, law of magnetism, 236.
Miles, Dr., on sparkling persons, 503.
Milesian doctrine, Theophrastus dissents from, 41; philosophy (see Thales).
Milton, reference to sun's "magnetic beam," 438.
Monconys, Balthasar, 388.
Mongols, 59, 62, 83.
Monmor, De, 378.
Moon, mag. effect of earth on, 292; mapped by Gilbert, 329.
Mortimer, Dr. Cromwell, on elec. fire, 505.
Motor, germ of electric, 49; Gordon invents electric, 507; Kinnersley's electric, 560; Peregrinus' magnetic, 167, 177, 182.
Mountains, magnetic, 96 *et seq.*
Musschenbrœck, Peter V., 517.
Mycenæ, amber at, 15.
Mysticism, 94.
Mythology, Greek, 31.

N.

Nature, Aristotelian use of term, 284.
Navigation, ancient Arab, 103; Chinese, 77; Egyptian, 58; Etruscan, 59; Greek, 54; Norse, 112; Phœnician, 15, 55; Solomon's, 55.
Navigator's supply, Barlowe's, 336.
Necho, voyages of, 58.
Neckam, Alexander, 120; and Peregrinus compared, 174.
Newton, Isaac, definition of ether, 511; disc'y of univ'l gravitation, 435; elec. exp'ts, 445; finds field of force in intervening medium, 440; laws of motion, 439; on lines of force, 442; on luminous bodies, 458; theory compared with Gilbert's, 435.
Nicander, legend of Magnes, 19.
Nichol, Prof., on inductive method, 330.
Nickel-in-slot machine, ancient, 87.
Nollet, Jean Antoine, 516; amplifies elec. theory, 554; denounces lightning-rods, 593; exp'ts on light, vegetables and animals, 527; on identity of elec'y and lightning, 573.
Nomads, the, 61.
Non-conductors, electrics become, 482.
Non-electrics, become conductors, 482; Gilbert's list of, 305.
Norman, Robert, 211, 213.
Normans, conquest by, 116.
Norse, legends of mag. rocks, 99.
Northmen, the, 112.
Nova Philosophia, Gilbert's, 318.
Novum Organum, criticisms of Gilbert in the, 328.

O.

Odyssey, supposed reference to compass in, 54.
Œneus of Euripides, 24.
Ophir, voyages to, 55.

608 INDEX.

Orb of coition, Gilbert's, 302.
Orb of virtue, Gilbert's, 272–291.
Orpheus, 23.
Orphic mysteries, 23.
Otiosi, Porta's society of the, 231.
Otocousticon, 430.
Oviedo on mag. rocks, 204.
Owen, epigram on Gilbert, 341.
Ozone, bleaching effect obs'd by Krüger, 508; odor obs'd by Boyle, 415; by Hausen, 494.

P.

Panaceas, magnetic, 25.
Paracelsus, 220; imitated by Rosicrucians, 372.
Paris, Matthew, 157.
Paschal, 378.
Peiresc, on condition of England, 334.
Penn, William, corresponds with Boyle, 425.
Pensieri of Fra Paolo, 226.
Pennsylvania Gazette, Kinnersley's adv't in, 585.
Pepys, account of R. Society, 407; receives terrella, 407.
Peregrinus, Peter, 165; and Neckam compared, 174; copied by Porta, 234; discoveries of, 184; field of force revealed by, 207; Gilbert refers to, 279; theories of, compared with those of Gilbert, 278.
Petty, Sir W., 404.
Phaeton, legend of, 16.
Philippe de Thaun, Bestiary of, 119.
Philosophers, Lærtius' lives of, 34.
Philosophy, at time of Socrates, 37; beginning of, 33; cosmical in middle ages, 163; of Paracelsus, 220; of Thales, 33; rise of Greek, 44; scholastic, 118.
Phœnicians, compass attributed to, 54; knowledge of magnet, 56; voyages for amber, 15; voyages in, 697 B. C., 55.
Phonograph, foreshadowed by Hooke, 429.
Phoronid, the, 22.
Phosphorus, 454; mercurial, 455.
Phrygia, first iron miners from, 22.
Physicians, mag. knowledge of English, 256.
Physiological Remains, Bacon's, 324.
Picard, Jean, observes barometer light, 453.
Pirorganon, Winkler's, 506.

Plato, on amber attraction, 35–36; on lodestone attraction, 24; on connection of lodestone and amber, 37.
Pliny, denounces Lyngurian stone as myth, 42; legend of amber attraction, 17; of Magnes, 19; on foundation of Magnesia, 26; on gagates, 43; on Heraclean stone, 27; on lodestone rings, 25; on lodestone roof, 44; on magnetic repulsion, 51.
Plutarch, on amber attraction, 50; on mag. repulsion, 51; on Manetho, 28.
Plus and minus electrification, 543.
Po, amber on shores of, 17.
Points, discharging effect of, obs'd by Franklin, 541; elec. effect of, obs'd by Guericke, 398.
Polarity, electrical, obs'd by Guericke, 398; magnetic, first suggestion of, 127; magnetic, supposed Egyptian knowledge of, 58; magnetic, supposed Etruscan knowledge of, 62; of lodestone, 19; Peregrinus on reversal of, 176; prehistoric knowledge of, 20.
Pole, magnetic, supposed places of, 204; sons of the, 164.
Poles, ancient ideas of heavenly, 164; confusion of magnet and earth's, 158; Gilbert's comparison of the, 277.
Porta, John Baptista, 230; first notion of telegraph, 239; Gilbert's attitude toward, 280; indebtedness of, to Peregrinus, 234; on magnetic sphere of virtue, 237; relations with Sarpi, 232.
Portuguese voyages, 194.
Power, confutation of Grandamicus, 405.
Priestley, Dr., on powerful elec. discharges, 505.
Prima Forma of Aristotle, 283.
Prime conductor, Bose invents, 496.
Primum Mobile, 163.
Principia, Descartes', 365.
Printing, invention of, 193.
Prolusiones Academicæ of Strada, 383.
Prometheus, legend of, 565.
Pseudodoxia epidemica, Browne's, 380.
Ptolemy on magnetic rocks, 203, 380.
Puritans, opposition to physical science, 371.

INDEX. 609

Pyramid, iron in great, 28; orientation of, 57.

Q.

Quakers, refuse to defend Phila, 546.
Queen Elizabeth, relations to Gilbert, 262-4.
Quelmalz, elec. theory of, 503.

R.

Races, peculiarities of, 61.
Réaumur, Musschenbrœck's letter to, 517.
Repulsion, electric, obs'd by Cabæus, 351; Guericke on, 397; laws of, 485; magnetic, 48, 51, 285.
Resinous electricity, Dufay on, 484.
Resistance, 360.
Respective point, Norman's, 216.
Rhea, 22.
Rhodes, mag. suspension at, 46.
Ridley, Mark, 338.
Ridotti, Italian, 342.
Rings, Samothracian, see Samothracian rings.
Roberval, 378.
Rocks, magnetic, 96.
Rolli, on spontaneous combustion, 503, 504.
Rose of the winds, 60, 187, 188, 191.
Rosicrucians, the, 371.
Royal Academy of Sciences, foundation of, 450; reorganized by de Ponchartrain, 452.
Royal Society, Berlin, founded by Frederick I., 490.
Royal Society, English, early exp'ts of, 410; first notice of elec'y, 402. foundation of, 406; influence on new phil'y, 409; insists on original research, 411; its opponents, 413; rejects Franklin's papers, 583; repeats Newton's exp'ts, 445.
Ruffinus, on magnet, 92; magnetic suspension, 45.
Runes, Finn, 138.
Rupert, Prince, inventions of, 406.

S.

Sagredo, Gian Francesco, 342-343; observes time changes in variation, 367.
Samothrace, Cabiri in, 23.
Samothracian rings, 23, 47, 87.
Sanconiathon, 56.
Saracens in Spain, 108.

Sarpi, Fra Paolo, 224; and Porta, 232; writes to Galileo about Gilbert, 343.
Saturation, magnetic, 290.
Sauveur, invents gambling system, 451.
Scaliger, on Laertius, 35; on Gilbert, 341.
Schilling, J. J., exp'ts of, 491.
Scholastic philosophy, 118.
Schott, Gasper, describes Guericke's air-pump, 414.
Schwenter, Daniel, compass telegraph, 382.
Sellers, produces artificial magnets, 446.
Semites, characteristics of, 61.
Septuagint, ligure disc'd in, 42.
Serapis, magnetic suspension in temple of, 45, 92.
Series, connection of elec. generators in, 557.
Sewall, Madam, sparkling frock of, 425.
S'Gravesande, W. J., 488, 516.
Shakespeare, William, on nature of lightning, 569; on St. Elmo's fire, 568; references to magnetism, 370.
Shoo King, the, 66.
Short circuiting, Descartes on, 361.
Siderites, 25.
Silk, filament suspension, 312; Gilbert's use of, for insulation, 308; Gray's use of, for insulation, 475.
Silver, attraction of, by lodestone, 281.
Similitudes, Neckam on, 123.
Simpson, Dr., on elec. sparkling, 503.
Sing, Philip, 540-544.
Siphon recorder, principle of, suggested, 499.
Societies, learned, in France, 378; names of Italian, 232.
Society, English Scientific in 1660, 404; Royal—See Royal Society.
Socrates, philosophy in time of, 37.
Solomon, voyages of, 55.
Somers, Lord, 463.
Sorcery, Finn, 138.
So-soung on mag. rocks, 98.
Soul in lodestone, Thales on, 33.
Sound due to electrification, 402.
South-pointing carts—See Carts, south-pointing.
Spain, compass in mediæval, 111; in time of Saracens, 107.

Spark, electric, discovered, 431, 494.
Sparkling phenomena of human body and apparel, 503, 504.
Speaking trumpet, inv'n of, 430.
Spence, Dr., elec. exp'ts of, 538.
Spider, Franklin's elec., 545.
Spindle, ancient use of amber as, 18.
Spontaneous combustion, 503.
Sprat, T., on learning in Elizabethan age, 334; on Royal Society exp'ts, 412.
St. Aldhelm on magnet, 115.
St. Amand, discoveries of, 192.
St. Ambrose on attraction, 90; mag. mountains, 98.
St. Augustine on mag. attraction, 87; on mag. suspension, 45; distinguishes between mag. and elec. effects, 89.
St. Brandaen, journey of, 99.
St. Elmo's fire, 567.
St. Gregory Nazianzenus on attraction, 90.
St. Gregory Nyssenus on magnet, 90.
St. Isidore, Etymologies of, 91; on magnet, 91; on mag. suspension, 45.
St. Jerome on attraction of magnet and amber, 90.
St. Paul's, Bacon's exp't at, 324; Gray proposes exp't at, 474.
St. Thomas Aquinas on Form, 283; on lightning protection, 592; on magnetic attraction, 281.
Statue, suspended by magnetism, 45.
Steam engine, inv'n of, by Hero, 44.
Stone of Heraclea, 27.
Stones, Theophrastus' hist'y of, 39.
Strada, Prolusiones of, 383.
Stubbe attacks R. Society, 413.
Stukely, Dr., notices Franklin's theories, 583.
Sun, name of amber derived from that of, 16.
Suspension, magnetic, 45.
Swift, Dean, parodies Boyle, 424.
Sympathy and Antipathy, 124.
Syrian women, observe amber attraction, 17.

T.

Taisnier, John, Gilbert on, 280; on mag. rocks, 101; plagiarisms of, 192.
Tartars, 59.
Tchoü dynasty in China, 68.
Telegraph, beginnings of, 239, 400;

Browne's exp'ts on, 386; mag. described by Strada, 383; by Addison, Akenside, Cabaeus, Galileo, Hakewill, 384, 385; disputed by de Boodt, 383; predictions of, by Beal and Glanvil, 387; Schwenter's compass, 382.
Temple, alleged lightning protection of, 564.
Terrella, Gilbert's lodestone, 277; Guericke's elec., 395; sent to Pepys, 407.
Tertullian on magnet, 90.
Thales of Miletus, 32; contrasted with Hero, 44; Grote on phil'y of, 37; Laertius on, 34; Theophrastus differs from, 41.
Theamedes, 5(.
Theodoritus on magnet, 90.
Theophrastus, 38 et seq.
Thermometer, elec., 515.
Thevenot, 378.
Timaeus of Locri, 37; Plato's, 35, 36.
Timochares, 44.
Tourmaline, 41.
Torpedo, shock of, noted by Bacon, 401.
Tubal Cain, 29.
Turanian family, 61.
Typhon, iron called bone of, 28.

U.

Ugric family, 61.
Umbrians, 59.
University of Alexandria, 44.
Universities, English in time of James I., 333.

V.

Vacuum, amber attraction due to, 50; Hauksbee's light in, 460.
Van Drebbel, see Drebbel.
Van Helmont, see Helmont.
Van Musschenbrœck, see Musschenbrœck.
Variation of compass, 196; alleged record of, by Andrea Blanco, 197; Burrowes, Bond, Gellibrand and Gunter on, 446; discovered by Chinese, 76; discovered by Columbus, 200; Gilbert's errors as to, 273; Halley on, 447; line of no, 200; time changes in, obs'd by Sagredo, 367.
Vasco da Gama, voyages of, 105, 205.
Vault, lodestone, 45.

INDEX. 611

Vegetables, electrifying, Nollet's exp'ts on, 527.
Vending machine in Egyptian temples, 87.
Venice and Genoa, rivalry of, 193.
Versorium, Gilbert's, 303.
Vesalius, 262.
Vincent de Beauvais, on mag. rocks, 101.
Virtue, Gilbert on, 272; Guericke's hypotheses of, 392; Neckam on attractive, 125; Peregrinus on magnetic, 177; Porta on magnetic, 235.
Vitreous electricity, Dufay on, 484.
Vitry, Cardinal de, on compass, 154.
Voltaire on Cartesian and Newtonian theories, 509.
Von Guericke, Otto, see Guericke.
Von Kleist, see Kleist.
Vortex theory, Descartes, 48, 357; Plutarch on, 51.

W.

Wall, Dr., on resemblance of elec'y and lightning, 469.
Wallis, John, 404; Dr., on lightning, 570.
Ward, Bishop, 404.
Water and earth circuit, Watson's, 550; circuit, Lemonniers', 582; electrification of running, 489–499; soul of Thales, 34.
Watson, Dr. Wm., 508; dealings with Franklin's theory, 552; with Lemonnier's theory, 548; determines velocity of elec'y to be instantaneous, 551; discovers water and earth circuit, 550; establishes elec. circuits across Thames and New rivers, 549; exposes Bose's beatification, 499; fires spirits by elec'y, 508; ignites gas by elec'y, 508; Leyden jar exp'ts, 532, 533, 554; makes circuit 12,276 ft. long, 555; provisional theory of elec'y, 51; publishes papers, 533; urges elec. research, 539.
Wheler, Granvile, exp'ts with Gray, 474.
William the Clerk, poems of, 149.
Wilkins, Bishop, 404, 430, 436.
Wilson, Benjamin, elec. theory of, 554.
Winkler, Johann Heinrich, 506; attempts to measure speed of elec'y, 506; elec. machine, 506; identity of lightning and elec'y, 572; Leyden jar exp'ts, 515, 524; theories of elec'y, 507, 511; Von Kleist describes Leyden jar to him, 513.
Winthrop, John, 424.
Wisbuy, 134, 146; laws of, 136.
Wright, Edward, 265, 274, 315, 335.
Wren, Sir Christopher, 411.
Worship of nature, Greek, 31.

X.

Xenophanes and Xenomanes, Bacon's comparison of Gilbert to, 328.

Z.

Zoroaster, legend of, 565.

HISTORY, PHILOSOPHY AND SOCIOLOGY OF SCIENCE

Classics, Staples and Precursors

An Arno Press Collection

Aliotta, [Antonio]. **The Idealistic Reaction Against Science.** 1914

Arago, [Dominique François Jean]. **Historical Eloge of James Watt.** 1839

Bavink, Bernhard. **The Natural Sciences.** 1932

Benjamin, Park. **A History of Electricity.** 1898

Bennett, Jesse Lee. **The Diffusion of Science.** 1942

[Bronfenbrenner], Ornstein, Martha. **The Role of Scientific Societies in the Seventeenth Century.** 1928

Bush, Vannevar. **Endless Horizons.** 1946

Campanella, Thomas. **The Defense of Galileo.** 1937

Carmichael, R. D. **The Logic of Discovery.** 1930

Caullery, Maurice. **French Science and its Principal Discoveries Since the Seventeenth Century.** [1934]

Caullery, Maurice. **Universities and Scientific Life in the United States.** 1922

Debates on the Decline of Science. 1975

de Beer, G. R. **Sir Hans Sloane and the British Museum.** 1953

Dissertations on the Progress of Knowledge. [1824].
    2 vols. in one

Euler, [Leonard]. **Letters of Euler.** 1833. 2 vols. in one

Flint, Robert. **Philosophy as Scientia Scientiarum and a History of Classifications of the Sciences.** 1904

Forke, Alfred. **The World-Conception of the Chinese.** 1925

Frank, Philipp. **Modern Science and its Philosophy.** 1949

The Freedom of Science. 1975

George, William H. **The Scientist in Action.** 1936

Goodfield, G. J. **The Growth of Scientific Physiology.** 1960

Graves, Robert Perceval. **Life of Sir William Rowan Hamilton.**
    3 vols. 1882

Haldane, J. B. S. **Science and Everyday Life.** 1940

Hall, Daniel, et al. **The Frustration of Science.** 1935

Halley, Edmond. **Correspondence and Papers of Edmond Halley.** 1932

Jones, Bence. **The Royal Institution.** 1871

Kaplan, Norman. **Science and Society.** 1965

Levy, H. **The Universe of Science.** 1933

Marchant, James. **Alfred Russel Wallace.** 1916

McKie, Douglas and Niels H. de V. Heathcote. **The Discovery of Specific and Latent Heats.** 1935

Montagu, M. F. Ashley. **Studies and Essays in the History of Science and Learning.** [1944]

Morgan, John. **A Discourse Upon the Institution of Medical Schools in America.** 1765

Mottelay, Paul Fleury. **Bibliographical History of Electricity and Magnetism Chronologically Arranged.** 1922

Muir, M. M. Pattison. **A History of Chemical Theories and Laws.** 1907

National Council of American-Soviet Friendship. **Science in Soviet Russia: Papers Presented at Congress of American-Soviet Friendship.** 1944

Needham, Joseph. **A History of Embryology.** 1959

Needham, Joseph and Walter Pagel. **Background to Modern Science.** 1940

Osborn, Henry Fairfield. **From the Greeks to Darwin.** 1929

Partington, J[ames] R[iddick]. **Origins and Development of Applied Chemistry.** 1935

Polanyi, M[ichael]. **The Contempt of Freedom.** 1940

Priestley, Joseph. **Disquisitions Relating to Matter and Spirit.** 1777

Ray, John. **The Correspondence of John Ray.** 1848

Richet, Charles. **The Natural History of a Savant.** 1927

Schuster, Arthur. **The Progress of Physics During 33 Years (1875-1908).** 1911

Science, Internationalism and War. 1975

Selye, Hans. **From Dream to Discovery: On Being a Scientist.** 1964

Singer, Charles. **Studies in the History and Method of Science.** 1917/1921. 2 vols. in one

Smith, Edward. **The Life of Sir Joseph Banks.** 1911

Snow, A. J. **Matter and Gravity in Newton's Physical Philosophy.** 1926

Somerville, Mary. **On the Connexion of the Physical Sciences.** 1846

Thomson, J. J. **Recollections and Reflections.** 1936

Thomson, Thomas. **The History of Chemistry.** 1830/31

Underwood, E. Ashworth. **Science, Medicine and History.** 2 vols. 1953

Visher, Stephen Sargent. **Scientists Starred 1903-1943 in American Men of Science.** 1947

Von Humboldt, Alexander. **Views of Nature: Or Contemplations on the Sublime Phenomena of Creation.** 1850

Von Meyer, Ernst. **A History of Chemistry from Earliest Times to the Present Day.** 1891

Walker, Helen M. **Studies in the History of Statistical Method.** 1929

Watson, David Lindsay. **Scientists Are Human.** 1938

Weld, Charles Richard. **A History of the Royal Society.** 1848. 2 vols. in one

Wilson, George. **The Life of the Honorable Henry Cavendish.** 1851

Randall Library – UNCW
QC507 .B45 1975
Benjamin / A history of electricity
NXWW

304900204299%